Advances in Intelligent Systems and Computing

Volume 1266

The series "Advances in Intelligent Systems and Computing" contains publications on theory, applications, and design methods of Intelligent Systems and Intelligent Computing. Virtually all disciplines such as engineering, natural sciences, computer and information science, ICT, economics, business, e-commerce, environment, healthcare, life science are covered. The list of topics spans all the areas of modern intelligent systems and computing such as: computational intelligence, soft computing including neural networks, fuzzy systems, evolutionary computing and the fusion of these paradigms, social intelligence, ambient intelligence, computational neuroscience, artificial life, virtual worlds and society, cognitive science and systems, Perception and Vision, DNA and immune based systems, self-organizing and adaptive systems, e-Learning and teaching, human-centered and human-centric computing, recommender systems, intelligent control, robotics and mechatronics including human-machine teaming, knowledge-based paradigms, learning paradigms, machine ethics, intelligent data analysis, knowledge management, intelligent agents, intelligent decision making and support, intelligent network security, trust management, interactive entertainment, Web intelligence and multimedia.

The publications within "Advances in Intelligent Systems and Computing" are primarily proceedings of important conferences, symposia and congresses. They cover significant recent developments in the field, both of a foundational and applicable character. An important characteristic feature of the series is the short publication time and world-wide distribution. This permits a rapid and broad dissemination of research results.

**** Indexing: The books of this series are submitted to ISI Proceedings, EI-Compendex, DBLP, SCOPUS, Google Scholar and Springerlink ****

More information about this series at http://www.springer.com/series/11156

Álvaro Herrero · Carlos Cambra ·
Daniel Urda · Javier Sedano ·
Héctor Quintián · Emilio Corchado
Editors

The 11th International Conference on EUropean Transnational Educational (ICEUTE 2020)

Springer

Editors
Álvaro Herrero
Grupo de Inteligencia Computacional Aplicada (GICAP), Departamento de Ingeniería Informática, Escuela Politécnica Superior
Universidad de Burgos
Burgos, Spain

Carlos Cambra
Grupo de Inteligencia Computacional Aplicada (GICAP), Departamento de Ingeniería Informática, Escuela Politécnica Superior
Universidad de Burgos
Burgos, Spain

Daniel Urda
Grupo de Inteligencia Computacional Aplicada (GICAP), Departamento de Ingeniería Informática, Escuela Politécnica Superior
Universidad de Burgos
Burgos, Spain

Javier Sedano
Technological Institute of Castilla y León
Burgos, Spain

Emilio Corchado
University of Salamanca
Salamanca, Spain

Héctor Quintián
Department of Industrial Engineering
University of A Coruña
La Coruña, Spain

ISSN 2194-5357 ISSN 2194-5365 (electronic)
Advances in Intelligent Systems and Computing
ISBN 978-3-030-57798-8 ISBN 978-3-030-57799-5 (eBook)
https://doi.org/10.1007/978-3-030-57799-5

This Springer imprint is published by the registered company Springer Nature Switzerland AG
The registered company address is: Gewerbestrasse 11, 6330 Cham, Switzerland

Preface

This volume of Advances in Intelligent and Soft Computing contains accepted papers presented at ICEUTE 2020 conference held in the beautiful and historic city of Burgos (Spain), in September 2020.

The 11th International Conference on EUropean Transnational Education (ICEUTE 2020) has been a meeting point for people working on transnational education within Europe. It has provided a stimulating and fruitful forum for presenting and discussing the latest works and advances on transnational education within European countries.

After a through peer-review process, the ICEUTE 2020 International Program Committee selected 44 papers which are published in these conference proceedings achieving an acceptance rate of 41%. Due to the COVID-19 outbreak, the ICEUTE 2020 edition was blended, combining on-site and on-line participation. In this relevant edition, a special emphasis was put on the organization of five special sessions related to relevant topics as: Role of English in Transnational Education and Teacher Training, Personalization and ICT: a Path to Educational Inclusion, Innovation and Research Findings in Engineering Higher Education, Practical Implementations of Novel Initiatives, Innovation in Computer Science Higher Education.

The selection of papers was extremely rigorous in order to maintain the high quality of ICEUTE conference editions and we would like to thank the members of the Program Committees for their hard work in the reviewing process. This is a crucial process to the creation of a high standard conference and the ICEUTE conference would not exist without their help.

ICEUTE 2020 has teamed up with "Research Papers in Education" (Taylor & Francis) and "European Journal of Education" (Wiley) for proposals of special issues including selected papers from ICEUTE 2020.

Particular thanks go as well to the conference main sponsors Startup Ole and the IEEE Systems, Man, and Cybernetics Society—Spanish, Portuguese, French, and Italian Chapters, who jointly contributed in an active and constructive manner to the success of this initiative.

We would like to thank all the special session organizers, contributing authors, as well as the members of the Program Committees and the Local Organizing Committee for their hard and highly valuable work. Their work has helped to contribute to the success of the ICEUTE 2020 event.

September 2020

Álvaro Herrero
Carlos Cambra
Daniel Urda
Javier Sedano
Héctor Quintián
Emilio Corchado

ICEUTE 2020

Organization

General Chair

Emilio Corchado

General Co-chair

Álvaro Herrero

International Advisory Committee

Veronika Peralta	Université Francois Rabelais de Tours/Blois, France
Carlos Pereira	Polytechnic Institute of Coimbra, Portugal
Zhen Ru Dai	Hamburg University of Applied Sciences, Germany
Gabriel Michel	University of Lorraine, France
Sorin Stratulat	University of Lorraine, France
Daniela Zaharie	West University of Timisoara, Romania
Paula Steinby	Turku University of Applied Sciences, Finland
Alessandra Raffaetta	University Ca' Foscari of Venice, Italy

Program Committee Chairs

Emilio Corchado	University of Salamanca, Spain
Álvaro Herrero	University of Burgos, Spain
Javier Sedano	Technological Institute of Castilla y León, Spain
Héctor Quintián	University of A Coruña, Spain

Program Committee

Agueda Gras-Velazquez	European Schoolnet, Belgium
Alessandra Raffaetta	University Ca' Foscari of Venice, Italy
Amaia Mesanza	University of the Basque Country, Spain
Ana Lara	University of Burgos, Spain
Ana Rosa Pereira Borges	Coimbra Institute of Engineering, Portugal
Andreea Vescan	Babes-Bolyai University, Romania
Angel Arroyo	University of Burgos, Spain
Bárbara de Aymerich	University of Burgos, Spain
Bruno Baruque	University of Burgos, Spain
Carlos Cambra	University of Burgos, Spain
Carlos Pereira	Polytechnic Institute of Coimbra, Portugal
Carmen Benavides	University of León, Spain
Colm O'Reilly	CTY Ireland, Ireland
Damián Fernández	University of Seville, Spain
Daniel Urda	University of Burgos, Spain
Daniela Zaharie	West University of Timisoara, Romania
David Aguilera	University of Granada, Spain
David Alvarez Leon	University of León, Spain
Deolinda Simões	Coimbra Institute of Engineering, Portugal
Dominique Laurent	Université Cergy-Pontoise, France
Dragan Simic	University of Novi Sad, Serbia
Eduardo Solteiro Pires	UTAD University, Portugal
Eloy Irigoyen	University of the Basque Country, Spain
Esteban Jove	UDC, Spain
Estibaliz Apiñaniz	University of the Basque Country, Spain
Federico Divina	Pablo de Olavide University, Spain
Fernanda Brito Correia	Polytechnic Institute of Coimbra, Portugal
Francisco Gómez-Vela	Pablo de Olavide University, Spain
Francisco Martínez-Álvarez	Pablo de Olavide University, Spain
Francisco Zayas Gato	University of A Coruña, Spain
Gabriel Michel	University of Lorraine, France
Ileana M. Greca	University of Burgos, Spain
Inmaculada Arnedillo-Sánchez	Trinity College Dublin, Ireland
Irene Arriassecq	UNICEN, Argentina
Isabela Dramnesc	West University of Timisoara, Romania
Isaias Garcia	University of León, Spain
J. David Nuñez-Gonzalez	University of the Basque Country, Spain
Jairo Ortiz-Revilla	University of Burgos, Spain
Jean-Yves Antoine	University of Tours, France
Jennifer Bruton	Dublin City University, Ireland
Jesús Chacón	Complutense University of Madrid, Spain
Jiri Dvorsky	VSB - Technical University of Ostrava, Czechia

Joaquín Barreiro García	University of León, Spain
Jorge Barbosa	Coimbra Institute of Engineering, Portugal
José Francisco Torres Maldonado	Pablo de Olavide University, Spain
Jose Luis Calvo-Rolle	University of A Coruña, Spain
José-Luis Casteleiro-Roca	University of A Coruña, Spain
José Manuel Galán	University of Burgos, Spain
Jose Manuel Gonzalez-Cava	University of La Laguna, Spain
Jose Manuel Lopez-Guede	University of the Basque Country, Spain
José Nuñez-Gonzalez	University of the Basque Country,
José-Lázaro Amaro-Mellado	University of Seville, Spain
Juan Pavón	Complutense University of Madrid, Spain
Julio Elias Normey-Rico	Federal University of Santa Catarina, Brazil
Laura Fernández-Robles	University of León, Spain
Lidia Sánchez-González	Universidad de León, Spain
Luis Alfonso Fernández Serantes	FH-Joanneum University of Applied Sciences, Spain
Manuel Castejón-Limas	University of León, Spain
Manuel Dominguez	University of León, Spain
Manuel Grana	University of the Basque Country, Spain
María Fernández-Raga	University of León, Spain
María Inmaculada González	University of León, Spain
Maria Jose Marcelino	University of Coimbra, Portugal
Marta Romero Ariza	University of Jaén, Spain
Matilde Santos	Complutense University of Madrid, Spain
Miguel Carriegos	University of León, Spain
Miguel García Torres	Pablo de Olavide University, Spain
Monika Ciesielkiewicz	Complutense University of Madrid, Spain
Natividad Duro	UNED, Spain
Nuno Ferreira	Coimbra Institute of Engineering, Portugal
Orlando Belo	University of Minho, Portugal
Paivi Oliva	Turku University of Applied Sciences, Finland
Paula Steinby	Turku University of Applied Sciences, Finland
Paulo Moura Oliveira	UTAD University, Portugal
Pedro Antonio Gutierrez	University of Cordoba, Spain
Rafael Corchuelo	University of Seville, Spain
Ramón-Ángel Fernández-Díaz	University of León, Spain
Raquel Dormido Canto	UNED, Spain
Richard Duro	University of A Coruña, Spain
Salvatore Orlando	Ca' Foscari University, Italy
Santiago Porras	University of Burgos, Spain
Sorin Stratulat	University of Lorraine, France
Unai Fernandez	University of the Basque Country, Spain
Vaclav Snasel	VSB: Technical University of Ostrava, Czechia

Verónika Peralta	University of Tours, France
Vicente Matellan	University of León, Spain
Victoria López	Complutense University of Madrid, Spain
Viorel Negru	West University of Timisoara, Romania
Zhen Ru Dai	Hamburg University of Applied Sciences, Germany

Special Sessions

Role of English in Transnational Education and Teacher Training

Special Session Organizers

María Amor Barros del Río	University of Burgos, Spain
Concetta Maria Sigona	University of Burgos, Spain
Ana Cunha	University of Lusófona, Portugal
Alina Doroch	Collegium Balticum, Poland
Onorina Botezat	Dimitrie Cantemir Christian University, Romania

Program Committee

Alina Doroch	SSW Collegium Balticum, Poland
Ana Cunha	Lusofona University, Portugal
Concetta Sigona	University of Burgos, Spain
Onorina Botezat	Dimitrie Cantemir Christian University, Romania
Paola Clara Leotta	University of Catania, Italy
Ramona Mihaila	Dimitrie Cantemir Christian University, Romania
Zut Koczalska	ZUT, SJO, Poland
Alina Doroch	SSW Collegium Balticum, Poland

Personalization and ICT: A Path to Educational Inclusion

Special Session Organizers

Beatriz Núñez Angulo	University of Burgos, Spain
Sonia Sapia	Universitá della Calabria, Italy

Program Committee

Miriam Lorenzo Bañuelos	University of Burgos, Spain
Natalia Muñoz-Rujas	University of Burgos, Spain
Pedro Luis Sánchez Ortega	University of Burgos, Spain
Wilco M.H. Verbeeten	University of Burgos, Spain

Innovation and Research Findings in Engineering Higher Education

Special Session Organizers

Eduardo Montero García	University of Burgos, Spain
Fatima E. M. Mhamdi Alaoui	University Chouaib Doukkali El Jadida, Morocco
María Consuelo Sáiz Manzanares	University of Burgos, Spain
Natalia Muñoz-Rujas	University of Burgos, Spain

Program Committee

Eduardo Montero García	University of Burgos, Spain
Fatima E. M. Mhamdi Alaoui	University Chouaib Doukkali El Jadida, Morocco
María Consuelo Sáiz Manzanares	University of Burgos, Spain
Miriam Lorenzo Bañuelos	University of Burgos, Spain
Natalia Muñoz-Rujas	University of Burgos, Spain

Practical Implementations of Novel Initiatives

Special Session Organizers

Jose Manuel Lopez-Guede	University of the Basque Country, Spain
Manuel Graña	University of the Basque Country, Spain
Julián Estévez	University of the Basque Country, Spain
Felipe Núñez	Pontificia Universidad Católica de Chile, Chile

Program Committee

Amaia Mesanza	University of the Basque Country, Spain
Ana Boyano	University of the Basque Country, Spain
Estibaliz Apiñaniz	University of the Basque Country, Spain
Felipe Nunez	Technological University of Chile INACAP, Chile
Francisco Martínez-Álvarez	Pablo de Olavide University, Spain

Julian Estevez	University of the Basque Country, Spain
Manuel Graña	University of the Basque Country, Spain

Innovation in Computer Science Higher Education

Special Session Organizers

Cristina Rubio Escudero	University of Seville, Spain
Guillermo Santamaría Bonfil	CONACYT-INEEL, Mexico
Federico Divina	Pablo de Olavide University, Spain
Isabel Sofía Brito	Polytechnic Institute of Beja, Portugal
Miguel Torres-García	Pablo de Olavide University, Spain
Maria Teresa Godinho	Polytechnic Institute of Beja, Portugal
Francisco Gómez-Vela	Pablo de Olavide University, Spain
Francisco Martínez-Álvarez	Pablo de Olavide University, Spain

Program Committee

Adela Has	University of Osijek, Croatia
Antonio Morales-Esteban	University of Seville, Spain
Jorge Reyes	NT2 Labs, Chile
Khawaja Asim	PIEAS, Pakistan
Laura Melgar-García	Pablo de Olavide University, Spain
María Victoria Requena García de la Cruz	University of Seville, Spain
Marinela Knežević	University of Osijek, Croatia

Organising Committee Chairs

Álvaro Herrero	University of Burgos, Spain
Javier Sedano	ITCL, Spain
Carlos Cambra	University of Burgos, Spain
Daniel Urda	University of Burgos, Spain

Organising Committee

Emilio Corchado	University of Salamanca, Spain
Héctor Quintián	University of A Coruña, Spain
Carlos Alonso de Armiño	University of Burgos, Spain
Ángel Arroyo	University of Burgos, Spain
Bruno Baruque	University of Burgos, Spain
Nuño Basurto	University of Burgos, Spain
Pedro Burgos	University of Burgos, Spain
David Caubilla	University of Burgos, Spain
Leticia Curiel	University of Burgos, Spain

Raquel Redondo	University of Burgos, Spain
Jesús Enrique Sierra	University of Burgos, Spain
Belén Vaquerizo	University of Burgos, Spain
Juan Vicente Martín	University of Burgos, Spain

Contents

Transnational Education at University Level

Special Session: Role of English in Transnational Education and Teacher Training

Transnational Education at Pre-university Level

Are Secondary Mathematics Student Teachers Ready for the Profession? A Multi-actor Perspective on Mathematics Student Teachers' Mastery of Related Competences

Laura Muñiz-Rodríguez[1(✉)], Pedro Alonso[2], Luis J. Rodríguez-Muñiz[1], and Martin Valcke[3]

[1] Department of Statistics and Operational Research and Mathematics Education, University of Oviedo, Oviedo, Spain
munizlaura@uniovi.es
[2] Department of Mathematics, University of Oviedo, Oviedo, Spain
[3] Department of Educational Studies, Ghent University, Ghent, Belgium

Abstract. Initial teacher education programs are expected to provide student teachers with the necessary competences to develop themselves as teachers. This article studies the extent to which initial teacher education programs achieve this purpose according to different context-, input-, process-, and output-indicators (CIPO) by adopting a multi-actor focus. We illustrate this approach based on initial teacher education programs for secondary mathematics student teachers in Spain. Data were gathered through an online survey administered to 315 participants, including student teachers (95), teacher educators (95), mentors (96), and graduate teachers (29). The results reflect that initial teacher education programs are rather moderately effective in view of preparing future secondary mathematics teachers for the profession. More specifically, the perceptions of student and graduate teachers indicate clear quality problems regarding future teachers' mastery of teaching competences. Implications and directions for future research are discussed.

Keywords: Initial teacher education · Mathematics · Teaching competences

1 Introduction

The quality of initial teacher education programs has been often determined by student teachers' readiness for the profession [21]. Attaining readiness for the profession is, in turn, connected to student teachers' competences [9,12]. Ensuring competent teachers are recruited for the teaching profession is, therefore, a primary goal for initial teacher education institutions.

Different factors influence the extent to which initial teacher education programs accomplish their purpose. The literature highlights the recruitment system [17], the content of the program [7], the theory-practice linkage [1], the existence of well-defined standards [6], or the preparation of teacher educators [20] as some of these factors.

Á. Herrero et al. (Eds.): ICEUTE 2020, AISC 1266, pp. 3–10, 2021.
https://doi.org/10.1007/978-3-030-57799-5_1

Worldwide, mathematics teacher education at this level is rarely studied. In 2008, seventeen countries participated in the first cross-national study providing data on the knowledge that future primary and lower-secondary mathematics teachers acquire during their initial teacher education: the TEDS-M [21]. However, that study focused on teacher education policies, institutional practices, and future mathematics teachers' content knowledge and pedagogical content knowledge. Some core pedagogical competences were not considered. In addition, the Spanish participation in the study was limited to future teachers at primary education due to difficulties in collecting data from dispersed and difficult-to-reach future teachers in secondary education.

This research study aims at exploring the quality of initial teacher education programs in terms of student teachers' readiness for the profession and mastery of related competences from a multi-actor perspective. To organize the study, we build on the CIPO approach [18] to structure a series of context-, input-, process-, and output-indicators. The study was conducted in the specific context of initial teacher education programs for secondary mathematics student teachers in Spain.

2 Theoretical Framework

The CIPO model [18] distinguishes four components that define education as a production system in which Inputs (I) are transferred into Outputs (O) through a Process (P) influenced by a Context (C). Therefore, this model serves as an analytic framework to identify educational quality indicators [5]. Considering the main research aim of the study, this paper focuses on the last two components (P and O).

The process component concerns the pedagogical activities determining the transition of inputs into outputs. In the case of initial teacher education, the process refers to student teachers' opportunities to develop professional teaching competences during the program. Although the structure of initial teacher education programs might vary between universities, they commonly incorporate a theoretical and a practical component. In this sense, the development of professional teaching competences is influenced by both the theoretical courses and the field experiences. In [7] the authors emphasize that the stronger the linkage between the later components, the larger the impact of initial teacher education programs. Previous studies stress how practical experiences significantly influence future teachers' beliefs and personal mission in view of teaching [4,10]. Research focusing on the efficacy of initial teacher education highlights this theory-practice connection and its balance [1]. Therefore, teaching competences should be measured from both a theoretical and a practical perspective.

Up to now, no established system is available to measure teaching competences. Previous studies relied on the perceptions of student teachers and teacher educators about the level of development of professional teaching competences during initial teacher education as quality process-indicators [2]. Building on this approach, in [8] the authors found significant correlations between future teachers' perceptions and their sense of self-efficacy, which is correlated with student achievement. Process-indicators measuring teaching competences are usually based on a framework reflecting a common understanding about the competences student teachers should master after completing their initial teacher education [9]. Despite the lack of international consensus, several validated frameworks are available in the academic literature [15].

The output component alludes to learner achievement in terms of competence development. In the context of initial teacher education, output applies to the extent initial teacher education programs accomplish their purpose, i.e., the actual acquisition of teaching competences [3,6]. Building on former research, two types of output-indicators are adopted. A first output-indicator builds on the perceptions of student and graduate teachers about their competence development level. As a second output-indicator, stakeholders' perceptions about the effectiveness of initial teacher education programs to prepare future teachers for the profession is used.

3 Method

This research study builds on an exploratory quantitative design to determine the level of pursuance and attainment of competences during initial teacher education programs for future secondary mathematics teachers. Two research questions are investigated: (1) To what extent are teaching competences pursued during an initial teacher education program for future secondary mathematics teachers? and (2) To what extent are initial teacher education programs effective in preparing future secondary mathematics teachers for the profession?

3.1 Participants

This study builds on data from student teachers, teacher educators, mentors, and graduate teachers from initial teacher education programs for future secondary mathematics teachers in Spain. The sample consisted of 315 individuals, distributed among four stakeholder groups: secondary student teachers enrolled in an initial teacher education program for future secondary mathematics teachers (95), mathematics teacher educators teaching in an initial teacher education program for future secondary mathematics teachers (95), mentors supporting field experiences in secondary education schools (96), and graduate secondary mathematics teachers (29). The reader notices graduate teachers represent the smallest subsample. Due to privacy reasons, no direct contact information of these individuals could be obtained from initial teacher education institutions. Participants represent 47 of the 55 Spanish universities – 36 public and 11 private – and 86 secondary education schools in Spain – 79 public and 7 private.

Student teachers participating in the study (95) were on average 29.35 years old (standard deviation = 7.33), and graduate teachers (29) were 31.28 years old (standard deviation = 8.38). Around 59% of student teachers and 62% of graduate teachers were female.

Data showed a slightly larger proportion (60%) of male teacher educators (95). The largest proportion of teacher educators was tenured (65.3%) at full professor or associate professor level holding a PhD. Other teacher educators were tenured as lecturer or got non-tenured jobs as assistant professor or teaching assistant. Average teaching experience was 23.49 years (standard deviation = 10.50) and 7.29 years (standard deviation = 7.25) as a mathematics teacher educator. Around 71.6% of respondents never received special preparation for training student teachers. Within those receiving training, 23.1% did before and 5.3% after becoming a teacher educator.

A gender-balanced distribution of female (51%) and male (49%) mentors (96) were involved in the study. Most mentors were secondary mathematics teachers, with an average experience of 22.67 years (standard deviation = 9.16). Average experience as a mentor was 4.35 years (standard deviation = 5.61). A large proportion of mentors (85.4%) never received special preparation for training student teachers. Within those receiving training, 11.5% did before becoming a mentor and 3.1% afterwards.

3.2 Data Collection and Data Analysis

Data were gathered through an online survey presenting a specific version of a questionnaire for each stakeholder group [13]. The online survey was designed and administered via LimeSurvey®. Respondents were asked to indicate their informed consent when submitting their replies. Anonymity and confidentiality were guaranteed.

All items were previously validated in a pilot study conducted for the purpose of this research [14]. To assess the level of pursuance and attainment of professional teaching competences, the questionnaire built on a framework of thirty-three competences, validated for the Spanish context and classified into twelve clusters [15]: Mathematical Content Knowledge (MCK), Mathematical Pedagogical Knowledge (MPK), Teaching and Learning Processes (TLP), Classroom Management (CM), Lesson Planning (LP), Assessment and Mentoring (AM), Developmental Psychology (DP), Inclusion and Diversity (ID), Technology Knowledge (TK), Communication Skills (CS), Contribution to School Organization (CSO), and Personal Commitment (PC). Respondents indicated the extent to which each competence was pursued or attained during the initial teacher education program.

Data analysis was performed using SPSS®. Seven-point Likert scales were consistently used throughout the questionnaire helping to establish larger reliability and validity. Based on the Likert scores, an average index was calculated for each competence cluster considering the four stakeholder perspectives.

4 Results

Table 1 summarizes the perceptions of each stakeholder group as to the extent each competence cluster is pursued during the initial teacher education program for future secondary mathematics teachers in Spain. For student and graduate teachers, a distinction is made between the theoretical and the practical component.

Participants perceive most competences are not being intensively pursued during their initial teacher education. Results reflect heterogeneous responses between stakeholders in all competence clusters, except in the mathematical content knowledge domain. Student and graduate teachers' answers are more critical than those of teacher educators and mentors. When focusing on the theory-practice gap, student and graduate teachers perceive higher pursuance levels in theory than in practice.

Table 1. Multi-stakeholder perspective about the pursuance level of each competence cluster.

	Student teachers		Teacher educators	Mentors	Graduate teachers	
	Theory M(SD)	Practice M(SD)	M(SD)	M(SD)	Theory M(SD)	Practice M(SD)
MCK	3.85(1.69)	3.26(1.60)	4.21(1.59)	3.66(1.46)	3.74 (1.70)	3.29(1.68)
MPK	3.79(1.53)	3.68(1.61)	4.58(1.27)	4.83(1.20)	4.28 (1.73)	3.97(1.74)
TLP	4.19(1.48)	3.88(1.64)	5.07(1.25)	4.89(1.25)	4.57 (1.63)	4.25(1.80)
CM	3.61(1.61)	3.50(1.69)	4.73(1.28)	5.13(1.32)	4.53 (1.70)	4.02(1.77)
LP	4.47(1.49)	4.15(1.64)	5.22(1.18)	5.17(1.25)	4.94 (1.58)	4.66(1.79)
AM	3.62(1.70)	3.29(1.68)	4.50(1.32)	4.75(1.44)	4.47 (1.86)	4.33(1.89)
DP	4.16(1.70)	3.56(1.77)	4.40(1.49)	4.79(1.40)	4.93 (1.66)	4.32(1.88)
ID	3.98(1.70)	3.34(1.78)	4.31(1.44)	4.71(1.50)	4.76 (1.75)	3.95(1.74)
TK	4.57(1.62)	4.02(1.98)	5.44(1.46)	4.96(1.38)	5.41 (1.72)	5.21(1.93)
CS	3.67(1.88)	3.47(1.96)	4.57(1.54)	4.92(1.51)	4.69 (1.63)	4.45(2.01)
CSO	3.23(1.67)	2.79(1.69)	3.83(1.49)	3.87(1.69)	4.00 (1.87)	3.67(1.97)
PC	4.02(1.75)	3.72(1.92)	4.87(1.52)	4.86(1.44)	4.71 (1.81)	4.45(2.02)

Note. Minimum score: 1. Maximum score: 7. M: Mean. SD: Standard Deviation.

Table 2 summarizes the perceptions of student and graduate teachers about the extent to which each competence cluster is attained during the initial teacher education program. The results show all competence clusters are rated significantly low.

Table 2. Student and graduate teachers' perceptions of the attainment level of each competence cluster.

Competence cluster	Student teachers	Graduate teachers
	M(SD)	M(SD)
MCK	3.46(1.79)	3.53(2.00)
MPK	3.67(1.59)	4.19(1.81)
TLP	4.06(1.59)	4.43(1.68)
CM	3.62(1.68)	4.09(1.68)
LP	4.29(1.61)	4.64(1.67)
AM	3.71(1.81)	4.26(1.85)
DP	3.96(1.76)	4.56(1.73)
ID	3.73(1.73)	4.40(1.71)
TK	4.31(1.85)	5.07(1.87)
CS	3.72(1.87)	4.21(1.80)
CSO	3.35(1.72)	3.74(1.69)
PC	3.79(1.76)	4.34(1.95)

Note. Minimum score: 1. Maximum score: 7. M: Mean. SD: Standard Deviation

Participants were asked to rate the effectiveness of the initial teacher education programs to prepare future secondary mathematics teachers for the profession. The average effectiveness rate is low from the four stakeholder perspectives: student teachers (mean = 3.52, standard deviation = 1.46), teacher educators (mean = 4.81, standard deviation = 1.40), mentors (mean = 4.75, standard deviation = 1.27), and graduate teachers (mean = 3.90, standard deviation = 1.59). Student and graduate teachers' perceptions are more critical than those of teacher educators and mentors.

5 Discussion

This research study aims at exploring the quality of initial teacher education programs in terms of student teachers' mastery of related competences from a multi-actor perspective. Building on student and graduate teachers' perceptions (see Table 1), all competences are insufficiently pursued; both in theory and in practice. The weak application of competences during field experiences explain the differences between the theoretical and the practical component. Previous research suggests establishing stronger school-university partnerships to bridge this theory-practice gap [19].

The low pursuance levels are clearly reflected in student and graduate teachers' perceived attainment of their teaching competences (see Table 2). The latter is not reassuring whether initial teacher education programs in Spain adequately train future secondary mathematics teachers and whether these are being prepared for the teaching career. The former also pushes the need to develop and implement alternative strategies to strengthen the full range of teaching competences in student teachers. In view of this, several authors propose innovative approaches to develop student teachers' competences for teaching mathematics in secondary education [16]. For instance, in [11] the authors suggest incorporating active learning and simulation approaches to engage student teachers in mathematical reasoning. Next to alternative learning strategies, the findings also reinforce the idea to define benchmark criteria in view of competence development in Spain. Though the above results were based on student teachers' perceptions, also other stakeholders admit student teachers' competence attainment is low. This emphasizes the critical voices stating initial teacher education hardly has an added value for future secondary mathematics teachers in Spain. The findings also raise concerns about the extent to which enough attention is paid to the continuous professional development of mathematics teachers in secondary education.

Teacher educators and mentors play an essential role in the preparation of future teachers. In line with previous national research [22], our data point at the limited experience and expertise of both stakeholder groups. It seems there are no specific requirements or professional trajectories to become teacher educator or mentor in an initial teacher education program in Spain. Authors explain that teacher educators and mentors are rather selected according to availability of teaching credits, instead of their career profile [23]. In this sense, Spain is not an exception. In most countries, teacher educators and mentors hardly receive formal training in view of their work. They rather learn through experience, with little institutional and professional support. Also [20] stresses that the career of teacher educators and mentors seems the only professional educational position not requiring formal training. Notwithstanding, some countries have already moved forward in the development of quality requirements or standards for teacher educators and mentors. In Spain, policymakers and educational leaders should urgently address the need to develop and establish a competence framework for teacher educators and mentors.

The present study contributes to the field of initial teacher education and teaching quality assessment in several ways. The theoretical relevance is reflected in the availability of validated instruments to assess initial teacher education programs [13]. Reliable quality indicators have been developed and implemented to assess the extent to which teaching competences are pursued and attained during initial teacher education programs, focusing on the mathematics specialty. They can serve as a starting point for (inter)national research in other contexts or specialties. From an empirical viewpoint, this study helps to explain the nature of initial teacher education for future secondary mathematics teachers in Spain. The results reveal a critical situation when looking at the mastery of teaching competences. Considering how initial teacher education influences the nature of future teachers' teaching practices and professional development [7], appropriate measures should be implemented. Finally, this study also contributes to educational practice and policy. The findings can be used as benchmark data to inform policymakers and educational leaders about the mastery of teaching competences, some of which need to be (re)addressed.

Developing a state-of-the-art initial teacher education system represents a complex enterprise. This complexity is largely explained by the demands of today's society, who requires teachers to demonstrate a full range of competences right from the start of their career. This research study provides evidence about different shortcomings regarding initial teacher education programs for future secondary mathematics teachers in Spain. The perceptions of student teachers, teacher educators, mentors, and graduate teachers are too important to be neglected. Future secondary mathematics teachers seem to be strongly motivated for teaching, but their initial teacher education falls short in providing them with the competences to become effective teachers. Therefore, improving initial teacher education becomes a major challenge for Spanish institutions, policymakers and educational leaders.

Acknowledgements. This research has been partially supported by the Spanish MINECO project (TIN2017-87600-P).

References

1. Allen, J.M., Wright, S.E.: Integrating theory and practice in the pre-service teacher education practicum. Teach. Teach. **20**(2), 136–151 (2013)
2. Bhargava, A., Pathy, M.: Perception of student teachers about teaching competencies. Am. Int. J. Contemp. Res. **1**(1), 77–81 (2011)
3. Buchberger, F., Campos, B.P., Kallos, D., Stephenson, J.: Green paper on teacher education in Europe. Thematic Network on Teacher Education in Europe, Umea (2000)
4. Caires, S., Almeida, L.: Teaching practice in initial teacher education. Its impact on student teachers' professional skills and development. J. Educ. Teach. **31**(2), 111–120 (2005)
5. Cuyvers, G.: Kwaliteitsontwikkeling in Het Onderwijs. Garant, Antwerpen (2002)

6. Darling-Hammond, L.: Constructing 21 st-century teacher education. J. Teach. Educ. **57**(3), 300–314 (2006)
7. Darling-Hammond, L., Hammerness, K., Grossman, P., Rust, F., Shulman, L.: The design of teacher education programs. In: Darling-Hammond, L., Hammerness, K. (eds.) Preparing Teacher for a Changing World: What Teachers Should Learn and Be Able to Do. Jossey Bass, San Francisco (2005)
8. Darling-Hammond, L., Newton, X., Wei, R.C.: Evaluating teacher education outcomes: a study of the Stanford Teacher Education Programme. J. Educ. Teach. Int. Res. Pedagog. **36** (4), 369–388 (2010)
9. European Commission: Supporting teaching competence development. European Commission, Brussels (2013)
10. Linden, W., Bakx, A., Ros, A., Beijaard, D., Bergh, L.: The development of student teachers' research knowledge, beliefs and attitude. J. Educ. Teach. **41**(1), 4–18 (2015)
11. Lovett, J.N., Lee, H.S.: New standards require teaching more statistics: are preservice secondary mathematics teachers ready? J. Teach. Educ. **68**(3), 299–311 (2017)
12. Mohamed, Z., Valcke, M., De Wever, B.: Are they ready to teach? Student teachers' readiness for the job with reference to teacher competence frameworks. J. Educ. Teach. **43** (2), 151–170 (2016)
13. Muñiz-Rodríguez, L.: Initial education of future secondary mathematics teachers in Spain. University of Oviedo, Oviedo (2017)
14. Muñiz-Rodríguez, L., Alonso, P., Rodríguez-Muñiz, L.J., Valcke, M.: Are future mathematics teachers ready for the profession? A pilot study in the Spanish framework. Procedia Soc. Behav. Sci. **16**, 735–745 (2016)
15. Muñiz-Rodríguez, L., Alonso, P., Rodríguez-Muñiz, L.J., Valcke, M.: Developing and validating a competence framework for secondary mathematics student teachers through a Delphi method. J. Educ. Teach. **43**(4), 383–399 (2017)
16. Muñiz-Rodríguez, L., Alonso, P., Rodríguez-Muñiz, L.J., De Coninck, K., Vanderlinde, R., Valcke, M.: Exploring the effectiveness of video-vignettes to develop mathematics student teachers' feedback competence. EURASIA J. Math. Sci. Technol. Educ. **14**(9), 1–17 (2018)
17. OECD (Organization for Economic Co-operation and Development): Teachers Matter: Attracting, Developing and Retaining Effective Teachers. OECD Publications, Paris (2005).
18. Scheerens, J.: School effectiveness research. In: International Encyclopedia of the Social & Behavioral Sciences, vol. 21, 2nd edn., pp. 80–85 (2015)
19. Schleicher, A.: Preparing Teachers and Developing School Leaders for the 21st Century: Lessons from Around the World. OECD Publishing, Paris (2012)
20. Smith, K.: So, what about the professional development of teacher educators. Eur. J. Teach. Educ. **26**(2), 201–215 (2003)
21. Tatto, M.T., Schwille, J., Senk, S.L., Ingvarson, L., Rowley, G., Peck, R., Bankov, K., Rodriguez, M.. Reckase, M.: Policy, Practice, and Readiness to Teach Primary and Secondary Mathematics in 17 Countries. Findings from the IEA Teacher Education and Development Study in Mathematics (TEDS-M). IEA, Amsterdam (2012)
22. Valdés, R., Bolívar, A.: La experiencia española de formación del profesorado: El Máster en Educación Secundaria. Ensino Em Re-Vista **21**(1), 159–173 (2014)
23. Vilches, A., Gil-Pérez, D.: Máster de Formación Inicial del Profesorado de Enseñanza Secundaria. Algunos análisis y propuestas. Revista Eureka sobre Enseñanza y Divulgación de las Ciencias **7**(3), 661–666 (2010)

From Trivium to Smart Education

Javier Teira-Lafuente[1], Ana Belén Gil-González[2(✉)], and Ana de Luis Reboredo[2]

[1] Faculty of Philosophy, University of Salamanca, Salamanca, Spain
teira@usal.es
[2] Computer Science and Automation Department, University of Salamanca, Plaza de los Caídos. s/n, 37008 Salamanca, Spain
{abg,adeluis}@usal.es

Abstract. Rethinking the classics for thinking the future. This could be the compendium of the present article, in which we propose a revision of the immediate future of education based on the classic project of the Trivium. We will analyze, first, the transformation of education in the perspective of *smart education,* determined by the impact of technology and by the reflection on the competences of the 21st century; secondly, we will review the strategic and methodological proposals in accordance with this transformation, based on the theory of generative learning; thirdly, from the point of view of contents, we will analyse the importance of core digital skills as programming and computational thinking.

On this basis, the paper offers a proposal from a dual perspective. Firstly, by rethinking the main issues of education in the light of the history of the Trivium and the epistemological principles that shaped it. Secondly, by proposing the recovery of the Trivium disciplines (Grammar, Rhetoric and Logic) having in mind the debate on the competences of the 21st century, as the best instrument to enrich the current educational systems, especially in view of the challenges of digitalization.

Keywords: Smart education · Trivium · Competences · Computational thinking

1 Introduction

The immediate future of education is determined by the exponential impact of technology. This transformation of education is leading to what the International Association of Smart Learning Environments (IASLE) has defined also as: "an emerging area alongside other related emerging areas such as smart technology, smart teaching, smart education, smart-e-learning, smart classrooms, smart universities, smart society. The challenging exploitation of smart environments for learning together with new technologies and approaches such as ubiquitous learning and mobile learning could be termed smart learning" [23].

In recent years, some voices have emerged calling for a look at the Trivium, both from the point of view of learning skills [16,20] and from that of interdisciplinary studies [5], as an inspirational source for facing the educational challenges

Á. Herrero et al. (Eds.): ICEUTE 2020, AISC 1266, pp. 11–20, 2021.
https://doi.org/10.1007/978-3-030-57799-5_2

of the 21st century. In this article we propose a revival of the classic Trivium project, both from the point of view of the philosophical reflections that emerge from its historical revision, as well as from the point of view of the content of the disciplines that compose it.

The term Trivium, whose first vestige of use dates from the 9th century [32], means "triple road" and includes the first three of the seven liberal arts, that is, Grammar, Rhetoric and Logic, which constituted the corpus of literary knowledge, today we would say humanities, from Antiquity to the Renaissance. Subsequently, these disciplines maintained an unequal presence in the faculties of arts [6] and, finally, when they were relocated at different rates within the framework of modern and contemporary education, they lost their central position. As a result of this process, Rhetoric and Logic are occupying today a residual place in the curricula, being accessible only under the specialized or complementary training modalities [36].

With this basic conceptualization, the reminder of this paper is organized as follow. In Sect. 2 we analyze the transformation of education in the perspective of Smart Education; in Sect. 3, we present some pedagogical and methodological issues of this transformation in the light of the theory of generative learning; at Sect. 4, we analyse the importance of computational and algorithmic thinking as well as programming, as key elements of digital skills; Sect. 5 is devoted to show how the classical Trivium concept can help us to face challenges in education today; Sect. 6 concludes setting out the main findings of this approach.

2 Smart Education and Education for the 21st Century

Zhu *et al.* [39], defined smart education this way: "The essence of smart education is to create intelligent environments by using smart technologies, so that smart pedagogies can be facilitated as to provide personalized learning services and empower learners, and thus talents of wisdom who have better value orientation, higher thinking quality, and stronger conduct ability could be fostered". According to Coccoli *et al.* [9], the environments of intelligent education are characterized by their richness, interactivity and flexibility, in order to be able to fulfill three objectives: (1) To take advantage of the devices availables in networks, (2) To enhance individual skills and competences (3) To reinforce collaborative work.

The consequences of this novelty take place on two levels, *objective* and *subjective*. On the *objective* level, it confronts individuals with situations for which, in principle, they lack tools and conceptual schemes. This objective novelty encompasses the personal sphere, as it is progressively immersed in an environment dominated by artificial intelligence; the work sphere, as it is constantly faced with a variation and complication of work profiles [30]; and the social sphere, as it is subjected to the dynamics of intense social mobility and forms part of the entrepreneurship competence [2], as well as of the new social and civic space defined by the elements of digital citizenship [13]. Education, therefore, must provide a high level of adaptability through polyhedral labor profiles, with a wide range of complementary skills and competences as citizens as well [35]. The model of education focused in reproducible protocols must be enriched with instruments to connect concepts and create knowledge adapted to new problems [25,26].

This brings us closer to the second level of consequences, that is, the *subjective* changes involved, since they produces a situation of insecurity and uncertainty that, in itself, constitutes an obstacle to the effectiveness of individual actions and decisions. This idea is emphasized by Segredo *et al.* [35]: "Citizens of the future must have full confidence in the tools and technologies involved in a smart environment". Thus, an adequate training must be an instrument of personal success also subjectively, favouring attitudes and feelings of self-confidence and security. This can be achieved by promoting key skills such as creativity or resilience at NMC Horizon Report [28] and all of these abilities, in short, that allow people to develop and "live effective at work and leisure time" (Trilling & Fadel [37], Zhu *et al.* [40]). Technology in education, therefore, is revealed in this context as fundamental, but not enough. Segredo *et al.* [35] provide us with a precise and illustrative synthesis of recent reflection on competencies for the 21st century, which outlines its main lines by determining dimensions, skill levels, main components, basic academic goals, and ICT skills. These works shows that technology should be a fundamental tool, but not the ultimate goal. Education must therefore serve to go further and develop a new digital citizenship, taking into account the concepts of social responsibility, quality of life and, ultimately, personal happiness. All of this has been object of the European report on social and emotional education conducted by Cefai *et al.* [12].

3 Methodology, Design and Educational Environment

With regard to smart education, Zhu *et al.* [40] identify three essential elements, that is, environments, pedagogy, and learners, while a main requirement: to provide higher thinking quality, and foster stronger conduct ability. According to this, Segredo *et al.* [35] highlight three methodological needs: (1) The design of learning processes according to the needs and preferences of the students. (2) The application of a generative learning model in which, instead of giving priority to the reception of transmitted content, the active role of the student is the main factor, who acts supported by the educational potential of the intelligent environment and (3) The design of intelligent environments according to a constructivist paradigm. This is also the main component of the third and fifth areas of competences of the European Framework for the Digital Competence of Educators [33].

Nonetheless, our tradition of teaching [21], in which personal presence and living word are primordial, together with the fact of the natural reticence to change of the educational systems, could be precisely at the basis of the "the consistent tendency of the educational system to preserve itself and its practices by the assimilation of new technologies into existing instructional practices", in such a way that technology is domesticated within the framework of the "prevailing educational philosophy of cultural transmission" [35].

4 Digital Skills, Computational Thinking and Educational Content

Since its first formulation at 2006 by Jeannette Wing [38], definitions of computational thinking have shown discrepancies about its basic components, but it exists a fundamental consensus that becomes visible, for example, in how the International Society for Technology in Education (ISTE) and, likewise, the Computer Science Teachers Association (CSTA), identify the core dimensions of computational thinking: 1. Formulate problems with a view to their solution by means of computers. 2. Organizing and analyzing data logically. 3. Representing data through abstractions, models, and simulations. 4. Automating solutions through algorithmic thinking, that is, through a series of steps ordered to those solutions. 5. Identify, analyze and implement efficient solutions. And finally 6. Generalize the solution process to a wide range of problems [35].

Bocconi *et al.* underline also the definition proposed by the Royal Society in 2012, in which "stresses that computation is not exclusively a human construct but is also present in nature": "Computational thinking is the process of recognising aspects of computation in the world that surrounds us, and applying tools and techniques from Computer Science to understand and reason about both natural and artificial systems and processes".

Algorithmic thinking, in turn, also used in many official policy documents to refer to CT [4], implies the following skills: 1. Analyze given problems. 2. Specifying or representing a problem accurately. 3. Finding the basic and appropriate operations (instructions) to solve a given problem. 4. Constructing an algorithm to solve the problem following the given sequence of actions. Think of all possible cases (special or not) of a given problem. Improve the efficiency of an algorithm [35].

"Fostering coding and programming" is one of the six main reasons to introduce computational thinking in curriculum [4]. But because provided that programming teaches thinking and this teaches to think in general, it results that programming and computational are mutually reinforcing strategies that benefits education.

Actually it has become clear that an effective way to improve educational content in general while digital skills also is by incorporating models of Computational Thinking (CT). The development of CT on this perspective goes along the thinking over these models, algorithmic thinking and programming [7].

Finally, among the reasons to introduce computational thinking in curriculum, stand out for us the two following: (1) CT foster logical skills and (2) CT foster 21'st skills as entrepreneurial skills, social and emotional skills [4].

5 Why a Revival of Classical Trivium to Face 21st Century Education Challenges?

5.1 Trivium Insights: Philosophical Reflections Emerging from History

At least five main blocks of questions emerge from the above considerations, in relation to which a retrospective look at the classical world [24] and the contents of the Trivium can help us to shed light:

The Need to Include Digital Skills in a Broad Sense in Educational Design. One of the most striking methodological problems facing contemporary education are operational and organizational difficulties produced by the amount of disciplines [4]. The need to introduce digital competencies in a broad sense and the lack of consensus aggravates the problem of the comparability of education systems. In the absence of a concept of unity of knowledge, it makes more visible the voluntarist nature of the requirements of interdisciplinary studies, that claim, for example, a naif substitution of STEM by STEAM (STEM + Arts) [22]. A careful look at the history and components of the Trivium can help us to shed light on these problems, pointing to an integral solution that starts from their common root: the idea of unity of science, nature and mind. The disciplines of the Trivium (grammar, rhetoric, logic) and those of the Quadrivium (music, astronomy, geometry and arithmetic) formed a coherent and complete whole ('enkiklios paidea', encyclopedic cycle of knowledge) with a common epistemological foundation: mathematics [21]. In this same direction, on the interweaving of mathematics and philosophy in the Quadrivium, the recent study of Sanna (2019) [34].

This fundamental idea of a coherent and organized whole is what emerges, for example, in *four moments* of prehistory and the history of the Trivium that we will now comment on.

The *first* is Plato's Philebus (18c), when Socrates invokes the number as the origin of the invention of Grammar, mythically attributed to the god Teuth [31]. The *second* is the commentary of the neo-Platonic Proclo (412–485) on the first book of Euclid's Elements [21]: "The importance and usefulness of mathematics for the other sciences and arts, we can learn it if we think how mathematics imposes perfection and order to theoretical sciences such as Rhetoric and to all those that are executed through discourse". The *third* is Book II of the *De Ordine* of St. Augustine (354–430), where the origin of all knowledge (and, therefore, also that which is consecrated to the study of "the meaning of words", that is, the three disciplines of the Trivium) is attributed to the activity of "reason" (which today we could translate as intelligence), which finds all its "strength" and "power" in "numbers" [1,21]. From this epistemological perspective, reinforced by the essential connection of programming languages with logic, a possible solution for the integration of digital literacy could be developed through the inclusion of transversal contents related to this epistemological unity, specially contents of Logic. The *fourth* takes us back to the 14th–15th centuries, in relation to the integration of humanistic and scientific knowledge carried out by Renaissance humanism and, in particular, by Salamanca humanists as Nebrija or the Brocense [15]. The Humanities do not represent certain human knowledge as opposed to others that are inhuman or dehumanized due to their scientific or technical nature. Thus, Ptolemy's Geography and Euclid's Elements of Geometry were as humanistic as Virgil's Aeneid or Plato's Banquet.

The Need to Include Communication and Critical Thinking Skills. As Segredo emphasizes [35], the majority of the approaches on 21st century skills stands out the importance of writing, critical and inventive thinking, communication, problem solving and teamwork skills. This highlights the importance of classics tools that includes Rhetoric (invention and arrangement, understanding emotions, argumentation, style an ornament, memory and delivery), what it can be found in all the literature, from classical antiquity to the present day [27].

Technology is Essential but it is Not Enough. The same idea emerges strongly in two of the most famous myths found in Plato's works, the myth of Prometheus and the myth of Theuth. "The theft of Prometheus is not enough to guarantee full human life. It only serves for human nutrition, so that man becomes a craftsman, a builder or a farmer, but not all professional arts together guarantee human coexistence" [17]. The insufficiency of specialized technical knowledge is also the theme of platonic version of Theuth's myth, whose moral is that not everyone who has been given something to discover, has also been given to understand the importance and convenience of their finding. In the same way that for Theuth the possibility of writing down science and history meant a revolutionary milestone of human consciousness and evolution, the present era hopes that the development of technology will make it possible, not just an advance without precedents of knowledge, but even a qualitative leap forward in the evolutionary history of the human species [14].

Previous Training of a Basic and Propaedeutic Nature, Including Literary and Humanistic Training, Is Necessary. This idea responds precisely to the idea of paideia as "general culture", basic and preparatory to any higher specialization, as well as necessary for the maximum development of the personality [5,21].

A Global and Comprehensive Design is Needed, Without Splits, Coherent, Defining a Progressive, Interdisciplinary and Complementary Skills Curriculum. The cycle of the seven liberal arts, the first part of which is the Trivium, was conceived as a whole of knowledge of a universal, coherent and internally connected interdisciplinary nature. A cycle of learning and training that is a direct continuation of the tradition of the *enkyklios paideía*, characteristic of Hellenism, which was joined by Cato (*Ad Marcum Filium*), Varrón (*Disciplinarum libri novem*) and, thereafter other authors as Macrobio, Boethius, Marciano Capela, Cassiodorus and Isidor of Seville [21].

5.2 Rationale of Rhetoric and Logic as Components of 21st Century Curriculums

Assuming the importance of digital skills like programming and computational thinking from the point of view of education in general (Sect. 4), is now a matter of exposing the reasons why the disciplines of the Trivium are decisive in promoting and optimizing the transformation of educational systems as proposed by Segredo *et al.* [35]: "that computational thinking may be used as a more general learning methodology, not uniquely devoted to those interested in a professional career in the field of Computing, but also *for every learner interested on training useful and promising skills*".

Our proposal to recover the scheme of the Trivium is based on the fact that the disciplines that compose it fulfil a double function: on the one hand, they reinforce the skills of computational thinking, algorithmic thinking and programming; and on the other hand, they develop the main skills of the 21st century that are not technological.

Given the fact that the study of Grammar and Literature has been preserved in current educational systems, our proposal is focused on Rhetoric and Logic, which today we would call formal and informal logic [10].

Rhetoric. As indicate above (Sect. 5.1), the five basic skills [27] that make up the instruments of classical Rhetoric cover almost all the competences of the 21st century by Segredo *et al.* [35]. We underline the following:

- *Invention*: reading, writing, thinking and problem solving, creativity, cognitive abilities or higher-order thinking skills innovation, information literacy, productivity and accountability.
- *Disposition*: master information, prioritizing and planning.
- *Elocution*: reading, writing, creativity, cross-cultural interaction
- *Memory*: cognitive abilities, productivity and accountability.
- *Action*: emotional skills, communication and collaboration, initiative and self-direction, leadership, responsibility, effective communication.

Next, we will explain in more detail the benefits of sequentially introducing Logic among educational contents in two specific versions: first, syllogistics logic [10], and second, the algebraic version of Fred Sommers' Aristotelian logic (TFL - Term Functor Logic) [18,19,29].

Syllogistics Logic. Its advantages from the perspective of current educational challenges [10,29], can be summarized as follows:

- Being a logic that uses natural language, facilitates learning, or otherwise reduces the cognitive load.
- Its basics operations opens access to the understanding of reality from abstract categories and, therefore, to the operations of formulation, organization, representation, abstraction and generalization typical of computational thinking.

- It is a logic that allows transition between natural language and the languages of mathematics (Set theory), electronic design (logic gates) and computer programming (TFL+).
- Its multidisciplinary nature makes it an irreplaceable methodological instrument for adaptability and interdisciplinary requirements.

Fred Sommers' Logic. Known as *Term Functor Logic* (TFL) [18,19,29], the system developed by Sommers and Englebretsen is a formal logical language easily assimilated into natural language. Among its advantages we can list:

- Based on the idea that natural language is the "genuine source of natural logic", TFL represents the categorical propositions using an arithmetic grammar (plus-minus algebra).
- Its algebraic representation t this plus-minus algebra offers a simple method of decision for syllogistic.
- Its visible "syntactic naturalness" and the simplicity of its reasoning rules, provide intuitively and immediately cognitively relevant information and make it a "logic of reasoning in natural language".
- Its direct usefulness from the point of view of logical programming languages [3]) and, in particular, through the programming language TFLPL+ [8,11].

6 Conclusions

The usual discourse about technological transformation of education suggests promoting computational thinking because it promotes 21st century skills that are fundamental and not necessarily technological. Our proposal, based on the concept of education by competences, consists of promoting the disciplines of the Trivium for four basic reasons:

- Because they directly promote 21st century skills (Sect. 2).
- Because they offer a sound philosophical framework to reflect on the various problems and conflicts that emerge from the transformations that education is undergoing today (Sect. 3 and Subsect. 5.1).
- Because they promote skills in computer thinking, algorithmic thinking and programming (Sect. 4 and Subsect. 5.2).
- Because they offer a solid base of interdisciplinarity from the epistemological unit of key humanistic disciplines (Trivium disciplines) and STEM disciplines (Sect. 1 and Sect. 5).

In short, what the Trivium offers is a compact disciplinary scheme that brings together the key skills of 21st century education and a solid epistemologically ground for an authentically interdisciplinary education that responds to the exposed requirements of smart education.

Acknowledgments. This research has been partially supported by the Department of Education of the JCyL and the project RTI2018-095390-B-C32 (MCIU/AEI/FEDER, UE).

References

1. Agustín. El Orden. En Obras de S. Agustín (vol. I). BAC, Madrid (1946). https://www.augustinus.it/spagnolo/ordine/index2.htm
2. Bacigalupo, M., Kampylis, P., Punie, Y., Van den Brande, G.: EntreComp: The Entrepreneurship Competence Framework. Publication Office of the European Union, Luxembourg, EUR 27939 EN (2016). https://doi.org/10.2791/593884
3. Bergin, T.J., Gibson, R.G. (eds.): History of Programming Languages II. ACM Press, New York (1996)
4. Bocconi, S., Chioccariello, A., Dettori, G., Ferrari, A., Engelhardt, K.: Developing computational thinking in compulsory education - Implications for policy and practice JRC Science for Policy Report. EUR 28295 EN (2016). https://doi.org/10.2791/792158. Edited by Kampylis, P., Punie, Y
5. Bugliarello, G.: A new trivium and quadrivium. Bull. Sci. Technol. Soc. **23**(2), 106–113 (2003)
6. Carabias Torres, A.M.: Evolución del concepto de Facultad de Artes en España: (siglos XIII al XVII). En A. Heredia Soriano (Coord.) Actas del IV Seminario de Historia de la Filosofía Española. Ediciones Universidad de Salamanca, Salamanca (1986)
7. Carretero, S.; Vuorikari, R., Punie, Y.: DigComp 2.1: The Digital Competence Framework for Citizens with Eight Proficiency Levels and Examples of Use, EUR 28558 EN (2017). https://doi.org/10.2760/38842
8. Castro-Manzano, J.M., Lozano-Cobos, L.I.: TFLPL: Programación con lógica de términos. Res. Comput. Sci. **147**, 265–283 (2018)
9. Coccoli, M., Guercio, A., Maresca, P., Stanganelli, L.: Smarter universities: a vision for the fast changing digital era. J. Vis. Lang. Comput. **25**(6), 1003–1011 (2014). https://doi.org/10.1016/j.jvlc.2014.09.007
10. Copi, I.: Introducción a la Lógica. Limusa, Mexico (2013)
11. Castro-Manzano, J.M., Lozano-Cobos, L.I., Reyes-Cardenas, P.O.: Programming with term logic. BRAIN Broad Res. Artif. Intell. Neurosci. **9**(3), 22–36 (2018)
12. Cefai, C., Bartolo P.A., Cavioni, V., Downes, P.: Strengthening social and emotional education as a core curricular area across the EU: a review of the international evidence (NESET II report). European Union, Luxembourg (2018)
13. Clever, S., et al.: Ethical analyses of smart city applications. Urban Sci. **2**, 96 (2018)
14. Diéguez, A.: Transhumanismo. Herder, Barcelona (2017)
15. Flórez, C., et al.: El humanismo científico. Caja de Ahorros y M. P. de Salamanca, Salamanca (1998)
16. Etzkowitz, H., Ranga, M., Dzisah, J.: The Novum Trivium and the transition from industrial to knowledge society. Soc. Sci. Inf. **51**(2), 143–164 (2012)
17. García Castillo, P.: Prometeo o la educación insuficiente. Campo abierto, n° 5 (1988)
18. Sommers, F.: The Logic of Natural Language. Clarendon Library of Logic and Philosophy. Clarendon Press/Oxford University Press, New York/Oxford (1982)
19. Sommers, F., Englebretsen, G.: An Invitation to Formal Reasoning: The Logic of Terms. Ashgate, Farnham (2000)
20. Gallar, M.: Trivium: una necesidad urgente y una responsabilidad compartida entre docentes y discentes. En XIV Jornadas de Redes de investigación en docencia universitaria. Investigación, innovación y enseñanza universitaria: enfoques pluridisciplinares, pp. 1581–1595. Universidad de Alicante. Instituto de Ciencias de la Educación, Alicante (2016). http://hdl.handle.net/10045/59691

21. Hadot, I.: Arst libéraux et philosophie dans la pensée antique. Études augustiniennes - CNRS, Paris (1984)
22. Hazelkorn, E., Ryan, Ch., et al.: Science Education for Responsible Citizenship. European Commision, Brussels (2015)
23. IASLE. (n.d.) Smart Learning. http://www.iasle.net/index.php/about-us/background. Accessed 10 Feb 2020
24. Jaeguer, W.: Paidea. Fondo de cultura económica, Madrid (1983)
25. Jonassen, D.H.: Learning to Solve Problems: A Handbook for Designing Problem-Solving Learning Environments. Taylor & Francis, New York (2010)
26. Kirschner, P., Wopereis, I.G.J.H.: Mindtools for teacher communities: a European perspective. Technol. Pedagogy Educ. **12**(1), 105–124 (2003). https://doi.org/10.1080/14759390300200148
27. Nance, J.B.: Fitting Words. Roman Roads Media, Moscow (2016)
28. NMC Horizon Report para la Educación Superior (2016). http://research.unir.net/wp-content/uploads/2016/05/2016-nmc-horizon-report-HE-ES.pdf
29. Oderberg, D.S.: The Old New Logic: Essays on the Philosophy of Fred Sommers. MIT Press, Cambridge (2005)
30. OECD Publishing: Economic Policy Reforms: Going for Growth. OECD (2009)
31. Platón: Filebo, en Platón III. Gredos, Madrid (2011)
32. Rajna, P.: Le denominazioni Trivium e Quadrivium, Studi Medievali, N.S.I., pp. 4–36 (1928)
33. Redecker, C.: European Framework for the Digital Competence of Educators: DigCompEdu. Punie, Y. (ed). EUR 28775 EN. Publications Office of the European Union, Luxembourg (2017). https://doi.org/10.2760/159770. ISBN 978-92-79-73494-6. JRC107466
34. Sanna, M.Ch.: La preistoria del quadriviumprima e dopo Platone. Universití degli Studi di Cagliari, Cagliari (2019)
35. Segredo, E., Miranda, G., León, C.: Towards the education of the future: computational thinking as a generative learning mechanism. In: EKS 2017, vol. 18, no. 2 (2017)
36. Téllez, D.: Gramática, lógica y retórica: un estudio histórico-filosófico del Trivium medieval, México: Ed. Johann Georg Hamann (2014)
37. Trilling, B., Fadel, C.: 21st Century Skills: Learning for Life in Our Times. Wiley, Hoboken (2009)
38. Wing, J.M.: Computational thinking. Commun. ACM **49**(3), 33–35 (2006). https://doi.org/10.1145/1118178.1118215
39. Zhu, Z.-T., He, B.: Smart education: new frontier of educational informatization. E-Educ. Res. **12**, 1–13 (2012)
40. Zhu, Z.-T., Yu, M.-H., Riezebos, P.: A research framework of smart education. Smart Learn. Environ. **3**(1). Article 4 (2016). https://doi.org/10.1186/s40561-016-0026-2

Analysis and Classification of Inappropriate Strategies Used by Students to Find the Winning Strategy in Catch the Frog and Daisy Games

Esther Lorenzo-Fernández[1(✉)] and Jordi Deulofeu[2]

[1] Department of Statistics and Operational Research and Mathematics Education, University of Oviedo, Oviedo, Spain
lorenzomaria@uniovi.es

[2] Department of Didactics of Mathematics and Experimental Sciences, Autonomous University of Barcelona, Barcelona, Spain

Abstract. This research is framed in the area of didactic mathematics, particularly within the so-called *small strategy games*. These games are intrinsically related to problem solving, because the different winning strategies used to solve a game are often similar to those applied in problem solving situations.

This paper analyses the responses provided by students while playing the games Frog Catcher and Daisy in pairs. This is an empirical study with 422 students in years 1 and 3 of Spanish Secondary Education (ESO) [years 8 and 10 of secondary education (IGCSE)]. The objective is to carry out a qualitative analysis of the different types of inappropriate strategies used by the students in their effort to find the winning heuristics in both games. In addition, a classification of these inappropriate strategies is established.

Keywords: Strategy games · Problems solving · Video games

1 Introduction and State of the Art

In the framework of a wider research [1], whose starting point is the investigation carried out by Corbalán [2], the study with *small strategy games* has been carried out as shown throughout this paper.

As a background to our study, we would like to mention the fact that there is evidence of games and recreational activities in the ancient Babylonian civilization, ancient Egypt, India, the Mayan and Etruscan cultures, as well as, of course, in ancient Greece or Rome. In fact, games appear, throughout most of history, integrated within the mathematical discipline. At the beginning of the 17th century, with the publication of the work *Problemes plaisants qui se font par les nombres* of Mezirac, recreational mathematics began to distinguish itself from the concept of mathematics in the current society.

In relation to games or recreational activities, we find numerous relevant figures in the world of mathematics such as the Italian Leonardo de Pisa (1170–1250), better known as Fibonacci, whose famous succession arises from a problem about the

Á. Herrero et al. (Eds.): ICEUTE 2020, AISC 1266, pp. 21–29, 2021.
https://doi.org/10.1007/978-3-030-57799-5_3

breeding of rabbits. For his part, Pierre de Fermat (1601–1665), in trying to solve the problem of points, related to games of chance, lays the foundations of the Theory of Probability in his correspondence with Pascal. The same applies to the Swiss Leonhard Euler (1707–1783), father of the Graph Theory, the branch of mathematics that was born when he tried to solve the problem of the Königsberg bridges, and to the mathematician Johan Von Neumann (1903–1957), of Hungarian origin, who marks the beginning of modern Game Theory.

With regard to board games, Beasley [3] establishes four groups: pure games of chance, mixed games of skill and luck, pure games of skill, and automatic games.

In relation to mathematical games, which we will henceforth simply refer to as games, the Oldfield classification [4] consists of twelve types. Some of them can belong to several categories at the same time.

In his book *Una recreación matemática: historias, juegos y problemas* [5], Deulofeu includes among the strategy games, in which there is no intervention of chance, the *small strategy games*.

In the context of the strategy games we have had access to several investigations, such as those developed by Corbalán, Edo, Mallart and Navarro [2,6–8], among others. These studies allow us to establish the existence of problem-solving methods that can be practiced through this type of game, since it is possible to establish a parallelism between the resolution phases of both.

Redeker et al. [9] note that formal education does not harness the potential of ICTs with the benefits and learning opportunities they could bring to the educational field, while the use of social media and the Internet continues to grow. Furthermore, they consider that the social media would be very useful in relation to the four challenges facing education and training policies in Europe in the period leading up to 2020:

1. Improve innovation and creativity.
2. Improving the quality and effectiveness of the tools and the results of the learning process.
3. Making lifelong learning and student mobility a reality.
4. Promote equity and active citizenship.

Before starting with the methodological framework, we would like to make it clear that the main objective of this article is to conduct a qualitative analysis of the inappropriate strategies that students use when they are trying to find the winning heuristics in these two games. In addition, a classification of the inappropriate strategies observed is established.

2 Methodological Framework

Firstly, we will make a general description of the two *small strategy games* used as instruments, used in [10]. Next to each game, the winning strategy foreseen in [11] to solve them is described:

Catch the Frog. This game is a simplified version of Nim, a classic Chinese game for two players. In this case, it is a game of ten pieces (or frogs), in which each of the two players must remove one or two pieces in turn, according to their own criteria. The player who removes the last file wins the game.

To find a way to always win in this game, the right strategy is to "start at the end", the key idea (or attack) to solve it is to realize that players can remove the files in threes. The starting player, as long as he leaves his opponent a number of checkers that is a multiple of three (nine, six and three pieces, respectively). In other words, the starting player wins the game, as long as he removes one piece, and in successive moves he removes a different number of his opponent's pieces, removing three pieces from between them each time. Consequently, it is a matter of making numerical considerations and performing the analysis from the end, forcing the opponent into a "fatal" situation, as the so called Corbalán [2], which occurs when he is left with three or a multiple of three pieces.

Daisy. This game is, like the game Catch the Frog, a *small strategy game*. In these games, a small variation of the rules can lead to the application of very different strategies. In this case, adding to a Nim type game the condition that, in the case of removing two pieces, these must be consecutive, implies that to win the game we must take into account not only the number of pieces (or petals), but also their position.

In this game, the strategy originally envisaged is the "use of symmetry", so that if we leave a symmetrical situation on the contrary, we win by replicating its movements in the opposite position.

In this case, it is the second player who has the winning strategy, as long as he moves again in the initial move a number of pieces opposite to those of his opponent (one if the first player chooses two and two if he chooses one). The checkers removed by the second player must be placed on the opposite side of the daisy, thus leaving two isolated and symmetrical groups of three petals or checkers each. In order to win, the second player only has to choose the symmetrical pieces in the following moves (Fig. 1).

Fig. 1. Winning situation for the second player

In addition to this strategy, Corbalán [2] describes another less obvious and less elegant one that can be obtained from a "systematic study of all cases".

On the other hand, the presentation of the rules of the games to the students has been carried out without oral explanations. The tests in technological format detected that many students, passed to the next screen without reading the rules, so it was decided to write the following instructions on the blackboards of the classrooms before starting the sessions:

- "The starting order in the first game is by draw and in the others, by turn".
- "The two players take turns making their moves".

In addition, in the Daisy game, the instruction was added:

- "You can play more comfortably by putting a piece in each petal at the beginning and removing a piece or two pieces that are together in each move".
- In the game Catch the Frog, the following instruction was added to the two initials:
- "Green chips are called frogs."

Two sessions were held with all the students. In each, students played one of the games, practicing the time needed to answer the questionnaires in pairs. The duration of each session was approximately one hour.

2.1 Instruments

The questionnaire to obtain the information has been designed from the one used by Corbalán [2]. It is designed to be answered by pairs of students and its format has evolved through different tests carried out during the months prior to the collection of data from the study. These tests were carried out with students from different ESO courses, belonging to different educational centers in Asturias (Spain). Each pair of students should describe what the winning strategy is in the two games.

Next, we will show the instruments of the two formats in which we have carried out the study: the non-technological format (board games) and the technological format (video games) (Fig. 2):

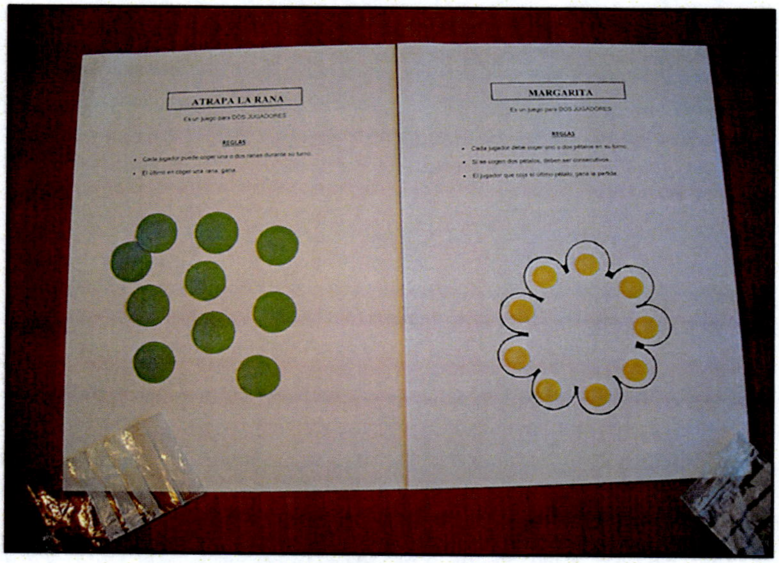

Fig. 2. Instruments in non-technological format

The games Catch the Frog and Daisy in technological format have been implemented in tablets so that the students participating in the study had access to the games in this format. Following the appropriate guidelines, the company WILDBIT STUDIOS S. L. (Madrid) has been in charge of their implementation in the android platform with HTML5 technology, as well as the graphic design aspects and other technical characteristics. The two resulting technology format games (or video games) are presented to students in a simple, intuitive and user-friendly format (Figs. 3 and 4):

Fig. 3. Screenshot of the Catch the Frog game

Fig. 4. Screenshot of the Daisy game, already started by one of the students

2.2 Population

The present research has been carried out with students of first and third year of ESO [years 8 and 10 of secondary education (IGCSE)], with ages of 12 and 14 years respectively, in most cases. The aim of this study was to collect data on the mode and mechanisms used by these students in their search for winning strategies in the games Catch the Frog and Daisy.

In total, a sample of 422 students from nine schools in Oviedo in a similar socio-cultural situation has been chosen. Of these, 224 were in their first year of secondary education and the remaining 198 in their third year. The data collection was carried out in two different sessions, one for each game, and it was carried out in two phases: the first one with 54 pairs of first-year students and 48 third-year students, who played both games in technological format; in the second phase 58 pairs of first-year students and 51 third-year students played both games in a non-technological format. In all, data were obtained with students from 19 different classes or groups, 8 classes for the games in technological format and 11 for the games in non-technological format. The first phase of data collection took place during the months of April, May and December 2016, and January and February 2017, while the second phase took place in the months of March and April 2017.

3 Analysis and Results

In those questionnaires where the pairs of students have not found the right winning strategy, we have found different types of answers:

- blank answer or explain that no strategy has been found;
- description of the objective or the rules of the game, explaining how to play, not the strategy followed;
- erroneous strategies, which do not lead to the resolution of the games. We have called this type of strategies as **inappropriate strategies** and have established a classification of them according to the different answers obtained in our research. Although some sections may overlap, they only do so partially and there are always nuances that differentiate them.

As we will see throughout this section, there are some differences between the answers obtained in Catch the Frog and Daisy, perhaps because they are games with different types of resolution strategies (arithmetic in the case of Catch the Frog and geometric in Daisy), so a specific classification has been developed for each game:

3.1 Catch the Frog

Next, we will describe the inappropriate strategies of the game Catch the Frog, with some examples of the responses of some couples:

1. **Parity.** An incorrect strategy is extracted based on the parity of the number of tiles. An example would be the following answer, obtained from the questionnaire of a

pair of students from first year of ESO: *You have to try to leave an odd number of frogs.*

2. **Conditioning the game of the opponent.** Setting how the opponent should play, thus conditioning his game. This occurs in answers like the following: *If you each touch two frogs, the first one wins. If you each touch one frog, the first one wins. If you each touch two frogs and when there are four frogs left, the second one touches only one, the second one wins.*
3. **Inappropriate Partial.** You set up the strategy from a situation that the other player imposes, that is, you depend on the opponent leaving a certain amount of pieces on the board. Let's see an example found among the answers of the students of third year of ESO: When there are four frogs left, touch only one, so that the opposite is true, you will win.
4. **Particular case that does not work.** It describes the strategy followed in a particular situation, in a wrong way. Here is the response of a first-year couple: *Whoever starts first and takes two all the time wins.*
5. **Others.** This section includes various strategies that are not very common. For example, in the following case there is a part of strategy, which is to choose which player you want to be: *Being the second player.*

3.2 Daisy

With respect to the game Daisy, because heuristics is, fundamentally, the "use of symmetry", of a geometric nature, we have contemplated some additional cases to those previously mentioned for the game Catch the Frog:

1. **Parity.** An erroneous strategy is formulated taking into account the parity of the number of counters. As in the game Catch the Frog, we find examples of this type in the answers of the pairs: *Let him always try to start first and to leave a number of petals, always odd.*
2. **Conditioning the game of the opponent.** Describes how the opponent should play, thus conditioning his game. This is the case of this answer, which belongs to a questionnaire of first year students of ESO: *We believe that if your partner starts, he plays two, then you follow, you give him one, the other two, you one, the other two and you one, therefore you win.*
3. **Strategy for the other player.** Describes the strategy for the opponent to win. As an example, we can mention the following situations:

   - Leave the opponent with 4 petals: 2 separate and 2 together, type 2-1-1;
   - Leaving the opponent with three consecutive petals;
   - Leaving the opponent with 3 consecutive petals, separated from 2 others, type 3-2.
   - The following response illustrates this type of inappropriate strategy: It is indifferent the player who starts the game, but what you have to do is try to leave the opponent with three petals that are consecutive at the end.

4. **Refer to quantity and not position.** Refers to the number of petals, regardless of the position they occupy. These are answers like the following one, taken from a

card of the first-year students: *Try to always have three petals of the flower left so you can win.*

5. **Inappropriate Partial.** The strategy is established from a situation that the other player imposes, that is to say, it depends on the opponent leaving a certain number or disposition of the pieces on the board. We have found answers like this:
   - "Let them leave three separate petals for me."
   - "Let them leave me three petals together and I'll take the one in the middle."
   - "Let them leave four petals together and I'll take the two in the middle."
   - "Let them leave two together and one apart, and I'll take one in a row."
   - "Let them leave me 4 petals: 2 separate and 2 together, type 2-1-1."

   We have chosen an example that refers to the latter case: *Eliminate all the petals until there is a pair and two others loose, in total four petals and you have to eliminate the pair, then the other will eliminate one and you eliminating the last one, you will win.*
6. **Particular case that does not work.** The strategy followed in a particular situation is described in a wrong way. As an example, the following answer is present in one of the questionnaires by pairs: *Do the opposite of what your partner does.*
7. **Others.** This section has been reserved for all those rare strategies that we have found throughout the research. The following is an unusual strategy different from the previous cases: *Try to keep the petals apart.*

4 Discussion and Conclusions

With regard to the "start at the end" strategy, students use it to win in the game Catch the Frog, as planned. They seem to have assumed its operation and application in this game with a winning heuristic of an arithmetical nature. Furthermore, the use of this strategy improves with age, as it is used to a greater extent by students in the third year of ESO.

However, they also make use of this same strategy when faced with games of a geometric nature, as they apply it in the game Daisy, which is not profitable, as with it they will only achieve partial strategies and not a complete winning strategy.

With regard to the strategy "use of symmetry", its application in the resolution of the game Daisy seems to indicate that geometric reasoning to solve the game is used with a much lower frequency than arithmetic reasoning. It may be that using this strategy, which requires consideration of the entire problem, will present greater difficulties for students.

As a conclusion, we can say that the number of couples who find the winning strategy in the games is much better than the number of couples who use wrong strategies. These data vary according to the game and the course, although the objective of this article is not to establish a quantitative analysis, but rather a qualitative one of the types of inappropriate strategies used by the students.

On the other hand, when conducting the qualitative analysis of inappropriate strategies, no notable differences have been observed between students who have used the technological format of the games and those who have played in a non-technological format.

Finally, as regards these inappropriate strategies used by couples who have not found the complete winning strategy, we have carried out a qualitative analysis of them and established a classification of the responses obtained in our research, as was our objective.

References

1. Lorenzo-Fernández, E.: Juegos de estrategia en formato tecnológico y resolución de problemas en la ESO. Dissertation, Universitat Autònoma de Barcelona (2018)
2. Corbalán, F.: Juegos de estrategia y resolución de problemas: análisis de estrategias y tipología de jugadores en el alumnado de secundaria. Dissertation, Universitat Autònoma de Barcelona (1997)
3. Beasley, J.D.: The Mathematics of Games. Oxford University Press, Oxford (1989)
4. Oldfield, B.J.: Games in the learning of mathematics. Math. Sch. **20**(1), 41–43 (1991)
5. Deulofeu, J.: Una recreación matemática: historias, juegos y problemas. Planeta, Barcelona (2001)
6. Edo, M.: Jocs, interacció I construcción de coneixements matemàtics. Dissertation, Universitat Autònoma de Barcelona (2002)
7. Mallart, A.: Estratègies de millora per a la resolució de problemes amb alumnes de segon d'ESO: ús de la matemática recreativa a les fases d'abordatge i de revisió. Dissertation, Universitat Autònoma de Barcelona (2008)
8. Navarro, A.: La influència de l'ús de jocs d'estratègia en l'aprenentatge de la resolució de problemes de matemàtiques a l'educació secundària. Dissertation, Universitat Autònoma de Barcelona (2013)
9. Redeker, C., Leis, M., Leendertse, M., Punie, Y., Gijsbers, G., Kirschner, P.A., Hoogveld, B.: The future of learning: preparing for change. Institute for prospective Technological Studies, Sevilla (2012)
10. Lorenzo-Fernández, E., Deulofeu, J., González, S.: Videojuegos de estrategia en la ESO. UNO **74**, 14–20 (2016)
11. Corbalán, F., Deulofeu, J.: Juegos manipulativos en la enseñanza de las matemáticas. UNO **3** (7), 71–80 (1996)

Implementation of an Integrated STEM Activity in Pre-primary Schools

Eva M. García Terceño[1(✉)], Ileana M. Greca[1], Andreas Redfors[2], and Marie Fridberg[2]

[1] Universidad de Burgos, 09001 Burgos, Spain
emgterceno@ubu.es
[2] Kristianstad University, Kristianstad, Sweden

Abstract. BotSTEM is an ERASMUS+ project. Its outputs are aimed to provide in- and pre-service teachers in Childhood and Primary Education with a didactical framework, research-based materials and best practices using integrated Science, Technology, Engineering, Mathematics (STEM) and robot-based approaches for enhancing scientific literacy in young children. Initial results from the implementation of activities following the proposed model in preschools in Spain are presented here. Despite the possible obstacles that preschool teachers initially expressed, the preliminary analysis indicates that the proposed STEM integrated framework, including inquiry teaching and engineering design methodologies, can be used with children as young as 4 y.o. In the case of a project about magnets, the children seem to have learnt some scientific ideas, applied these ideas to design a magnetic toy and learnt about spatial orientation using robots.

Keywords: Early childhood education · Integrated STEM education · Inquiry-based teaching · Variation theory · Design-based research

1 Introduction

Robotics and STEM education for children and primary schools is an ERASMUS+ project with partners in Spain (coordinators), Sweden, Italy and Cyprus, that aims to develop a new didactical framework for integrated STEM education, understood as a combination of STEM disciplines that allows a holistic teaching-learning process based on solving real problems. This approach includes robotics and coding, into education curricula for childhood and primary schools (4–8 y.o.). The project outputs are specifically aimed to provide in- and pre-service teachers in Childhood and Primary Education with a didactical framework and materials, based on the framework, for enhancing scientific literacy in young children.

STEM in early childhood education should be preferably holistic, child centred and project and problem based. As have been shown, meaningful hands-on STEM experiences for early childhood and elementary school-age children positively affect their perceptions and dispositions towards STEM [1–3], when these activities are integrated [4, 5]. To reach this integration, two methodologies, inquiry teaching and engineering design, seem to be appropriate to intertwine the different fields in STEM through real

Á. Herrero et al. (Eds.): ICEUTE 2020, AISC 1266, pp. 30–39, 2021.
https://doi.org/10.1007/978-3-030-57799-5_4

world problems. Working with inquiry and engineering based STEM activities provide children with opportunities to practice skills such as reasoning, reflection, questioning, modelling, justifying decisions and communicating. These problems should revolve around certain key ideas indispensable for children to understand, enjoy and marvel at the natural world [6]. Although research suggests that children can follow both methodologies, some adjustments must be made for very young children, especially with the first part of both cycles. For example, inquiry is about questions, but it′s difficult for children to ask questions about something they have not seen, touched or experienced. Also, young children need time to explore, create, and innovate [7]. Therefore, for young children it is very important first to engage, notice, wonder and question [8]. That is, to give time to play in a rich science environment. As many of the emergent questions may not be possible to investigate, children need teachers to focus observation and clarify questions. After that, children can, with teacher support, follow the other stages. So, inspired in Chalufour and Worth's inquiry cycle [8], the activities developed at botSTEM project are designed to follow the cycle that appears in Fig. 1.

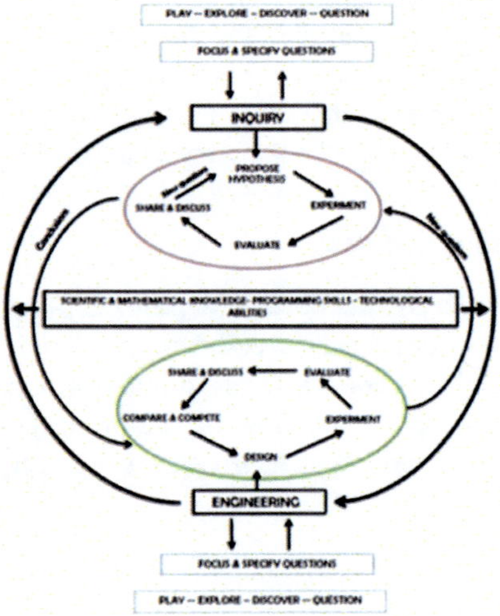

Fig. 1. Phases of inquiry teaching and engineering design for young children. Inspired in Chalufour and Worth (p. 74) diagram [8].

Related to computational thinking, included because of its potential to teach logical thinking, problem solving and digital competence, we consider that it should be introduced at early childhood through the use of scaffolding devices. In the activities developed, 4 year-old children begin to work with simple robots that are programmed with buttons on the back. For older children the complexity increases because the programming of robots with tablets is added, the use of Scratch and finally, the programming of measurement instruments with a physical device like the BBC-microbit.

Integrated STEM approaches imply several challenges for teachers. The long history of studies on the STEM approach has not allowed teachers to incorporate approaches to teaching-learning processes that integrate the four disciplines significantly [9]. Making cross-cutting STEM connections is not an easy task, requiring that teachers prepare lessons that allow students to understand how STEM knowledge is applied to real-world problems. To make these connections, teachers need to address both content and pedagogy, but as Dare et al. [10] acknowledge, identifying exactly what content and pedagogy is complex. Nevertheless, Fleer et al. [11] show that with a 'sciencing attitude', teachers have unique possibilities to teach science in preschool.

As Spanish preschool teachers are not trained in any of these aspects (integrated STEM approaches; inquiry and engineering design methodologies and robotics) we follow a design-based implementation [12] of the activities, guided by variation theory [13] in order to implement this proposal. One of the basic ideas of variation theory is that learning is always directed to something (phenomenon, object, skill, aspect of reality). This something is called the object of learning and 'learning' entails a qualitative change in the way of experiencing the object of learning – ways of acting originate from ways of experiencing [14]. Experiencing an object of learning requires that the learner becomes aware of its different aspects, and is provided the opportunity to discern these aspects simultaneously. Other relevant aspect of this theory is the dynamic nature of the object of learning. The intended object of learning planned by the teacher may not be the same as the enacted object of learning that the teacher implements in complex classroom situations, and what the students actually experience (the lived object of learning) may not be the same as the enacted object of learning [15].

1.1 Aim and Research Questions

The overarching aim of the 3-year botSTEM project is to, through a design-based research approach, develop and analyse collaborative inquiry teaching and learning STEM activities scaffolded by robotics in pre-primary school. The research question guiding the analysis presented here is: How do pre-primary school teachers and children deal with the proposed didactical approach and the objects of learning in their classrooms?

2 Method

With the objective of reducing the gap between the theoretical studies produced and the educational practice the botSTEM project is committed to use a design-based research framework that combines "the joint reflection of researchers and professionals in educational practices" (p. 217) [16]. This paper presents an extract of this process focused on the implementation of a first prototype of a teaching sequence on magnetism through a multiple-case study because this "design provides rich descriptions and interpretations of teachers' experiences, and by examining multiple cases, this information provides a broader description of their experiences relating to STEM integration" (p. 4) [17].

2.1 Participants

At the beginning of the project, the Spanish botSTEM partner established collaboration with four different schools in Spain. Three of them are in Burgos and one in Algeciras. In all, 15 pre-primary teachers and two researchers are involved in five different work groups. During the 2018–2019 academic year, eight teachers with around 20 children each, decided to implement an activity related to magnets, while the others preferred to work with other topics, such as changes of matter, parachutes or inclined planes.

2.2 Instruments

In order to achieve the aim of the study, two working groups were created with the teachers who decided to implement the STEM magnet activity, and they were interviewed before and after the implementation of the activity. The former interview aimed to make teachers reflect about two main questions related to the planning of the activity: (1) what knowledge (about STEM and programming) do you want the children to develop? (2) How can you help them to develop that knowledge? The latter interview had the purpose of comparing what they planned (is it the intended teaching) before the activity and what and how they finally did it: (1) did you follow the steps initially planned? (2) What did you finally do? (3) At the end of the project, what have children learnt? During this second meeting teachers also shared what they think about science and its role in Early Childhood Education. In addition, the teachers video-recorded the development of some activities. In these videos, children's attitudes and changes of reasoning during the implementation could be observed.

2.3 Procedure

The complete investigation process, from which this case study is extracted, is planned following the stages proposed by Plomp [18] for the design-based research: a preliminary research, prototyping phase and assessment phase. In the preliminary research, the botSTEM project developed the theoretical framework discussed above that guided the development of different activities. After analysing the possibilities offered, 8 teachers decided to implement a specific activity related to magnets. Originally, this activity focuses on encouraging children to design and build a magnetic toy. For achieving this objective, students develop scientific knowledge about magnets following a guided inquiry, and design a prototype, thinking critically and creatively as engineers do. During this process, children are supposed to apply mathematical knowledge – related to series, classifications, counting – and programming skills to program a simple robot, used as a consolidation tool of the new knowledge. The proposals included in the botSTEM toolkit offer general ideas, following the didactic model, so teachers, assisted by researchers, fine-tune the activity selected to their educational context. In order to adapt and adjust the activity, teachers were encouraged to reflect about what they wanted the children to learn (Intended object of learning) and how they would manage to achieve that goal. After the implementation of the activities, the working groups met again with the purpose of analyzing the teaching-learning process followed during the implementation (Enacted object of learning) and reflect about the initial goals and procedures (Intended object of learning) and what and how children finally learn (Lived object of learning).

3 Data Analysis

The analysis of the collected data has been carried out using the NVIVO software, after the transcription of the interviews and videos. The information gathered was categorized, also taking into account the teachers' attitudes during the interviews and the observations extracted from the videos recorded during the implementation of the activities, in order to have a comprehensive view of the process. The categorization was carried out independently by two of the researchers, without finding significant differences between the two contributions. After a joint reflection, agreement was reached on a final set of categories.

4 Results and Discussion

Intended Object of Learning
During the meetings with both working groups, the intended object of learning was identified. Magnetism, as a scientific concept, was acknowledged as the direct object of learning. But the other STEM disciplines, mathematics, technology or engineering and programming were not considered by the teachers, as well as neither the phases of inquiry teaching and the engineering design (Fig. 1), the methodologies to be used to achieve the object of learning, were identified as indirect aspects to be learnt. Occasionally, teachers are not familiar with what the object of learning entails. In this case, the development of scientific and engineering skills that would be used during the teaching-learning process as well as the mathematical knowledge, the programming skills and the technological abilities were obviated. Therefore, from the teachers' perspective, the intended object of learning focused specifically on developing scientific concepts about magnets (magnetic materials, force and polarity), about their history, their applications, even about their influence on some animals. However, none of these aspects were connected with the development of scientific and engineering competences or with the nature of science, which could be related to a misperception of teachers about the inability of children to understand how science works [19]. Related to programming, the teachers were more hesitant, and they only talked about the use of robots to consolidate the knowledge generated during the activity, as the toolkit proposes, never considering it as an intended object of learning.
Enacted Object of Learning
Once the teachers had planned the intended object of learning, they implemented the activity in their classrooms, that is, the enacted object of learning. Some of the teachers decided to carry out the sequence with the whole group, about 20 children at a time, while others chose to separate them into two groups, *We always work with half of the class, with 25 children it would have been impossible* (Teacher 5). These teachers sometimes brought the whole group together only to draw conclusions and reflect about

what they did and learnt. In the end, although with different intensity and investment of time, most of the teachers put into practice the following steps:

- Playground and experimental areas where children could discover different properties of magnets with the guidance of the teacher
- Definition of hypothesis for simple experiments
- Experimentation
- Documentation of data
- Drawing conclusions
- Creation of a magnetic toy

Although teachers did not consider the understanding of the methodologies used as an object of learning, they followed the steps of an inquiry method during the process, but not in an explicit or reflective way. Moreover, the engineering design methodology was not used, although all ended the project with the development of a magnetic toy.

In the final interview, the teachers commented that during the first phase, it was necessary to support the children in the playground areas because they did not focus their attention on the critical aspects of the intended object of learning. They were aware that magnets stuck on the lid (made of a magnetic metal), but they seemed not be able to think about anything else. *If you really want kids to observe that there is an attraction of the iron shavings through the water (and the glass), they need some guidance so as not to get stuck* (Teacher 2). Moreover, the teachers introduced new experimental moments in order to address the new hypothesis children proposed during the activities and therefore reinforced the development of the enacted object of learning, *They shared very curious hypothesis, they said that cold things were attracted by magnets, so I decided to bring an ice cube* (Teacher 5). These playground areas enabled teachers to scaffold learning through the experience of the children, in accordance with Rahm [20].

When teachers were asked about how they introduced robots into the teaching-learning process, most of them felt a lot more enthusiastic, in comparison with the first interview. They realized that, in addition to strengthening knowledge about magnets, children had worked on: spatial orientation, sequence of movements, reducing impulsivity, counting or working memory. It is worth stressing that although mathematical ideas were not included in teachers' intended object of learning, during the enacted object of learning, some mathematical ideas were addressed. They also asserted that although programming a robot implies a complex task for young children, especially when they had to deal with "turns", they did not get bored because they experienced these activities as "play situations".

Lived Object of Learning

As we have seen, the enacted object of learning sometimes can differ from what the children actually learn unless the teachers know how to draw their attention to what they want them to learn. With the intention of analyzing what the children achieved in terms of knowledge and skills, it is important to detect what they initially knew about magnetism and programming. This prior knowledge was identified by teachers at the beginning of the implementation, when the children played and explored the magnets through three different questions: What kind of things are attracted by magnets? Can magnets attract each other? Do magnets have different powers? With this knowledge, the teachers were able to create specific situations to scaffold them through the experimentation, changing their intended object of learning.

Initially, the children associated the temperature and the weight of objects with their magnetic properties, in such a way that cold and heavy things were attracted by magnets. It is interesting to note that, from a sensory point of view, these characteristics can be associated with many metallic objects with which the children interact every day. After the experimentation, the children modified their thoughts in two ways: they understood that "the cold and heavy things" are made of metals and that not all metal objects are attracted by magnets, in spite of the fact that they were not able to distinguish between magnetic and non-magnetic metals. Table 1 shows the new understandings achieved, in opposition to their previous ideas. Finally, they applied these ideas to design a magnetic toy. However, although the teachers generated situations in which children were able to play and discover properties of magnets and redirected what they had planned in order to address the new hypothesis, which is actually STEM in early childhood [21], the teachers were concerned about the lack of ability of the students to generalize what they have learnt.

Table 1. Children's ideas about magnetism before and after the implementation

Children's ideas previous the STEM activity	Children's ideas after the STEM activity
What kind of things are attracted by magnets?	
Objects that weigh little are not attracted by magnets *Cold things are attracted by magnets*	Not all metal objects are attracted by magnets (But they were not able to identify the ones that were attracted)
Can magnets attract each other?	
Some little faces can be joined together but others cannot stick together *The S with the S do not come together, they repel each other* *They do not meet because there is air between them*	Magnets attract and repel each other Magnets can attract each other even if there is a surface between them
Do magnets have different powers?	
Yes, big magnets take more clips and those which are small only take one or two	The power of a magnet is not related to its size

When it comes to programming, before the implementation not all the students had experience with robots. However, most of the children found them fun once the robots were introduced. They considered robots as a game and teachers took advantage of it to develop and improve their competences and abilities. Above all, teachers observed significant improvement in their ability to sequence steps and spatial orientation. *First of all, we worked the spatial orientation with our bodies, and then we started with the robots* (Teacher 5). Occasionally, although children verbalised that robots are controlled by people, when they made a mistake and the robot did not go where they wanted, they pointed to the robot as being responsible for the error. In other occasions, the children felt confident to try the same pattern of movements over and over again until they achieved their objective. This was because they did not perceive the error as a failure, but as a new opportunity: *This robot is crazy, where does it go? We tell them where they have to go. If we make a mistake, nothing happens. You can do it again* (Different children of 4 years).

Related to the attitudes/motivation during the STEM activity, both, teachers and children enjoyed it. *We took photos and we left them in the library. They are looking at them all the time. They loved it* (the activity), *we drew the attention to few clear concepts, no more and they enjoyed it a lot* (Teacher 3). In our data, children discussed, cooperated and physically tried out skills to understand how magnets behave and how to use them to create a racetrack toy.

Although all teachers expressed their satisfaction about the results achieved with the implementation, after the interviews, it was possible to categorize them in three groups taking into account their willingness to implement integrated STEM approach in their classrooms: Still reluctant due to teachers' limitations (number of children per class; lack of scientific knowledge; lack of knowledge in didactics of STEM; and lack of time); Entirely favourable (boost children curiosity; great opportunity to develop a meaningful learning from their personal experiences; exhaustive scientific knowledge is not required; the comments of the children have a lot of coherence); Still reluctant due to children's limitations (children are not able to pose scientific questions, to make generalizations and their prior conceptions are scarce).

5 Conclusions

The enacted and the lived object of learning observed for this implementation seem to show that both teachers and children could deal with the methodological proposal, adapting it to the realities of their classrooms, despite the drawbacks that teachers initially expressed. It's worth stressing that although not necessarily consciously, the teachers implemented, in practice, an integrated STEM approach, integrating the S, T and M parts along with robotics, in spite of focusing their efforts almost exclusively in science when discussing their intended object of learning. The reason could be their holistic view of education and the fact that pre-primary teachers usually work with projects. For the children, the activities, modified by the teachers for their specific settings, were useful for improving their knowledge and skills. However, in order to make teachers aware of the integration of the STEM disciplines and help children to understand these connections, it is necessary to highlight the specific content of each discipline and the links between them [22].

Even though Spanish pre-primary school teachers are not used to or trained to teach STEM subjects integrated and by experimental means, the teachers working within the botSTEM activity seem to have grasped key aspects of the framework and successfully applied them. Nevertheless, some of the teachers, notwithstanding this success, are still reluctant to integrate this approach in their teaching practice, perceiving it like isolated projects, although they note that the students' competences improve significantly. This may be related to their feelings and perceptions of self-efficacy about STEM, an aspect that has to be studied in depth.

In summary, the activities developed within the methodological framework, associated with the design based implementation seem to be useful to improve STEM teaching and learning at pre-primary school, although more implementations are needed to improve the model.

Acknowledgements. botSTEM project is funded by the European Union and the SEPIE Spanish National Agency under the ERASMUS+ KA2 Strategic Partnerships for School Education European programme. We want to thank the teachers who collaborate with dedication in this project, sharing their knowledge and experiences.

References

1. Bagiati, A., Yoon, S.Y., Evangelou, D., Ngambeki, I.: Engineering curricula in early education: describing the landscape of open resources. Early Child. Res. Pract. **12**(2), 1–15 (2010)
2. Bybee, R.W., Fuchs, B.: Preparing the 21st century workforce: a new reform in science and technology education. J. Res. Sci. Teach. **43**(4), 349–352 (2006)
3. DeJarnette, N.K.: America's children: providing early exposure to STEM (Science, Technology, Engineering and Math) initiatives. Education **133**(1), 77–83 (2012)
4. Toma, R.B., Greca, I.M.: The effect of integrative STEM instruction on elementary students' attitudes toward science. Eurasia J. Math. Sci. Technol. Educ. **14**(4), 1383–1395 (2018)
5. Kermani, H., Aldemir, J.: Preparing children for success: integrating science, math, and technology in early childhood classroom. Early Child Dev. Care **185**(9), 1504–1527 (2015)
6. Harlen, W. (ed.): Principles and Big Ideas of Science Education. Association for Science Education, Hatfield (2010)
7. DeJarnette, N.K.: Implementing STEAM in the early childhood classroom. Eur. J. STEM Educ. **3**(3), 18, 1–9 (2018)
8. Chalufour, I., Worth, K.: Building Structures with Young Children. Redleaf Press, St Paul (2004)
9. Kelley, T.R., Knowles, J.G.: A conceptual framework for integrated STEM education. Int. J. STEM Educ. **3**(1), 1–11 (2016)
10. Dare, E.A., Ring-Whalen, E.A., Roehrig, G.H.: Creating a continuum of STEM models: exploring how K-12 science teachers conceptualize STEM education. Int. J. Sci. Educ. **41** (12), 1701–1720 (2019)
11. Fleer, M., Gomes, J., March, S.: Science learning affordances in preschool environments. Austr. J. Early Child. **39**(1), 38–48 (2014)
12. Barab, S.A., Squire, K.: Design-based research: putting a stake in the ground. J. Learn. Sci. **13**(1), 1–14 (2004)
13. Marton, F., Booth, S.: Learning and Awareness. Lawrence Erlbaum Association, Mahwah (1997)
14. Marton, F., Tsui, A.B.M. (eds.): Classroom Discourse and the Space of Learning. Routledge, New York (2004)
15. Marton, F., Runesson, U., Tsui, A.B.M.: The space of learning. In: Marton, F., Tsui, A.B.M. (eds.) Classroom Discourse and the Space of Learning, pp. 3–40. Routledge, New York (2004)

16. Ortiz-Revilla, J., Greca, I.M., Meneses-Villagrá, J.Á.: La investigación de diseño en el desarrollo de propuestas didácticas STEAM. In: Membiela, P., Cebreiros, M.I., Vidal, M. (eds.) Nuevos retos en la enseñanza de las ciencias, pp. 217–222. Educación Editora, Ourense (2019)
17. Dare, E.A., Ellis, J.A., Roehrig, G.H.: Understanding science teachers' implementations of integrated STEM curricular units through a phenomenological multiple case study. Int. J. STEM Educ. **5**(4), 1–19 (2018)
18. Plomp, T.: Educational design research: an introduction. In: Plomp, T., Nieveen, N. (eds.) An Introduction to Educational Design Research, pp. 9–35. SLO, Enschede (2007)
19. Akerson, V.L., Carter, I., Pongsanon, K., Nargund-Joshi, V.: Teaching and learning nature of science in elementary classrooms. Sci. Educ. **28**(3–5), 391–411 (2019)
20. Rahm, J.: Reframing research on informal teaching and learning in science: comments and commentary at the heart of a new vision for the field. J. Res. Sci. Teach. **51**(3), 395–406 (2014)
21. Van Keulen, H.: STEM in early childhood education. Eur. J. STEM Educ. **3**(3), 1–3 (2018)
22. Martín-Páez, T., Aguilera, D., Perales-Palacios, F.J., Vílchez-González, J.M.: What are we talking about when we talk about STEM education? A review of literature. Sci. Educ. **103** (4), 799–822 (2019)

Intentions Towards Following Science and Engineering Studies Among Primary Education Pupils Participating in Integrated STEAM Activities

Jairo Ortiz-Revilla(✉) and Ileana M. Greca

Department of Specific Didactics, Faculty of Education, University of Burgos, c/Villadiego, 1, 09001 Burgos, Spain
jortizr@ubu.es

Abstract. The implementation of an integrated Science, Technology, Engineering, Arts and Mathematics (STEAM) didactic unit and its effects among primary education pupils is reported. The unit, designed to help pupils improve their competencies, revolves around the resolution of an engineering problem and addresses the Spanish curriculum for the sixth grade of primary education in the areas of Natural Sciences, Mathematics and Arts. Compared to the control group, the pupils following the STEAM didactic unit expressed stronger intentions to pursue science and engineering careers, despite the short—two week—intervention. Although the inter-group analysis showed non-significant differences, an effect size was noted for the group that had followed the didactic unit.

Keywords: Attitudes towards science and engineering · Intentions towards science and engineering · Integrated STEAM education · Primary education

1 Introduction

Policymakers from almost all developed countries have expressed concern about citizen scientific-technological literacy and also the fall of the number of students studying science and engineering. The Organisation for Economic Co-operation and Development (OECD) has highlighted that the proportion of students from OECD countries choosing to enter all Science, Technology, Engineering and Mathematics (STEM) fields has been dropping since the mid-1990s [1]. An additional concern is the lack of diversity among engineers [2], in as much as women are under-represented in most nations. In the search for alternatives, many researchers and national educational programs have in recent years opted for integrated STEM/STEAM—STEM + Arts—education. In those approaches, the intentional integration of STEM disciplines is supposed to improve the competences and the interests of the public towards science and technology through the resolution of problems that form part of the real world [3]. Specifically, integrated science, technology, engineering, arts, and math (STEAM) education has been broadly defined as a transdisciplinary approach the goal of which is to prepare students to solve the world's pressing issues through innovation, creativity,

Á. Herrero et al. (Eds.): ICEUTE 2020, AISC 1266, pp. 40–49, 2021.
https://doi.org/10.1007/978-3-030-57799-5_5

critical thinking, effective communication, collaboration, and ultimately new knowledge [4]. These proposals are introduced from childhood, as it is known that many engineers and scientists form their career choices even before adolescence [5, 6]. Interest in and positive attitudes towards science among children has generally been observed at quite high levels among both girls and boys younger than ten years old, although it has also been observed to lessen during early adolescence [7].

Furthermore, most available measurements of STEM attitudes, found in almost all STEM research studies, have focused on student attitudes toward science, as well as their intentions towards the future study of science and careers in STEM fields, although few studies have addressed student attitudes and intentions towards engineering [8], in particular following a STEM intervention.

During 2017–2018, the authors of this study implemented an integrated Science, Technology, Engineering, Art and Mathematics (STEAM) intervention directed at primary school pupils that made explicit use of inquiry teaching and engineering design methodologies, with the intention of both improving the competencies of pupils and fostering positive attitudes and intentions towards science and engineering in STEAM areas. In this paper, we examine the effects of this implementation, placing special emphasis on the intentions of pupils towards science and engineering studies. To do so, in this study we used an instrument, based on the Reasoned Action Model, which measures several dimensions, among which the intention to follow different careers, including science-related professions. The underlying assumption of the Reasoned Action Model is that people's attitudes, subjective norms and perceptions of control are a result of their beliefs, influencing their intentions and behavior [9].

2 Didactic Proposal

The planning of transdisciplinary didactic proposals and their implementation are complex tasks that are both useful and viable in natural contexts of Primary Education when committed to the empowerment of pupils, so that they can develop their competences. The study we present was developed in the context of an integrated STEAM teaching unit (STEAM UNIT) [10] implemented through a transdisciplinary approach, transcending individual disciplines and placing the focus of interest on a field of knowledge related to the real world [11]. In this case, the topic is the artificial lighting of a room. The main problem is presented as follows: How to design a prototype for lighting the room in which I study? All the didactic activities, which lasted two weeks, were designed around this problem, and the different experiments, results, conclusions and prototypes developed by the students were recorded in their field notebooks.

The specific integral model underlying this proposal addresses the contents of three subjects of the Primary Education stage, specifically linked to Natural Sciences, Arts and Mathematics, using inquiry teaching and engineering design as the methodologies in a precise and integral manner. Figure 1 shows the didactic model that has been developed, explaining how the methodologies were used to address the contents of the three subjects to solve the initial problem.

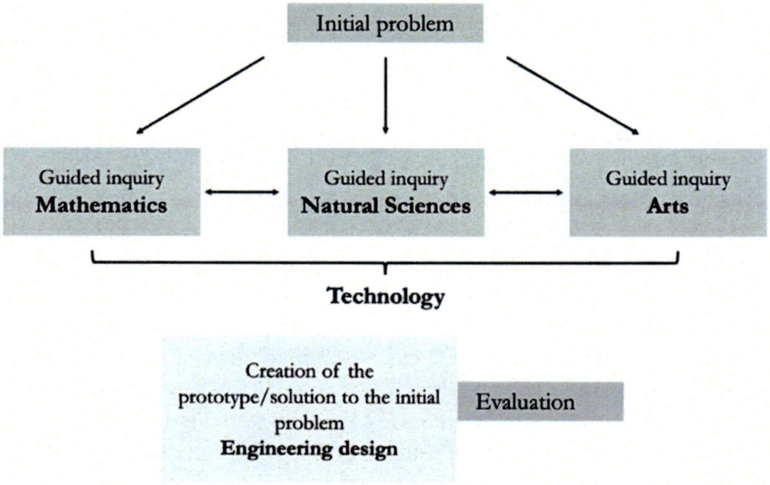

Fig. 1. Transdisciplinary model for integrated STEAM education. Adapted from [12].

3 Method

3.1 Design

In this study, a quasi-experimental cohort design with a post-test [13] was used in order to study the intentions of primary education pupils towards science and engineering after the implementation of an integrated STEAM unit. We compared the results obtained with an experimental group of sixth-grade pupils with the results of a control group of fifth-grade—at the end of their school year—, being the socio-cultural and demographic characteristics of both groups similar. So, it was assumed that the control group was equivalent to the experimental group before the intervention. It was decided to opt for this design, seeking as little interference as possible in the usual development of the ordinary classes of the experimental group before treatment.

3.2 Data Collection

The BRAINS test [14], validated as an instrument for assessing student attitudes towards science, was applied. For its transcultural adaptation, the original version of the instrument in English was translated into Spanish, with a direct and a reverse translation [15]. In addition to a series of interviews with pupils on their understanding of the wording of the instrument, a pilot version was used to check for any differences between the use of the term *ciencias*—science in its original version—and *Ciencias de la Naturaleza*—the equivalent term used in Spanish primary schools as a synonym for science—. No differences between the use of both terms were found. Moreover, given that no instruments were found in the literature for the joint evaluation of attitudes towards science and engineering, some new items relating to attitudes towards engineering were drafted and structured in a similar way to the existing science-related items. The translation into Spanish of the complete instrument that incorporates these new items is undergoing a validation process with a larger group of subjects. In this paper, we present the results that refer to the dimension 'Intentions towards science'—items 4, 11, 13, 16, 20 and 28 of the original instrument—, and the construct 'Intentions towards engineering' that was drafted along similar lines. These constructs are focused on studying the intention of continuing studies related to science or engineering, whether through future subjects or university careers. Table 1 shows the list of items for both constructs.

Table 1. Constructs and items under analysis.

Construct	Items in English	Items in Spanish
Intentions towards science	I will study science if I get into a university	Si voy a la universidad estudiaré ciencias
	I will not pursue a science-related career in the future	No voy a estudiar una carrera relacionada con las ciencias en el futuro
	I will become a scientist in the future	Llegaré a ser un científico en el futuro
	I will continue studying science after I leave school	Continuaré estudiando ciencias cuando termine el colegio
	I would enjoy working in a science-related career	Disfrutaría trabajando en una carrera relacionada con las ciencias
	I will take additional science courses in the future	Estudiaré más asignaturas de ciencias en el futuro
Intentions towards engineering	I will study science if I get into a university	Si voy a la universidad estudiaré ciencias
	I will not pursue a engineering-related career in the future	No voy a estudiar una carrera relacionada con la ingeniería e el futuro
	I will become an engineering in the future	Llegaré a ser un ingeniero en el futuro
	I will continue studying science after I leave school	Continuaré estudiando ciencias cuando termine el colegio
	I would enjoy working in a engineering-related career	Disfrutaría trabajando en una carrera relacionada con la ingeniería
	I will take additional science courses in the future	Estudiaré más asignaturas de ciencias en el futuro

In three of the items related to the Intentions towards Engineering construct, engineering replaced the terms *science* and *scientific*. The two items "I will study science if I get into a university" and "I will take additional science courses in the future" were not modified, because the option of taking engineering subjects falls within the scope of studying science in the Spanish context. Regarding the item "I will continue studying science after I leave school", we chose to keep the original wording, because there is no possibility in Spain of studying engineering at the end of the primary education stage and in the first year of secondary education. Considering the complete sample of pupils to whom the instrument was administered, the reliability analysis reported a Cronbach's Alpha of .818 and of .701 for the constructs intentions towards science and intentions towards engineering, respectively.

3.3 Context

The proposal was implemented during the 2017/18 academic year at Fernando de Rojas School, a state school located in the city of Burgos, Spain. The approval of the Provincial Director of Education of Burgos had previously been sought for the intervention, as it involved changes to the scheduled evaluations at the center.

3.4 Participants

A total of 258 school children divided into two groups participated in the study: the experimental group with 121 children—54.5% boys—, attending the sixth grade of primary education—$M = 11.33$ years, $SD = .52$—and the control group with 137 students—54.7% boys—from the fifth grade of primary education—$M = 10.62$ years, $SD = .59$—. Non-probabilistic convenience sampling was applied to select the final sample.

3.5 Data Analysis

The data from both constructs—Intentions towards science and Intentions towards engineering—were treated using descriptive and inferential statistics, with the 26.0 IBM SPSS Statistics package. The Kolmogorov-Smirnov test with Lilliefors significance correction was used to evaluate the normality of the data distribution. It yielded significant p-value results—<.05—for both constructs, which indicated a violation of the assumption of normality. The non-parametric Mann-Whitney U test and Rosenthal's r coefficient were therefore used, the former to compare both groups, and the latter to determine the effect size of the differences. The formula for the calculation of "r" is shown below:

$$r = \frac{Z}{\sqrt{N}}$$

It should be specified that we use the terms proposed by [16] to interpret the magnitude of the effect on the statistical values of "r". In the case of the separate samples from the fifth and the sixth grade, the Kolmogorov-Smirnov test returned a p-value of <.05, which was significant in both constructs, indicating a violation of the assumption of normality, so the Mann-Whitney U test was applied for the gender comparison within each group. The Kolmogorov-Smirnov test was also used to evaluate the normality of the data distribution of the separate samples of boys and girls within in the total sample, which again yielded a significant—<.05—p-value for both constructs, indicating a violation of the assumption of normality, so the non-parametric Mann-Whitney U test was used to compare courses by gender.

4 Results

Table 2 shows the descriptive statistics of both constructs under analysis.

Table 2. Descriptive statistics of the constructs under analysis.

Grade		Intentions towards science	Intentions towards engineering
5°	n	137	137
	M	3,049	2,970
	SD	,743	,654
6°	n	117	117
	M	3,169	3,151
	SD	,840	,714

* n = sample; M = mean; SD = standard deviation

These values imply the existence of high and positive intentions towards science and engineering, all of which, as can be seen, are higher for both the first construct and the values of the experimental group. In relation to the two grades—fifth and sixth year —, the Mann-Whitney U test revealed a lack of significant differences between the two constructs under analysis for this variable—Table 3—.

Table 3. Mann-Whitney U test for grade in the total sample.

	Intentions towards science	Intentions towards engineering
Mann-Whitney U	7498,500	7123,500
Wilcoxon W	16951,500	16576,500
Z	−,888	−1,534
Asymptotic Sig. (bilateral)	,375	,125

a. Grouping Variable: Grade

However, the ranges of this same test, although not significant, show that there were differences in the two constructs in favor of the year sixth grade. With these values, the effect size was analyzed, in order to examine the differences between both groups, by means of the Rosenthal r coefficient:

- Intentions towards science: r = 0.055
- Intentions towards engineering: r = 0.096.

These results show the presence of a small effect size for both constructs, which is associated with the implementation of the STEAM UNIT, as the sixth grade participated in no activities other than traditional science teaching, in addition to the activities developed in this STEAM UNIT.

Regarding the gender grouping within the control group, the Mann-Whitney U test revealed a lack of significant differences between both of the constructs under analysis —Table 4—.

Table 4. Mann Whitney U test for gender in the control group.

	Intentions towards science	Intentions towards engineering
Mann-Whitney U	2141,000	2221,000
Wilcoxon W	4992,000	4174,000
Z	−,794	−,452
Asymptotic Sig. (bilateral)	,427	,651

a. Grouping Variable: Gender

The Mann-Whitney U test for gender in the experimental group revealed little or no significant differences for the intentions towards science. However, significant differences were detected for the construction of intentions towards engineering—Table 5—.

Table 5. Mann Whitney U test for gender in the experimental group.

	Intentions towards science	Intentions towards engineering
Mann-Whitney U	1584,000	1321,000
Wilcoxon W	2962,500	2699,000
Z	−,583	−2,036
Asymptotic Sig. (bilateral)	,560	,042

a. Grouping Variable: Gender

It can be seen from Table 6 that the mean values for the boys are higher than for the girls.

Table 6. Descriptive statistics for intentions towards engineering by gender.

Gender		Intentions towards engineering
Boy	N	65
	Me	3,166
	SD	,667
	M	3,259
Girl	N	52
	Me	3,000
	SD	,753
	M	3,016

* N = total sample; Me = median; SD = standard deviation; M = mean

Finally, to verify to what extent these results were due to the implementation of the STEAM UNIT, the behavior of the group by grades for each gender was observed. In this case, two Mann-Whitney U tests revealed a lack of significant differences for the two constructs under analysis—Table 7—.

Table 7. Mann-Whitney U-test for grade by gender.

		Intentions towards science	Intentions towards engineering
Boy	Mann-Whitney U	2140,500	2002,500
	Wilcoxon W	4990,500	4852,500
	Z	−1,248	−1,829
	Asymptotic Sig. (bilateral)	,212	,067
Girl	Mann-Whitney U	1611,500	1581,500
	Wilcoxon W	3564,500	3534,500
	Z	−,003	−,174
	Asymptotic Sig. (bilateral)	,998	,862

a. Grouping Variable: Grade

However, the range table—Table 8—of those same tests shows that, although statistically not significant, there were differences in the two constructs in favor of the sixth grade.

Table 8. Table of Mann-Whitney U test ranges for grade by gender.

		Grade	n	Average range
Boy	Intentions towards science	5°	75	66,54
		6°	65	75,07
		Total	140	
	Intentions towards engineering	5°	75	64,70
		6°	65	77,19
		Total	140	
Girl	Intentions towards science	5°	62	57,49
		6°	52	57,51
		Total	114	
	Intentions towards engineering	5°	62	57,01
		6°	52	58,09
		Total	114	

* n = sample

In this case, Rosenthal's r, used to analyze the effect size, reported the following values:

- Intentions towards science among boys: r = 0.105
- Intentions towards engineering among boys: r = 0.154
- Intentions towards science among girls: r = 0.000
- Intentions towards engineering among girls: r = 0.016.

These results show the presence of an effect size for the two constructs among the group of boys and an effect size for the intentions towards the engineering construct among the group of girls, associated with the implementation of the STEAM UNIT. These effect sizes are small on all occasions, although it is much greater in the case of boys for both constructs.

5 Discussions and Conclusions

As proposed in the Reasoned Action Model, the intentions people express towards a specific behavior—in our case, continuing studies related to science or engineering, whether through future subjects or university careers—are determined by their attitudes, subjective norms and perceptions of control. These intentions are therefore quite complex to change. In this study of a two-weeks didactic intervention that kept to a very precise design, following a transdisciplinary STEAM approach, the intentions expressed by the students were seen to change, within a very short period. A result that suggests quite a promising approach. This results seems to support the conclusion achieved by [17] who indicate, after their study of the perception of science and technology in teenager students from Madrid, that exposing students to a social, scientific, multifaceted and humanitarian view of engineering might have a greater impact if performed with younger students.

However, it is worth stressing that although both genres showed greater intentions than those of the control group, the gains observed were greater among the boys than among the girls, particularly in the case of intentions towards engineering. Hence the need for successive studies, in order to inquire into the possible reasons behind those intentions. We guess that this could be related with the themes chosen for the problems proposed, since some studies relate girls' interest in STEM areas to the perceived social relevance of the addressed topics [18]. As it has been reported that males seem to be more interested in physics sciences, while females preferer those relating to health careers, the environment, and other "helping" professions [19, 20], we will study this hypothesis using engineering problems related to life sciences.

The results presented here may contribute to understand how the STEAM education movement in the K-12 instructional space influence students' attitudes about these STEM practices, and their interest in future STEM careers. However, these results must be interpreted taking into account the limitations of the study, in terms of samples, being necessary more researches for more robust results as well as greater attention to gender issues in the didactic proposals.

References

1. Organisation for Economic Co-operation and Development: Encouraging Student Interest in Science and Technology Studies. OECD, Paris (2008)
2. Buccheri, G., Gürber, N.A., Brühwiler, C.: The impact of gender on interest in science topics and the choice of scientific and technical vocations. Int. J. Sci. Educ. **33**(1), 159–178 (2011)
3. European Commission: Science education for responsible citizenship. European Union, Brussels (2015)
4. Quigley, C.F., Herro, D.: "Finding the joy in the unknown": implementation of STEAM teaching practices in middle school science and math classrooms. J. Sci. Educ. Technol. **25** (3), 410–426 (2016)
5. Maltese, A.V., Tai, R.H.: Eyeballs in the fridge: sources of early interest in science. Int. J. Sci. Educ. **32**(5), 669–685 (2010)
6. The Royal Society: Taking a Leading Role. The Royal Society, London (2004)
7. Murphy, C., Beggs, J.: Children's perceptions of school science. Sch. Sci. Rev. **84**(308), 109–116 (2003)
8. Wiebe, E., Unfried, A., Faber, M.: The relationship of STEM attitudes and career interest. Eurasia J. Math. Sci. Technol. Educ. **14**(10), 1–17 (2018)
9. Fishbein, M., Ajzen, I.: Predicting and Changing Behavior: The Reasoned Action Approach. Taylor & Francis Group, New York (2010)
10. Ortiz-Revilla, J.: ¿Cómo diseñar un prototipo de iluminación para mi sala de estudio? In: Greca, I.M., Meneses Villagrá, J.Á. (eds.) Proyectos STEAM para la Educación Primaria. Fundamentos y aplicaciones prácticas, pp. 163–93. Dextra, Madrid (2018)
11. Gresnigt, R., Taconis, R., van Keulen, H., Gravemeijer, K., Baartman, L.: Promoting science and technology in primary education: a review of integrated curricula. Stud. Sci. Educ. **50** (1), 47–84 (2014)
12. Ortiz-Revilla, J.: El desarrollo competencial en la Educación Primaria: efectos de una propuesta STEAM integrada [Doctoral dissertation]. Burgos, Universidad de Burgos (2019)
13. Creswell, J.W., Guetterman, T.C.: Educational Research: Planning, Conducting, and Evaluating Quantitative and Qualitative Research, 6th edn. Pearson, New York (2019)
14. Summers, R., Abd-El-Khalick, F.: Development and validation of an instrument to assess student attitudes toward science across grades 5 through 10. J. Res. Sci. Teach. **55**(2), 172–205 (2018)
15. Borsa, J.C., Damàsio, B.F., Bandeira, D.R.: Cross-cultural adaptation and validation of psychological instruments: some considerations. Paidéia **22**(53), 423–432 (2012)
16. Rosenthal, R.: Meta-Analytic Procedures for Social Research, 2nd edn. SAGE, Newbury Park (1991)
17. Pérez Sedeño, E., Sánchez Balmaseda, I., Miranda Suárez, M.J., García Dauder, S.: Perception of science and technology in teenage students from Madrid. Arbor **184**(733), 949–966 (2008)
18. Blanchard Kyte, S., Riegle-Crumb, C.: Perceptions of the social relevance of science: exploring the implications for gendered patterns in expectations of majoring in STEM fields. Soc. Sci. **6**(1), 1–17 (2017)
19. Jones, M.G., Howe, A., Rua, M.J.: Gender differences in students' experiences, interests, and attitudes toward science and scientists. Sci. Educ. **84**(2), 180–192 (2000)
20. Lachapelle, C.P., Brennan, R.T.: An instrument for examining elementary students' engineering interests and attitudes. Int. J. Educ. Math. Sci. Technol. **6**(3), 221–240 (2018)

Engagement and Learning from a Team-Based Mini-Project in Mechatronic Engineering

Noel Murphy(✉) and Jennifer Bruton

School of Electronic Engineering, Dublin City University, Dublin 9, Ireland
{noel.murphy, jennifer.bruton}@dcu.ie

Abstract. We outline our experiences with hidden and unsignposted learning by us and by our students arising from a team-based project activity in a 3rd-year undergraduate engineering module in the general Mechatronics area. We discuss the hidden learning achieved in areas such as team communications, team management, problem-solving skills, and communication through the media of student-produced video and presentations, as well as technical engineering reports. We describe the enablement of student reflection on their learning and its benefits and use these reflections to evidence various aspects of their learning. The work is situated within the literature on innovations and quality of STEM education.

Keywords: Engagement · Engineering project · Student-centered learning

1 Introduction

The results described in this paper were not derived as part of a research project that set out to answer particular questions about collaborative, team-based and peer-based learning. Instead, the developments were driven by challenges to innovate, to collect and act on feedback, and to aim for continuous improvement [1]. The means offered for recognizing the learning needs and delivering improvement were, and are, engagement with students, flipping the classroom, searching for motivation, teachers as partners in learning, and putting learning into the context of the students' future needs.

However, in terms of helping other teachers to innovate, it is useful to contextualize the un-anticipated learning-for-teaching that happened as a part of the aim for continuous improvement, in the context of the literature on collaborative teaching methods for improving student learning outcomes. Ralston *et al.* [2], for example, set out to investigate why, although the literature shows that almost any active or collaborative approach will improve learning outcomes as compared to traditional lectures, these methods have not been universally, or even widely, embraced. They cite the reasons for this as

- lack of pedagogical support from institution and colleagues,
- losing control of content coverage,
- student attitudes, expectations and resistance to change,
- lack of ongoing professional development,
- personal beliefs and expectations about teaching and learning.

Á. Herrero et al. (Eds.): ICEUTE 2020, AISC 1266, pp. 50–60, 2021.
https://doi.org/10.1007/978-3-030-57799-5_6

The outcome of the developments described in this paper is that many of these barriers can be overcome, not by assuming collaborative learning is the 'solution' and promoting ways of adopting it, but by setting *continuous improvement* as the objective, with listening and innovation (by the teacher) as the methods. As continuous improvement is actually about *quality of learning*, we see the progression as being one from quality to student engagement. Then, one natural expression of student engagement is collaborative learning – this now being an outcome and conclusion rather than a starting point.

Although Wittek and Kvernbekk [3] point out that 'quality of learning' is an elusive concept in higher education, assessment of teaching and learning is often considered as the means for 'achieving' it [4], and the enhancement of quality of learning can be linked to the quality of the student experience [5]. Klemenčič [6] focused on student-centered learning (SCL) as being a sort of 'litmus test' for quality, noting that student-centered learning features in the 2015 version of the European Standards and Guidelines for Quality Assurance in Europe (ESG), which in turn, strongly influences national and institutional policies. This is a reflection of the fact that policies for quality have shifted from student satisfaction to *student engagement*, and research on student engagement has been the cornerstone upon which policies and practices have been positioned over the last 10 years or so.

Indicators for student engagement are 'groupings of related items designed … to measure the extent to which an institution's environment promotes effective educational practices' [8]. For the Irish Survey of Student Engagement, there are nine such engagement indicators: higher-order learning, reflective and integrative learning, quantitative reasoning, learning strategies, *collaborative learning*, student-faculty interaction, effective teaching practices, quality of interactions and supportive environment [9]. Focusing on the collaborative learning element, Ralston *et al.* note as benefits: improvements in student achievement (content learning and understanding), the quality of interpersonal interactions (communication and social skills), self-esteem, self-confidence and student retention. Collaborative learning – which could be considered as any structured form of small group interaction [10] – is generally associated with active learning in the literature.

To situate the change process that led to the learning described here, it is useful to consider Borrego and Henderson's categorization of innovations in STEM teaching and learning into eight change strategies. In their terms, this work arose from individual 'reflective teachers' forming informal 'reflective communities', cross-fertilizing and supporting each other's innovations. However, there are also elements of the 'shared vision' strategy with its subcategories of 'learning organizations' and 'complexity leadership'. In the complex and disruptive educational environment in which we now find ourselves, these categories can help us to structure our efforts so that they support the agility and innovation that will characterize successful educators in the future.

2 Module Overview

Electromechanical Systems is a module in the first semester of the third year of related four-year (Bachelor) programmes in Electronic and Computer Engineering and in Mechatronic Engineering in DCU. This is where five years of post-second-level engineering education to Masters level is required to satisfy the educational requirements for Professional/Chartered Engineer status. This 5-ECTS credit module focusses on the sensors, actuators/motors and interfacing electronics that are key elements in a mechatronics system. Typical class sizes are in the range of 45–60 students.

Notwithstanding earlier modules on digital and analogue electronics and circuit theory, including associated labs, on commencing this module, the majority of the students are not competent or confident in designing and implementing straightforward interfaces between general analogue or digital devices and machines. Hence these learning outcomes are targeted by the inclusion of a team mini-project in the second half of the semester. Here we describe the nature of this project, and discuss the hidden and unsignposted learnings that the students encounter, evidenced by their own reflections.

The unsupervised team mini-project is carried out by groups of three with a standard kit (comprising components and tools), available to be taken off-campus if the team wishes. It is worth 17% of the module mark and is assessed on the basis of a team presentation to their peers, a short video produced by the team, and a written team technical report – all evaluated by rubric-based marking.

There is a marked tendency for students in the class to consider earlier years in their programme as 'hurdles' to be overcome and left behind, achieving superficial levels of retained knowledge, and not solid platforms on which to build further learning. Hence, there is guidance for student revision at the start of the module. This also supports student-centered design and implementation throughout the practical learning. To action this revision, specific local material and links to general external references (such as [12]) are provided to the class by way of the VLE. The individual student's mastery of this material is encouraged by a combined diagnostic and competency MCQ test. This test may be taken multiple times, as it has a pass mark of 70%, and any subsequent CA mark awarded to the student is zeroed until the diagnostic/competency test is passed.

3 The Team-Based Mini-Project

The team-based mini-project is timetabled for six 3-h lab sessions in the second half of semester, but teams can work outside this time if they wish, as all components are selected to be safely used without supervision. The teams spend this time on constructing, debugging, testing and operating a microcontroller-based 2-wheeled robot from a kit of components, with technical officer support during the scheduled lab times. The teams are usually self-selected, with the stipulation that each of the two programmes must be represented on a team. In the last week of the teaching part of the semester, each team makes a presentation to the class, and shows a short video that they have produced about the building and operation of the 2-wheeled robot. About a week or so later, they also need to submit a single agreed technical report per team.

One of the explicit aims of the mini-project is to move the students to the 'understanding', 'application' and 'analysis' levels of learning in Bloom's taxonomy, and this helps to prepare them for their fourth-year individual engineering project where they should be achieving at these, along with 'evaluation' and 'creation' levels of learning.

The MSP432P401R LaunchPad™ development board was chosen as the target embedded microcontroller to avoid the 'path-of-least-resistance' associated with the Arduino. The same processor is used for a more challenging project-based module on Mobile Robotics in the second semester, so the students benefit from the time and effort invested in learning about the MSP432 during the first semester.

The (returnable) kit is distributed without charge to each team. It is up to each team to decide what is done with the available components. The main objective is that they demonstrate a capacity to interface sensors and actuators/motors to the microcontroller for '*some constructive outcome*'. Marks are awarded (with rubric-based marking) for

- functionality and achievement of the objectives of the prototype;
- quality, structure and technical content of the written (team) report;
- quality and content of a video;
- quality and content of a team presentation.

The teams are asked to use at least 80% of the supplied components as a core part of their design, with up to 10% of the marks for the project being awarded for a team successfully interfacing to an additional, unspecified sensor that does not feature in the original kit. It turns out that a very positive and constructive direction to the students is that they should "*Feel free to surprise me here*". A 'Getting started guide', along with all necessary datasheets are provided through the VLE. The students are encouraged to use digital oscilloscopes for debugging.

The projects are assessed on a per-team basis by default. However, the students are asked to give a "clear and detailed" description of their personal contributions in the joint technical report, with advice at the start that they keep individual or team logs of contributions and progress throughout the project, including possible means for doing this. Follow-up action can be and is taken when a discrepancy in the extent of contribution within the team becomes apparent, either from the assessment materials or from direct inputs by students themselves.

Feedback on the mini-project consists of the selected assessment rubric levels in each category, along with additional textual descriptions of both positive and negative issues that are not captured by the rubric text. In addition, the technical reports are corrected within the VLE with detailed corrections visible to the team. The feedback includes notes on the quality of the presentations and the video, including advice on how to perform more effectively in front of an audience, and a class-wide message with generic areas for improvement that were common across many of the technical reports.

3.1 Sources of Innovation on This Module

The idea of getting the teams to present their work through the medium of a short video, along with other aspects of how the Electromechanical Systems module operates, came from a presentation given in DCU by Marc Prensky in Sept 2011, called "*Engage me or Enrage me*"[1]. In this challenging and provocative presentation, Prensky talked *inter alia* about engagement, listening, balancing top-down with bottom-up, flipping the classroom, peer-to-peer learning, mutual respect, motivation, partnering, and module coordinators not as lecturers but as coaches, guides and partners in learning: asking questions, adding quality and rigor, and putting learning into context. He pointed out that while yesterday's literate person might write a letter, report or essay, and the 2011 literate person might write an email, make a slide presentation or create a blog post, tomorrow's literate person would work in a virtual community, make a video or write a computer program. He challenged his audience with questions such as:

"Have I innovated today?"
"Did I get feedback and act on it?"
"Do I have a continuous improvement plan for my teaching?"
"Am I sharing the innovations that work with my colleagues?"

One of the reasons for writing this paper is to share the lessons learned from adopting and innovating around these ideas in one particular module delivered over more than a decade. The implementation of such innovations can make the educator a more effective fosterer of learning and can indeed make their work easier rather than harder.

4 Student Evidence of Their Learning Experiences

A notable aspect of the technical reports submitted on the project was the number of teams that undertook extensive introspection and self-analysis of their work and their experience under headings such as Learning Outcomes, Project Management, Team Operation and Conclusions and Evaluation. They understood that the relatively unstructured nature of the mini-project and the requirement to operate as a team were aspects that were foundational for their learning experiences, and could not easily have been 'taught' in other ways. Several teams came to understand that problems, hurdles, obstacles, mistakes and unforeseen issues were actually valuable learning opportunities, rather than impediments to learning. In the following sections, we outline some of the experiences described in the reports.

4.1 Team Communications

Some of the teams discussed the need for, and operation of, communications within the team. For example, Team M say:

[1] The content of the presentation is no longer available online, but sources containing similar advice are described on [2].

Since the lab was open the whole week, we were able to work out times that best suited each other to work outside of the lab time slot. We had to make sure we clearly communicated to each other to ensure we were meeting deadlines and that everyone attended any extra sessions that we had ... Since each member would work on a different aspect of the project during each session, we would have a quick debrief at the end of these sessions. This was when each member talked about what they had done that day. This allowed each member to understand every aspect of the project even if they weren't directly involved in the building of a specific aspect.

Team R report that:

The importance of adequate communication was evident during the project. Soft skills such as communication are invaluable in any industry and this project provided students with the opportunity to develop these skills in various ways, such as the presentation, or just simply communicating as a team in general.

Not everything was 'rosy', however, in this team:

There were some issues with communication and as a result it was difficult to allow the team to perform to its utmost potential, but regardless, the significance of communication and team-work were evident throughout the project.

4.2 Consolidation of Learning for Targeted Outcomes

Team G say in their report,

The mini project was a great introduction to microcontrollers and interfacing with various sensors and actuators. Exposure to electronics in a practical way was informative and useful when the results are able to be seen firsthand, giving a sense of familiarity. Having gained this experience, the team feels confident heading into the next semester: Mobile Robotics.

Team N conclude in their report that,

Overall, the team greatly benefitted from this project. Each team member acquired new knowledge and skills throughout the entire development of the project.

Team J give good descriptions of the problems that they encountered and how they debugged them. In the report, they say:

Not only did these problems help us to better understand the equipment and components, but [they] also taught us problem-solving, critical thinking and how to work together as a team.

Another student from Team J outlines her experience as follows:

I also wired up the H-Bridge ... [which] gave me the knowledge and information ... that I needed to fully understand them, why we use them and how to use them. This also gave me experience in reading data sheets and circuit diagrams as I had to use them both to figure out which wire needed to be connected to where.

The same student mentions about learning to use a multimeter:

Doing this enhanced my knowledge on how to use a multimeter correctly – which may seem like a trivial skill, however, this is a piece of equipment we will always need to use and need to know how to use properly.

This reminds the lecturer that no matter how mundane or trivial a fact or skill seems to us, if a student doesn't know it, then it's an impediment to their further learning and thus of crucial importance that they acquire it. A student from Team Q shows that notwithstanding the relatively unstructured and independent nature of the mini-project, even the straightforward engineering objectives of the module were achieved for her:

> *This project deepened my knowledge of basic electronic components and circuitry ... My participation in this project proved extremely beneficial towards my individual knowledge of the use and applications of microcontrollers ... Overall, the challenge was difficult at times but thoroughly engaging and eye-opening to the applications of microcontrollers and circuits working simultaneously.*

4.3 Learning Problem-Solving Skills

The effective practice of problem-solving is notoriously difficult to teach well. However, if we can put our students in the way of having 'light bulb moments', then it seems that we are facilitating the right type of learning. A Team J student says:

> *In the final weeks we kept running into issues and we could not figure out what the problem was. After reading through the lecture notes again and through vigorous problem solving, I realised that the cause to our problem was likely due to noise caused by ground loops ... I have now learned a valuable lesson: to always double check with the oscilloscope for noise or interference in the circuit when the circuit suddenly stops working for what seems like no reason.*

A female student also from Team Q outlined her revelation:

> *The main point of learning throughout this project although challenging, was troubleshooting! At the time of the problems I felt them unfortunate but once solving these issues I knew it was quite the opposite. From rewiring the entire bread board to rewriting the code, although these activities were troublesome, it was where I learned the most. I thoroughly enjoyed this project, and look forward to using microcontrollers again in the future.*

Comments like these illustrate the learning benefits of innovating in pedagogy in order to positively engage students. Team R report that

> *... the ability to problem-solve and debug errors is a crucial skill to have as an engineer, and this project greatly enhanced that ability. The available equipment (oscilloscopes and multimeters) was especially useful for this.*

4.4 Communication Through the Medium of Video

Team M programmed the car to do the "Cha Cha Cha", shown to impressive effect in their video, with a really excellent soundtrack that reflected the mood of different stages of development. They talked in the report about "keeping the video entertaining" and "motivating people to take up engineering", which certainly constitute aspects outside the scope of the original module descriptor.

The members of Team G mention, in relation to their excellently produced, well-edited and very entertaining video that its style was based on

> *... an 'old-school' over the-top infomercial, using the iconic black and white filter to highlight negative elements and an upbeat tune.*

This choice of design indicates an expression of motivation and talent that is far outside anything normally envisaged for an electromechanical systems module. The video for Team L was quite consciously directed and edited, and they used a voice actor from outside the team working to the team's script to give it a strong storyline, which was very effective.

4.5 Communication Through the Medium of a Technical Report

Team H mentioned that they divided up the writing of the report according to what each team member focused on during the build, and after combining all the sections they read through each other's sections to ensure no grammatical mistakes were made and all statements were factual. It was clear from other reports that this wasn't universally the case. Technically, and in terms of the care taken with the writing and presentation, the report from Team C was one of the best reports that I received. At the other extreme, Team K received quite negative feedback for the language and structure of the report, which had a level of informality entirely unsuitable for a technical report. These two extremes remind us that where a learning outcome activity is challenge-based and less structured, the learning can be quite inconsistent across teams. Where there are very clear standards expected, such as for a written technical report, we need to provide explicit guidance, templates, and minimum standards to be achieved universally.

4.6 Students' Own Reflection on Their Learning

Team N referenced Bloom's Taxonomy and understood that one aim of the project was to move to the 'apply' and 'analyze' stages of learning and be able to understand the material at a deeper level than before. They understood the need to recognize and learn from mistakes, to exploit knowledge gained in earlier modules, and to recognize and work around their shortcomings. Team J recalled the stated Learning Outcomes of the mini-project, and claimed to have achieved these. They also mentioned that it was.

> *important to note that other valuable learning outcomes ... not categorically outlined at the beginning of the project, were achieved.*

Then, after listing the levels of Bloom's taxonomy, they pointed out that

> *this project helped the students to remember and understand (level 2: Understanding) what they learned [in lectures], apply this knowledge to the project (level 3: Application), analyse any errors that occurred (level 4: Analysis), evaluate how best to fix these errors (level 5: Evaluation) and finally to create the wheeled robot (level 6: Creation).*

This may not be a perfect mapping, but it is interesting how giving students tools to understand the nature of learning, as well as appropriate opportunities, can be strongly enabling. Team J also pointed to 'soft' skills gained by the team members throughout the project, and they explicitly call out interpersonal skills, teamwork, time management, problem solving, creative thinking, critical thinking and decision-making.

4.7 Team Management and Dysfunctional Teams

Team F describe in their report that they

> *decided on who will manage the team and ensure all the deadlines for the various tasks are complete. This meant that the project leader has to conduct five processes to successfully manage the project which are initiating, planning, executing, monitoring and closing, the main objectives were to ensure that the team achieves each and every task effectively.*

From information received after the fact we believe that this team had an issue with one member dominating the others. It is not hard to imagine which student may have written the above text. A later paragraph talks about how [one student] "was lead in the role of the robot's hardware and software", and the other students "assisted" that student, while they were responsible for non-technical aspects such as the presentation and the video. In the conclusion, reported as written by one of the non-dominant students, we read that the project "... enhanced my learning in interfacing electromechanical systems, and it turned out to be a project worth doing", which is some distance from the 'glowing' reports of students in other teams of their learning experience. There is an opportunity to 'repair' the lack of learning opportunities for students in this team in the subsequent Mobile Robotics module, but more importantly, this occurrence points to the need to provide students with the skills necessary to manage such situations as they arise.

Team N were very clear in the report when one team member didn't contribute, and they didn't try to cover up the fact out of a misplaced sense of loyalty. One of the students in Team J reports that:

> *I helped my teammates practice for the presentation by giving constructive feedback about their body language and the importance of projection of their voices and organising a choreography so that each member who spoke would be standing in the center of the room to command attention and the following person was ready to take over, with the last person controlling the slides.*

Interestingly, it was also clear from the Team J report that this team had a member who contributed significantly less than the others. This was not explicitly stated, but there is little doubt that the students knew what they were implicitly communicating.

For Team K,

> *Teamwork was a core soft skill learned throughout this project. Working in a team can often be quite challenging and difficult due to conflicting ideas or someone not pulling their weight. The team learned that good project management skills are ... vital in making sure that everyone in the team is on the same page and that everyone knows what needs to be done.*

5 Conclusions

It has been a very useful exercise to reflect on the results of the learning innovations introduced to this module. These were significantly inspired by Mark Prensky's presentation, but also derive from a project-based and student-centered learning ethos in our School that has existed for at least the last two decades. We have learned that given the right motivations, challenges, freedoms and supports, that our students can achieve in ways that we could never have imagined, and that different students will achieve in ways that are particular to them. If we needed to be reminded of the Confucian truism: "I do and I understand", we can see it reflected here in the reports of many of these teams. We can see the benefits of giving them the tools to enable introspection, to enable students themselves to give names to the types and natures of their learning experiences. We can see the unintended lessons or the hidden lessons that contribute to the broader education, maturity and professionalization of the students.

It is not all positive, however. While dysfunction within teams can provide an opportunity for learning, without appropriate preparation or structures or framework guidelines for teamwork, it can also be a significant impediment to effective learning.

Critical thinking may have been developed, but these students have only just set out on this journey. They have much more to learn in this area. While problems encountered and, in many cases, overcome have encouraged the students to become more careful, systematic, orderly, and patient, these qualities are still not universal. The logical, rational, rigorous thinking needed to be an effective debugger or problem-solver still needs more structured opportunities, more experiences and more or more appropriate targeted interventions to develop this skill and expertise to its full extent. The production of videos may allow creative talents to emerge and be encouraged, but creativity alone does not help to improve the standard of technical writing. This will need more specific guidelines and more explicit standards to be set so that students can learn in a consistent way what is required to be effective and professional in this area.

We feel that the lessons on hidden learning that we have uncovered will benefit our teaching, our students' learning and the wider academic community into the future.

References

1. See www.marcprencky.com
2. Ralston, P.A.S., Tretter, T.R., Kendall-Brown, M.: Implementing collaborative learning across the engineering curriculum. J. Sch. Teach. Learn. **17**(3), 89–108 (2017). https://doi.org/10.14434/v17i3.21323
3. Wittek, L., Kvernbekk, T.: On the problems of asking for a definition of quality in education. Scand. J. Educ. Res. **55**, 671–684 (2011). https://doi.org/10.1080/00313831.2011.594618
4. Altbach, P.G., Hazelkorn, E.: Measuring education quality in global rankings: what's the likelihood? Int. High. Educ. (95), 12–14 (2018). https://doi.org/10.6017/ihe.2018.95.10721
5. Goff, L.: University administrators' conceptions of quality and approaches to quality assurance. High. Educ. **74**, 179–195 (2017). https://doi.org/10.1007/s10734-016-0042-8
6. Klemenčič, M.: From student engagement to student agency: conceptual considerations of European policies on student-centered learning in higher education. High. Educ. Policy **30** (1), 69–85 (2017). https://doi.org/10.1057/s41307-016-0034-4
7. Standards and Guidelines for Quality Assurance in the European Higher Education Area (ESG), Brussels, Belgium (2015)
8. Fosnacht, K., Gonyea, R.M.: The dependability of the updated NSSE: a generalizability study. Res. Pract. Assess. **13**, 62–73 (2018). https://eric.ed.gov/?id=EJ1203503. Accessed 9 Apr 2020

9. StudentSurvey.ie: Irish Survey of Student Engagement National Report, p. 80. Higher Education Authority, Dublin, Ireland (2019). https://hea.ie/assets/uploads/2018/11/ISSE-Report-2018-TEXT-Tag-A.pdf
10. Barkley, C., Barkley, E.F., Cross, K.P., Major, C.H.: Collaborative Learning Techniques: A Handbook for College Faculty. Jossey Bass, San Francisco (2005)
11. Borrego, M. Henderson, C., Increasing the use of evidence-based teaching in STEM higher education: a comparison of eight change strategies. J. Eng. Educ. **103**(2) (2014). https://doi.org/10.1002/jee.20040
12. https://www.allaboutcircuits.com/textbook/. Accessed 30 Jan 2020

Educational-Oriented Mobile Robot: Hidden Lessons

Francisco Manuel García-Álvarez[1] and Matilde Santos[2](✉)

[1] Universidad Nacional de Educación a Distancia, Juan del Rosal, 16, 28040 Madrid, Spain
fgarcia1917@alumno.uned.es
[2] Complutense University of Madrid, 28040 Madrid, Spain
msantos@ucm.es

Abstract. Autonomous robots are a potential learning platform for degrees related to mathematics, programming, signal processing, electronics, control, artificial intelligence and robotics, among others. Furthermore, working with these real robotic systems involves not only the knowledge of many different disciplines but it is also an ideal platform to acquire other transversal skills such as to concentrate on manual tasks, patience, team work, imagination and creativity, inquiring, etc. In this work, some aspects of the design and implementation of an autonomous robot are presented. This mobile robot is educational-oriented, and it serves as a platform for the application of different techniques to navigate in a maze. It also helps to compare theoretical and practical results. It is proved that this type of robotic platform is a good tool for teaching and learning different disciplines and that there are many hidden lessons beyond the theory that can be learned using this type of real experiences.

Keywords: Educational robotics · Autonomous robots · Learning platform · Maze

1 Introduction

Nowadays there are many autonomous vehicles that are used for very different purposes. Students know well these technological advanced systems and are aware that working with them can help them to further develop in their professional career [1, 2].

Robotic systems are a way to get closer to real world. Even more, introducing these real systems in the classroom engages students more deeply in the learning process and make them active learners [3–7].

Autonomous robots are a potential learning platform for academic degrees related to mathematics, programming, signal processing, electronics, control, artificial intelligence and robotics, among others. Furthermore, working with these real robotic systems involves not only the knowledge of many different disciplines but it is also an ideal platform to acquire other transversal skills such as to concentrate on manual tasks, patience, team work, imagination and creativity, inquiring, etc. [8].

In this work, a twofold contribution is pursued. On the one hand, the assembly of an autonomous robotic system can teach several hidden lessons to the students, besides

Á. Herrero et al. (Eds.): ICEUTE 2020, AISC 1266, pp. 61–71, 2021.
https://doi.org/10.1007/978-3-030-57799-5_7

the creativity and initiative the design and its construction require. On the other hand, it can be used as a teaching platform where different navigation strategies can be applied and tested, drawing some conclusion from it.

The wheeled autonomous robot construction is proposed with materials and resources that are easy to find and affordable. The design, selection of the components and the implementation of the physical navigation scenario are addressed.

The scenario for the robot navigation is a maze. It can be used as an ideal way to train and improve the ability of autonomous robots to move around the environment. To obtain an optimal route, decision-making algorithms adapted to that specific situation are used, whether they take into account distance, location, energy efficiency or process power, to give some examples [9, 10].

Some works found in the literature use this scenario in educational robotics [11] but they are mainly more focused on the efficiency of the maze navigation algorithms. On the contrary, our approach is to take advantage of the lessons students can learn when working with a real system in this complex scenario and to foster their critical thinking and other transversal skills.

Besides, this educational-oriented robotic platform allows the student to compare theoretical and practical results. Thus, this real robotic system is not only a good tool for teaching and learning different disciplines, but there are many hidden lessons beyond the theory that can be learned using this type of real experiences.

The structure of the paper is as follows. In Sect. 2 the robot assembly and hidden lessons that can be learnt from the robot building are presented. Section 3 describes the robot navigation scenario and the algorithms that are going to be tested. Results are discussed in Sect. 4, and hidden lessons derived from these experiments are presented. The paper ends with the conclusions and future works.

2 Robot Assembly

The proposed learning project involves the construction of the mobile robot. To assembly the wheeled robot, a commercial kit has been used, the Elegoo Smart Car Kit [12]. The parts are standard, easily replaceable and economical. The main components of the robot are the following:

1. Arduino Uno Rev3 board.
2. Ultrasonic and infrared sensors.
3. SG90 servo motor.
4. Micro-controller for DC motors.

To assembly the mobile robot, the students are suggested to follow the flow chart of Fig. 1, paying attention to the connections. The expected result is something similar to the robot shown in Fig. 2, though students can assemble the pieces in different ways in order to foster their creativity.

In this phase, it is especially important to place and calibrate the sensors well, since the navigation will depend on the measures these instruments give.

Besides, a PID controller has been designed, tuned and implemented on the hardware of the robot board. This PID regulator controls the trajectory tracking. The

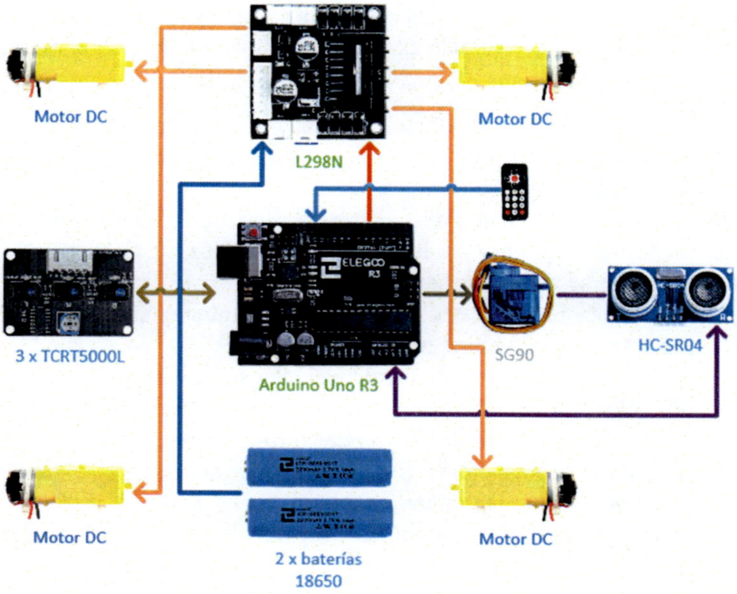

Fig. 1. Block diagram

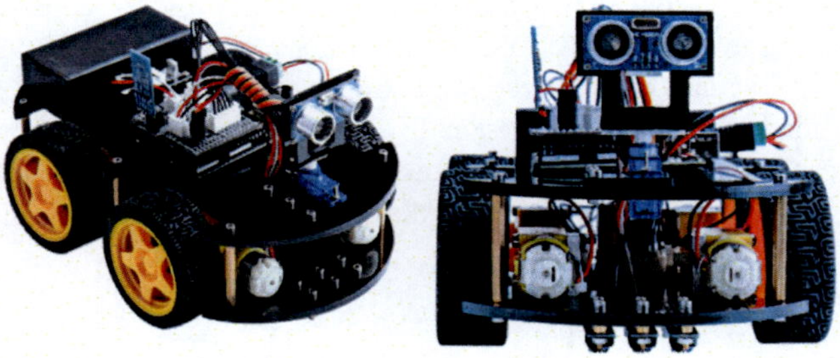

Fig. 2. Side and front view of the mobile robot

input is the error, i.e., the difference between the desired trajectory, measure by the infrared sensors using the lines on the scenario surface (see Fig. 4), and the real trajectory followed by the robot. The control output is the power of the left and right motors.

2.1 Hidden Lessons from the Robot Building

Once the mobile robot has been built by work teams, it would be good to make the students think about the lessons learned during this task, to develop a very important skill, critical thinking. It is quite common that students of technical degrees, mainly STEM (science, technology, engineering and mathematics) only see the correctness of the execution of the task without going beyond.

Some of the possible conclusions they could reach, and that would lead them to consider lessons beyond the simple physical assembly of the robot, could be the following:

- If the board is supplied with less than 7 V, it may be unstable. If more than 12 V are used, the voltage regulator may overheat and damage it. So 7–12 V is the recommended interval although the technical characteristic says [5, 15] V.
- The kit comes with only one ultrasonic sensor, making it difficult for the robot to follow the desired trajectory on its own. To solve this, at least three ultrasonic sensors would be necessary, at both sides and in the front of the robot.
- A fourth ultrasonic sensor located at the rear of the robot could help the system to change directions faster, without making a turn.
- The connections to Arduino board should be better soldered to avoid them to break or slip, and cables protected. Errors derived from these defects are difficult to detect.
- Similarly, the wheels are a bit loosely coupled to the motors.
- The memory of the Arduino board, 2 Kbytes, may be not enough for running some computational demanding algorithms.
- The servos are not ideal, they do not have a perfectly symmetric behavior.

3 Labyrinth and Navigation

The second part of the educational project with the mobile robot consists of autonomously navigating in a labyrinth. The designed maze is shown in Fig. 3 (left), with dimensions 5 × 5 square cells, where the entrance is at the bottom left corner (1, 1) (blue X) and the output is in the center (3, 3) (red S). The maze shown in Fig. 3 (middle) corresponds to the one used by the flood-fill algorithm. The optimal path is shown in Fig. 3 (right). It goes through 8 tiles, with three turns, to reach the target.

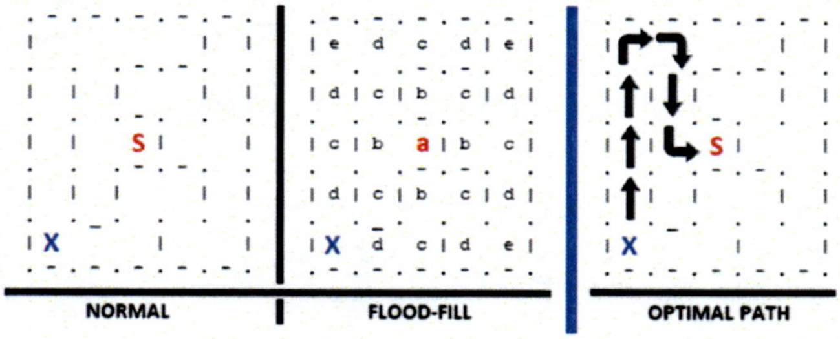

Fig. 3. Designed maze (left), flood-fill maze (middle) and optimal path (right)

The maze proposed in Fig. 3 has been implemented:

- Virtually in the Arduino board, the simulated robot goes through the virtual maze.

- Physically (Fig. 4, right): the real maze has been built with cardboard; lines have been added to mark the path for the infrared sensors. Friction due to surface materials, real engines performance, etc. must be taken into account.

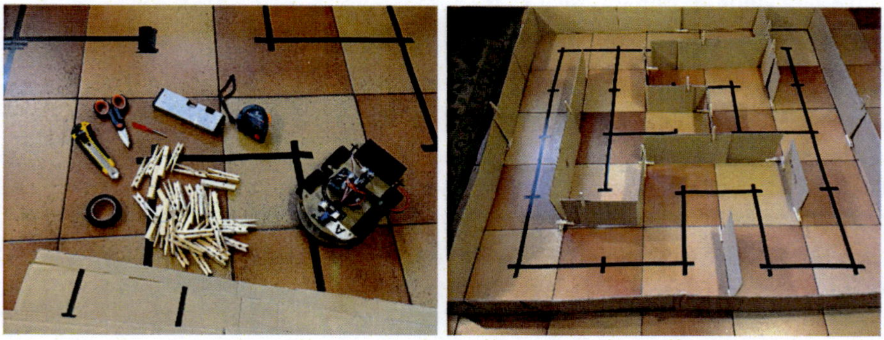

Fig. 4. Materials used and maze built with tracking lines

The selected algorithms to navigate the maze are relatively easy to implement in the Arduino board, and they do not demand too much memory or computational complexity. Different algorithms have been first simulated in the virtual environment and the most efficient ones have been chosen to be implemented on the hardware.

The goal of this part of the project is to have the students code various maze solving algorithms and compare the results in terms of efficiency and computational time.

3.1 Uninformed Algorithms

Uninformed algorithms, without any prior knowledge, explore the search space as they navigate the maze. Among them, two simple and easy to understand ones have been selected, in order to get the students analyze their advantages and limitations [13].

Random or "Mouse" Algorithm
Its sequence of steps is as follows:

1. The robot leaves the starting point of the maze and moves straight ahead.
2. At a junction, it makes a random decision about the next direction to follow.
3. It continues advancing this way until it reaches the goal.

The robot will always eventually reach the goal, but this blind search method is usually very slow and not predictable, it can take the wrong path over and over again.

Wall-Follower
The robot must keep one "hand" (left or right) in contact with one wall of the maze. This way the robot is guaranteed not to get lost and will reach a different exit if there is one; otherwise, the algorithm will return to the entrance having traversed every corridor next to that connected section of walls at least once. The maze must be simply-

connected because if there are passage loops it does not work. Besides it is quite inefficient for large mazes and there may be exits closer to the current position that are ignored.

As part of the tasks left to the students' free choice, other algorithms can be suggested within this group, such as the Pledge, or Trémaux's [13].

3.2 Informed Algorithms

With the informed algorithms, the robot uses domain-specific knowledge, for instance, it is possible to estimate how close a state is to the goal, where the obstacles are, etc. Therefore, these maze solver algorithms are faster than those without additional information as they have to explore a smaller search space.

The robot navigates through the maze first virtually, in simulation, and once the best path has been found, it is implemented in the Arduino board.

Again, two informed maze algorithms are proposed to the students to implement them in the real system and to draw conclusions [13].

Optimized Wall-Follower
It is basically the wall follower algorithm but with the information obtained once the goal has been reached in the simulation scenario. So the optimal path (shortest) is reproduced, avoiding unnecessary direction changes.

Using the left hand strategy (LHR), the optimization rules for the robot will be (symmetrical for RHR) in this particular case:

$$LTR = RTL = T \quad LTS = SCL = R \quad SCS = T \quad LTL = S$$

Where Lis left, R means right, S is straight and T is turn.

Flood-Fill
This algorithm is based on assigning an initial heuristic value to each cell, which will be updated by exploration at each iteration, taking into account that its final value should take the robot to the goal by the shortest route. It requires information on the dimensions of the maze and the position of the goal. The cost of the entrance X is 0, the cost of the adjacent cells in any horizontal or vertical direction is b (first level), at a second level the cost is c, and so on (Fig. 3, center). If there is a wall between cells they are not adjacent. The algorithm has three phases, the first two ones are carried out in simulation and the last one with the real robot:

1) Labyrinth exploration, to assign values to each cell by trial and error.
2) Once the path is found, the robot goes backward from the goal to the entrance, updating the values of each cell to give higher priority to the optimal route.
3) The real robot now navigates through the maze.

4 Discussion of the Results

Those four search algorithms have been tested on the same maze, with the obstacles fixed and being the optimal path 8 cells (Fig. 3, right). For both, the virtual experiment (simulation on the Arduino board) and real case with the robot, the results are given in terms of:

- Iterations during the executions of each phase/algorithm = Cells explored until goal.
- Time = time spent on the iterations.

4.1 Results

Random or "Mouse" Algorithm
It is undoubtedly the worst algorithm. Its randomness implies that it can vary between obtaining an optimal route or following the worst possible. Its coding is quite simple. The number of iterations and computational time are much higher in the real case due to the time the robot require to physically navigate the maze (Fig. 5).

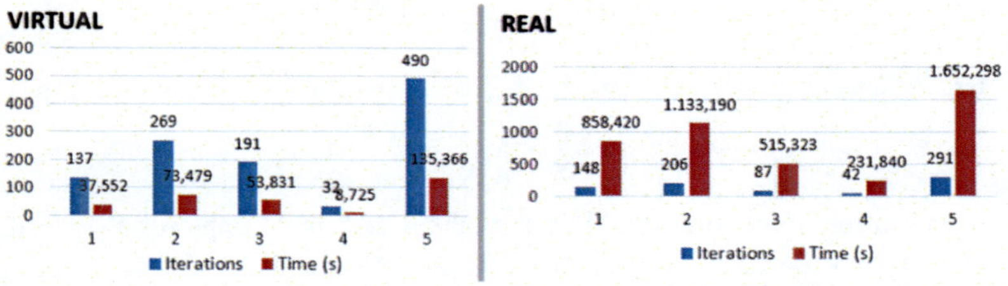

Fig. 5. Random mouse results, virtual maze (left) and real maze (right)

Wall-Follower
Its strength lies in the fact that it will always reach the goal, generally requiring an average time that can be estimated.

Results (time and iterations) (Fig. 6) are much better than with Random algorithm. The robot does not navigate randomly through the maze but with a preferential direction set in advance (in this case, the left hand), which avoids errors in the direction more than once and thus reach the goal much sooner.

Due to the specific design of the labyrinth in this project, the LHR strategy provides better results than the RHR. In the latter case, half of the maze would be explored unnecessarily.

Optimized Wall-Follower
It requires two phases, first to optimize and save the path (memory consumption) and then to implement the solving maze algorithm (with the rules already learnt).

The results (Fig. 7) of the robot in the virtual and the real maze are very different, since in the real scenario the algorithm already knows where the goal is and how to

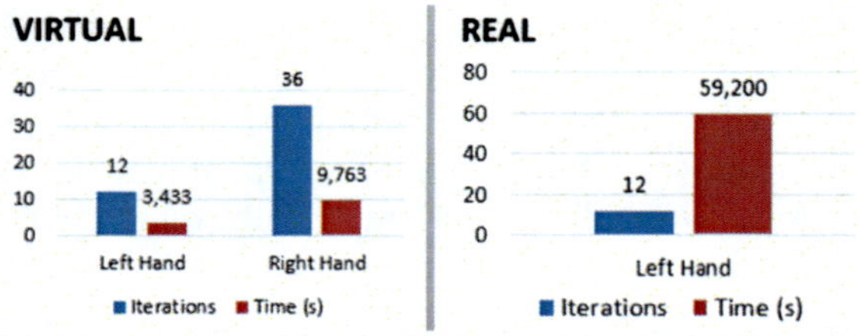

Fig. 6. Wall-follower (RMI) results, virtual maze (left) and real maze (right)

reach it in an optimal way. Besides, if the robot knew where the obstacles are and it could navigate at its maximum speed, times would be significantly shortened.

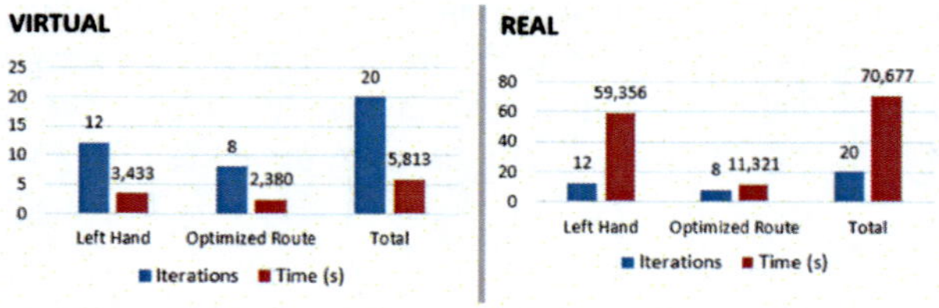

Fig. 7. Optimized Wall-follower results, virtual maze (left) and real maze (right)

Flood-Fill

This algorithm requires more time than the others because it explores all possibilities before finding the optimal path (see Sect. 3.2). The results of the first two phases (Fig. 8) are the same for the virtual and the real maze. The navigation through the maze in the virtual case is immediate, and in the real maze more time is required due to the delay of the sensors, turns it takes, etc.

However, it is a very efficient method; the time to explore the labyrinth is much shorter than in the LHR case, if some information on the maze is available.

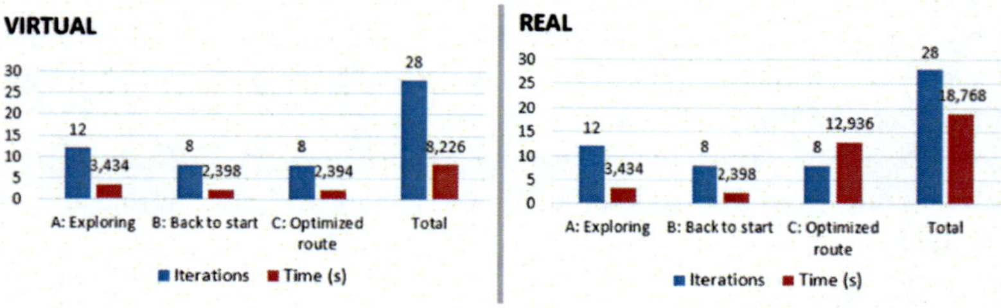

Fig. 8. Flood-fill results, virtual maze (left) and real maze (right)

4.2 Comparison

The results of the maze solving algorithms (average values in the case of the mouse algorithm) in both the virtual and the real maze are shown in Table 1.

Table 1. Results of applying the maze solving algorithms

Algorithm	Virtual maze		Real maze	
	Iterations	Time (s)	Iterations	Time (s)
Random	223,8	61,79	154,8	831,84
Wall follower LHR	12	3,433	12	59,20
Optimized LHR	12 + 8 (opt)	3,433 + 2,38	12 + 8 (opt)	59,35 + 11,32
Flood-Fill	12 + 8 + 8	3,43 + 2,39 + 2,39	12 + 8 + 8	3,43 + 2,39 + 12,93

From these results the following conclusions can be drawn. The "mouse" algorithm is, by far, the worst of all. Being completely random makes it very inefficient. The Wall-follower is very predictable, with the same number of iterations in both the virtual and the real case. However, the navigation time is longer in the real case due to the delays of the sensors, the friction of the wheels, etc. The Optimized Wall-follower will always be better than the not-optimized one for the same maze if it is used more than once. Finally, the Flood-Fill is the most efficient search algorithm but it should be noted that global time would increase if the time spent assigning values to the cells is considered, as it has been done manually in our case.

5 Hidden Lessons from the Simulated and Real Experiences with the Mobile Robot in the Maze

There are several aspects that can be pointed out to awaken the critical thinking of the students and to make them think and draw conclusions about the experiences.

First, regarding the operation of the robot:

- Memory management is essential in Arduino, because it is limited.
- It is necessary to calibrate the line sensors.
- The rotors are controlled used the PWM modulation.
- To avoid deviations and defects of rotation, the servo requires a previous calibration.
- The ultrasonic sensor is very accurate.
- Line sensors are controlled digitally, not with PWM pulses.
- Once tuned, the PID controller has worked very satisfactorily.

Regarding the algorithms:

- The randomness of the "Mouse" algorithm may drain the batteries completely even before reaching the goal.
- On the contrary, the Flood-Fill algorithm does not move the robot until the optimal path is known, which saves time and batteries.

The following possible improvements for the robot can be explored:

- An Arduino Mega board instead of the One would increase memory, faster speed, more connection possibilities, more timers for more devices, etc.
- Three ultrasound sensors would be very useful. This way the line guidance would not be required neither any added servo.
- For optimal line guidance, seven line sensors should be needed to detect deviations and junctions more accurately.
- To make the turns more accurately and to better position the robot, encoders for the wheels and a digital compass could be used.
- An LCD screen could display informational messages about the robot.
- The connection and power of the batteries should be better and more reliable.

6 Conclusions and Future Works

A mobile robot is a learning platform not only to learn lessons closely related to robotics or other technological subjects, but also to develop students' learning capacity and critical thinking. Working with a real system allows to learn lessons that go beyond theory.

In this work it has been proven how the mobile robot allows students to face real problems, and may foster their creativity to solve these problems.

In further works, the environment can also be used to test more optimization algorithms and even event-based control strategies for the robot, in order to compare with classical control systems, not only in terms of iteration and time of execution but also in terms of communications efficiency. Also, it is proposed to use the robot as a platform to implement intelligent strategies to navigate the maze, with online learning [14].

References

1. Cañas, J.M., Martín, A., Perdices, E., Rivas, F., Calvo, R.: Entorno docente universitario para la programación de los robots. Revista Iberoamericana de Automática e Informática Industrial **15**(4), 404–415 (2018)
2. Vesikivi, P., Lakkala, M., Holvikivi, J., Muukkonen, H.: The impact of project-based learning curriculum on first-year retention, study experiences, and knowledge work competence. Res. Pap. Educ. **35**, 64–81 (2019, online)

3. Ramos, J.J.M., Pineda, J.L.L., Moctezuma, R.A.G., Zavala, J.G.C.: A teaching methodology based on an educational experimental platform. IEEE Lat. Am. Trans. **17**(8), 1363–1370 (2019)
4. Bruder, S., Wedeward, K.: Robotics in the classroom. IEEE Robot. Autom. Mag. **10**(3), 25–29 (2003)
5. Barradas, R., Lencastre, J. A., Soares, S., Valente, A. Usability evaluation of an educational robot for STEM areas. In: 11th Proceedings of Conference on Computer Supported Education (CSEDU 2019), Crete, vol. 2, pp. 218–225 (2019)
6. Santos, M., Farias, G.: Laboratorios virtuales de procesamiento de señales. Revista Iberoamericana de Automática e Informática Industrial **7**(1), 91–100 (2010)
7. Verner, I.M., Alhgren, D.J.: Robot contest as a laboratory for experiential engineering education. J. Educ. Resour. Comput. **4**, 2 (2004)
8. Acevedo-Londoño, J.A., Caicedo-Bravo, E., Castillo-García, J.F.: Playful robotic environment for rehabilitation therapies of pediatric patients with upper limb disability. Revista Iberoamericana De Automática E Informática Industrial **15**(2), 203–210 (2018)
9. Su, J., Cai, X., Lee, C., Chen, C.: The development of a half-size micromouse and its application in mobile robot education. In: Int. Conference on Advanced Robotics and Intelligent Systems (ARIS), pp. 1–6 (2016).
10. Peñaranda, C., Palanca, J., Julian, V., Botti, V.: A flexible and dynamic mobile robot localization approach. Log. J. IGPL **28**, 197–210 (2018)
11. Junior, J.E.B., Queiroz, P.G.G., de Lima, R.W.: A study of the publications of educational robotics: a systematic review of literature. IEEE Lat. Am. Trans. **16**(4), 1193–1199 (2018)
12. Elegoo. https://www.elegoo.com. Accessed 01 Feb 2020
13. Gupta, B., Sehgal, S.: Survey on techniques used in autonomous maze solving robot. In: 5th International Conference - Confluence Noida, pp. 323–328 (2014).
14. Santos, M.: An applied approach of intelligent control. Revista Iberoamericana de Automática e Informática Industrial **8**(4), 283–296 (2011)

Transnational Education at University Level

The Acquisition of Competences in Transnational Education Through the ePortfolio

Monika Ciesielkiewicz[1], Claire Frances Bonilla[2(✉)], and Matilde Santos Peñas[1]

[1] Complutense University of Madrid, 28040 Madrid, Spain
[2] UDIMA – Madrid Open University, 28400 Collado Villalba, Madrid, Spain
clairefbonilla@ieee.org

Abstract. An educational transformation, characterized by ICTs, programs that transcend national borders, and progress measured by acquisition of competences, is burgeoning. In this study, Spanish university students in a Preschool and Primary School Education Degree Program participated in an international service-learning project and used an electronic portfolio (ePortfolio) to document and assess the acquisition of the competences specified in the undergraduate education degree program. Participants completed a survey in which they evaluated how the use of ePortfolio helped them to acquire these competences. Though responses to the survey were consistently positive across all competences, a principal component analysis identified the following three dimensions, listed in order of eigenvalue: acquisition and use of ICTs, problem solving, and global considerations. Our results seem to indicate that both the key competences and subject-specific competences prescribed by the degree program in which the participants were enrolled, were indeed acquired. It is interesting to note, however, that the gains were in a proximal distal pattern, with global considerations being the component that accounted for the least variance among the three dimensions identified, though it reflects the transnational nature of the learning.

Keywords: Competence-based education · ePortfolio · Transnational education

1 Introduction

The first principle of the European Pillar of Social Rights states that "everyone has the right to quality and inclusive education, training and lifelong learning in order to maintain and acquire skills that enable them to participate fully in society and manage successfully transitions in the labour market" [1]. The Bologna Process made the Europe-wide integrated educational system a necessity. In the higher education context, this meant ensuring that degree programs would be "comparable and compatible, to facilitate transparency and academic recognition" [2]. In 2006, the Council of the European Union formally adopted a framework of key competences for lifelong learning, in support of this goal [3].

Á. Herrero et al. (Eds.): ICEUTE 2020, AISC 1266, pp. 75–83, 2021.
https://doi.org/10.1007/978-3-030-57799-5_8

Meanwhile, interest in the ePortfolio is growing [4–6]. The ePortfolio can be a catalyst for the shift from teacher-centered to student-centered learning [7] that competence-based education requires. The ePortfolio was added to the list of educational high impact practices (HIPs) in 2017 [8, 9] and has even been considered as a meta HIP as it supports the use of other HIPs [8].

Given these considerations, it seems likely that the use of ePortfolio within the context of competence-based education would be a beneficial practice. This study sets out to test the effectiveness of using ePortfolio in conjunction with an international service-learning project in the acquisition of the discrete competences.

The paper is organized as follows: Sect. 2 summarizes some related works, both regarding competence-based education and the ePortfolio; the methodology is described in Sect. 3, and results are presented and discussed in Sect. 4. The paper ends with conclusions and suggestions for future research.

2 Literature Review

2.1 Competence-Based Education

Competences became crucial in the European context as a result of the Bologna Declaration in 1999, which called for the establishment of the European Higher Education Area (EHEA) which should ensure education to be "coherent, compatible and competitive" for EU students and students from other continents by 2010 [9].

The Tuning project was begun the next year by a group of universities in order to begin the work of creating a system that would produce comparable higher education degrees across Europe, establishing a common system of undergraduate degrees and graduate level degrees, and a system of college credits that would be transferable throughout the participating countries [2]. Competences were the mainstay of the work of the Tuning project and became the *lingua franca* of the stakeholders in the nascent learner-centered and competence-based education system, which was being created as a direct result of the Bologna process [10]. However, the Tuning project clearly stated that competences were to be used as points of reference and not constraints for curriculum design and evaluation [2]. A distinction between learning outcomes and competences was also made by the Tuning project. The former was to be devised by academic staff, preferably with the consultation of students and representatives from the labor market, while the latter were to be developed by the student during the course of study [10].

In 2007, the European Council formally defined the key competences as "a combination of knowledge, skills and attitudes… which all individuals need for personal fulfilment and development, active citizenship, social inclusion and employment" [3, p. 3]. In this document, each key competence is considered as essential, and none more crucial than another, and it is acknowledged that there is some overlap between the competences [11]. Subject-specific competences are those that correspond to a particular program or degree. In order to create quality educational degree programs, the degree profile would have to be built around the acquisition of both key and subject-specific competences, as well as the course content [2].

In 2018, the Council of the European Union adopted a revision of the 2006 framework of key competences, in which some competences were better defined, terminology was adapted, and the names of several competences were modified to better reflect their content. Figure 1 displays the key competences as adopted in 2018.

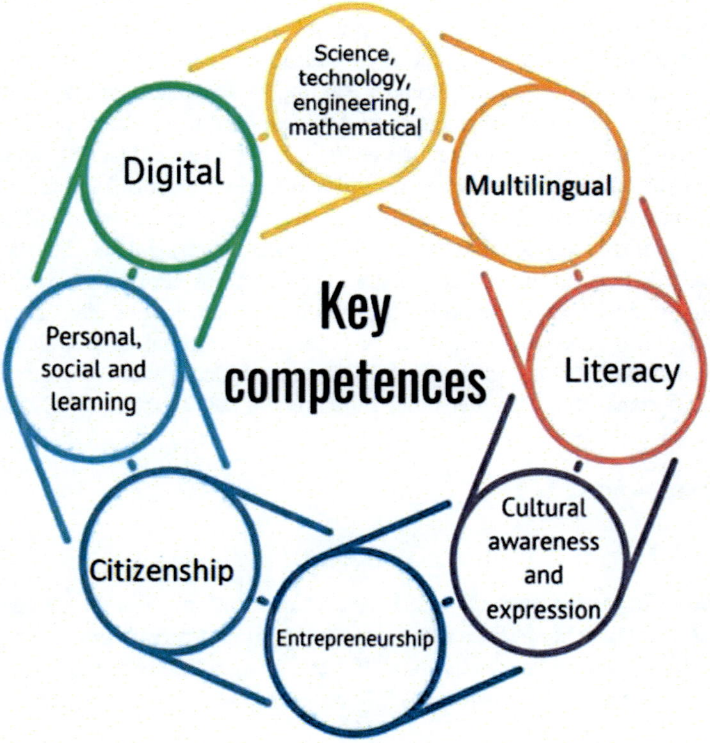

Fig. 1. Key competences as adopted by the council of the European Union in 2018 [12].

An important part of competence-based education is determining whether a student has acquired a given competence, be it a key competence or a subject-specific one. Assessing acquisition of competences can be difficult and demanding, but certainly not impossible as these assessments are made quite frequently in the business world by directors of human resources. However, in order to adequately assess acquisition of competences, it is helpful to have tools to aid in the evaluation [13]. One of these tools can be the ePortfolio, which is shared by both the educational and professional domains.

2.2 The ePortfolio

The popularity of and interest in electronic portfolios has been expanding in the field of higher education in recent years [4, 5]. However, the main focus of past research has been centered on the ePortfolio as a formative and summative evaluation tool in higher education [14]. Chen and Black [15] assert that the ePortfolio is much more than an evaluation tool and due to its multifaceted nature, it can be used as a pedagogical mechanism, a technological platform, a process and a product depending on its purpose and application.

Cochrane et al. [7] pinpoint that ePortfolios can be a catalyst for a shift from teacher-centered to student-centered learning. Watty et al. [5] also agree that the ePortfolio can prompt a powerful shift in education in general. It fosters students' autonomy, responsibility and self-directed learning by helping them take ownership of their learning, set goals, and engage in self-directed learning. Additionally, Ramirez [16] highlights that ePortfolio can generate a multi-centric learning space whilst Rowley [17] describes it as a student-owned and student-managed space.

Recent literature unveils the ePortfolio's increasing use and transformative potential to improve student learning and outcomes [6]. Numerous researchers identified a range of competences that can be developed through the use of ePortfolios, such as active learning skills, goal setting for both education and career, independent learning, collaborative learning, cross-curricular competences, interpersonal communication skills, self-assessment skills, self-regulating skills, computer skills, work readiness, lifelong learning, self-management, higher order thinking and self-awareness [18–26].

3 Methodology

3.1 Sample

In this study, the participants (N = 39) were in their second year of a four-year preschool and primary school education degree program in a private university in Spain. All the participants were between 18 and 20 years of age. Eighty five percent (85%) of them were female and fifteen percent (15%) were male. All students participated in one

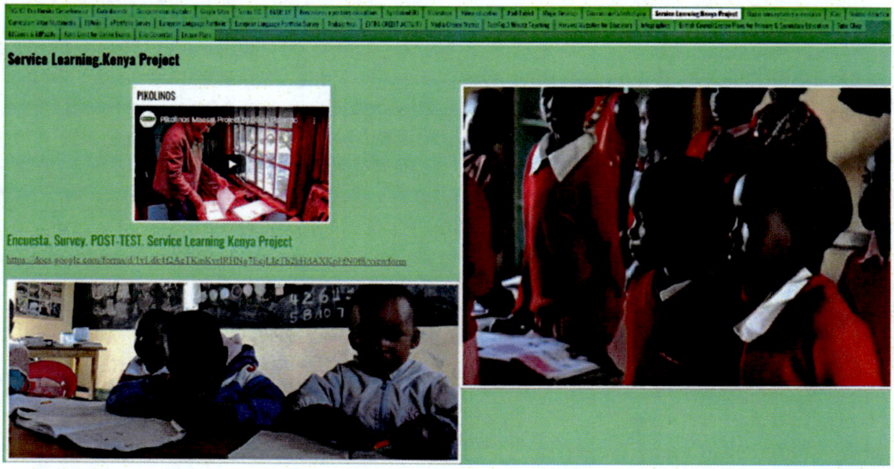

Fig. 2. Service-learning project website as part of the ICT in education course platform.

of three asynchronous undergraduate sections of an elective course on ICTs in education. This course was taught onsite by the same professor, during the second semester of the academic year, which lasted approximately 14 weeks. This course combined the use of an ePortfolio with a Service-Learning initiative to create educational materials to help educators in Kenya teach computer skills. The ePortfolios were used to document and assess the acquisition of the competences specified in the undergraduate education degree program. Figure 2 displays a graphical example of an ePortfolio.

Students designed a textbook and two activity books: basic and advanced, which covered computer skills such as using a computer, Windows, PowerPoint, Paint and Excel. They also recorded video tutorials that accompanied the textbook. These materials were revised and edited by American students from the University of Nebraska who were participating in an exchange program in Madrid. Then all these didactic materials were uploaded and shared on an ePortfolio platform (Fig. 3).

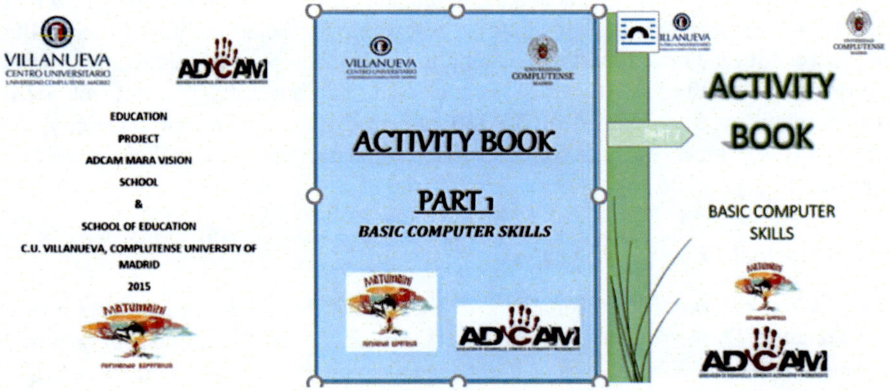

Fig. 3. Cover pages of the two activity books and text book designed by students who participated in the project.

3.2 Instrument

For the purposes of this study the authors used an online, quantitative survey, administered through Google Forms. The data are answers from this 13-item survey which used a five-point Likert-type scale (1-Strongly disagree to 5-Strongly agree).

Each item asked the respondent to assess how effective the combination of the international service-learning activity combined with the use of ePortfolio was in the acquisition of the discrete competences, both transversal in nature, as well as those specific to the undergraduate degree program in which the participants were enrolled. Students signed a consent allowing the researcher to use the data for research purposes.

4 Results

4.1 Data Screening

There were neither univariate outliers nor missing values in the data. The internal reliability was determined to be good, with a Cronbach's Alpha of .854. The responses to the items were, without exception, negatively skewed, indicating that the answers were generally strongly positive responses, which suggested that respondents agreed that they did acquire the specific competences contained in the survey.

In order to determine the factors that influenced the development of the competences, a factor analysis would be necessary. However, the authors had to determine if the data were suitable for this analysis. The correlation matrix showed that all items had at least five correlations with other items greater than 0.3, which would suggest suitability for principal component analysis [27] and that an oblique rotation would be indicated for the execution of said analysis [28]. Furthermore, all communalities were above 0.4, and 70% of the items had a communality greater than 0.6, which according to [29] indicates that the data set would be acceptable for factor analysis. Additionally, the Kaiser-Meyer-Olkin Measure of Sampling Adequacy was calculated. The KMO test returned a result of 0.748, which is considered middling suitability, but well above the minimum acceptable level of 0.5 [30]. Finally, Bartlett's test of sphericity was statistically significant ($p < .005$), confirming the suitability for principal component analysis.

4.2 Principal Component Analysis

Once having determined the suitability of the data, a Principal Component Analysis was executed. Of the oblique rotations that Version 25 of SPSS offered, Promax with Kaiser normalization was chosen as it obtained a simple structure, as defined by [31] in six iterations. This analysis detected three factors with an eigenvalue greater than one. These three factors accounted for 57% of the total variance. Table 1 shows the results of the principal component analysis.

The first factor identified, and the one that accounted for the most variance, grouped competences together that shared ITCs at their core, with elements such as learning ICTs and putting them into practice.

The second factor grouped elements together that dealt with problem solving. Competences in this group included understanding problem solving when creating the materials, learning to work with different cultures and languages and general problem-solving skills.

Finally, the last factor identified which was very close in Eigenvalue to the previous factor, was global considerations. Elements in this factor included understanding gender inequality prevalent in some cultures, as well as the value of the culture of peace and democratic values.

It is worth noting that the competences seemed to be acquired by the students in a proximal-distal pattern, meaning that the component closest to the participant, i.e. those centered around acquisition and use of ICTs, accounted for the most variance in the sample, while global considerations accounted for the least variance. This is of interest,

Table 1. Result of principal component analysis: components, elements and eigenvalues.

Component	Elements	Eigen value	% of variance	Cumulative % of variance
Using and acquiring skills, especially ICTs	• Put knowledge of ICTs into practice • Acquire knowledge of ICTs in education • Apply knowledge of ICTs in the classroom, design and select appropriate ICT activity • Develop skills for teamwork, working autonomously, and self-reflection	4.822	37.090	37.090
Problem solving	• Address situations of bilingualism, cultural diversity, especially cultural differences • Solve problems that emerged during creation of didactic materials • Develop problem solving skills, self-discipline and perseverance	1.488	11.449	48.539
Global considerations	• Promote fundamental rights and equality between men and women • Promote culture of peace and democratic values • Promote human rights and universal access to education	1.119	8.607	57.146

as the project was transnational, designed to give students the opportunity to interact with a vastly different educational system, with distinct needs and challenges.

5 Conclusions

The participants in this investigation were asked to self-evaluate their acquisition of thirteen different competences, both transversal (key) and those specific to their degree program. Quality educational degree programs should include key and subject-specific competences, as well as the course content [2]. The above research project followed these indications. The strong positive responses to each item indicate that the students believed that they had indeed acquired these competences through participating in this class that used the ePortfolio as an e-learning platform and the service-learning project in Kenya.

The findings in this research are in line with numerous studies that indicate that diverse competences can be developed through the use of ePortfolios [18–26]. Additionally, Hussain [32] states that transnational education can serve diverse purposes in different countries. However, it always benefits from the development and implementation of information and communication technologies. Transnational education offers new and unique opportunities for collaboration and cooperation to educational institutions, as well as to students who can broaden their knowledge and participate in educational experiences that they would not have otherwise [32]. The Service-Learning collaboration project studied in this paper immersed students in a very distinct culture and educational system. They had to use ICTs and adapt all the materials to the needs and cultural context of the teachers and students in Kenya. They also collaborated with students from the University of Nebraska on the elaboration of these materials. These experiences, which were possible thanks to transnational education, helped them to develop competences related to problem solving, ICTs, as well as global considerations, including cultural awareness, expression, and multilingual skills, which have been identified by the Council of the European Union as key competences and described as essential for the personal, political, and professional fulfillment of its citizens [3].

One must also take into consideration the positive impact of the pedagogical materials provided to teachers and students in Kenya through this project. As noted in previous research, transnational education can be particularly beneficial for developing countries since they often face lack of resources and quality education [32].

The limitations present in this investigation include the limited number of participants, and that the data were acquired through self-reporting. Future research may consider adding a more objective method of evaluation of competence acquisition.

References

1. The European Pillar of Social Rights in 20 Principles. https://ec.europa.eu/commission/priorities/deeper-and-fairer-economic-and-monetary-union/european-pillar-social-rights/european-pillar-social-rights-20-principles_en. Accessed 25 Oct 2019
2. González, J., Wagenaar, R. (eds.): Tuning Educational Structures in Europe: Universities' Contribution to the Bologna Process: Final Report: Pilot Project: Phase 2. Universidad de Deusto (2005)
3. European Commission: Key Competences for Lifelong Learning: European Reference Framework. Office for Official Publications of the European Communities, Luxembourg (2007)
4. Ciesielkiewicz, M., Coca, D.M.: The electronic language portfolio as a tool for lifelong learning. In: Conference Proceedings ICT for Language Learning, pp. 464–467 (2013)
5. Watty, K., McKay, J., Leanne Ngo, L., Holt, D., McGuigan, N., Leitch, S., Kavanagh, M.: Eportfolios in business education. A National Study of ePortfolio Implementation (2013)
6. Ciesielkiewicz, M.: Education for employability: the ePortfolio from school principals' perspective. Horizon **27**(1), 46–56 (2019)
7. Cochrane, T.: Secrets of learning failures: confronting reality. Res. Learn. Technol. **20**, 123–134 (2012)
8. Eynon, B., Gambino, L.: High-Impact ePortfolio Practice: A Catalyst for Student, Faculty, and Institutional Learning. Stylus, Sterling (2017)
9. González, J., Wagenaar, R. (eds.): Tuning Educational Structures in Europe. University of Deusto, Bilbao (2003)

10. Edwards, M., Sánchez-Ruiz, L.M., Sánchez-Díaz, C.: Achieving competence-based curriculum in engineering education in Spain. Proc. IEEE **97**(10), 1727–1736 (2009)
11. Halász, G., Michel, A.: Key competences in Europe: interpretation, policy formulation and implementation. Eur. J. Educ. **46**(3), 289–306 (2011)
12. Blazic, A.: EU key competences 2018. http://traveloteacher.blogspot.com/2018/03/. Accessed 1 Mar 2018
13. Lievens, R.: A proposal: mitigating effects of the economic crisis with career ePortfolios. Int. J. ePortfolio **4**(2), 157–168 (2014)
14. Wuetherick, B., Dickinson, J.: Why ePortfolios? student perceptions of ePortfolio use in continuing education learning environments. Int. J. ePortfolio **5**(1), 39–53 (2015)
15. Chen, H.L., Black, T.C.: Using e-portfolios to support an undergraduate learning career: an experiment with academic advising. Educ. Q. **33**(4) (2010)
16. Ramirez, K.: ePerformance: crafting, rehearsing, and presenting the ePortfolio persona. Int. J. ePortfolio **1**(1), 1–9 (2011)
17. Rowley, J.: ePortfolios in Australian Universities. Springer, Singapore (2017)
18. Goldsmith, D.J.: Enhancing learning and assessment through e-portfolios: a collaborative effort in Connecticut. New Dir. Stud. Serv. **119**, 31–42 (2007)
19. Kabilan, M.K., Khan, M.A.: Assessing pre-service English language teachers' learning using e-portfolios: benefits, challenges and competencies gained. Comput. Educ. **58**(4), 1007–1020 (2012)
20. Meyer, E.J., Abrami, P.C., Wade, A., Scherzer, R.: Electronic portfolios in the classroom: factors impacting teachers' integration of new technologies and new pedagogies. Technol. Pedagog. Educ. **20**(2), 191–207 (2011)
21. Lowenthal, P., White, J.W., Cooley, K.: Remake/remodel: using eportfolios and a system of gates to improve student assessment and program evaluation. Int. J. ePortfolios **1**(1), 61–70 (2011)
22. Ciesielkiewicz, M.: El portfolio electrónico como recurso educativo y su impacto en la búsqueda de trabajo. Revista Tecnología, Ciencia y Educación **2**, 83–99 (2015)
23. Mummalaneni, V.: Reflective essay and e-portfolio to promote and assess student learning in a capstone marketing course. Marketing Education Review **24**(1), 43–46 (2014)
24. Barrett, H.C.: Researching electronic portfolios and learner engagement: the REFLECT initiative. J. Adolesc. Adult Lit. **50**(6), 436–449 (2007)
25. Zellers, M., Mudrey, R.: Electronic portfolios and metacognition: a phenomenological examination of the implementation of e-Portfolios from the instructors' perspective. Int. J. Instr. Media **34**(4), 419–430 (2007)
26. Ciesielkiewicz, M.: The use of e-portfolios in higher education: from the students' perspective. Issues Educ. Res. **29**(3), 649–667 (2019)
27. Tabachnick, B.G., Fidell, L.S.: Using Multivariate Statistics, 5th edn. Pearson Allyn & Bacon, Upper Saddle River (2007)
28. Bravo-Agapito, J., Bonilla, C.F., Seoane, I.: Data mining in foreign language learning. Wiley Interdiscip. Rev.: Data Min. Knowl. Discov. **10**(1), e1287 (2020)
29. MacCallum, R.C., Windaman, K.F., Zhang, S., Hong, S.: Sample size factor analysis. Psychol. Methods **4**, 84–99 (1999)
30. Kaiser, H.F.: A second generation little jiffy. Psychometrika **35**(4), 401–415 (1970)
31. Thurstone, L.L.: Multiple Factor Analysis: A Development and Expansion of Vectors of the Mind. University of Chicago, Chicago (1947)
32. Hussain, I.: Transnational education: concept and methods. Turk. Online J. Distance Educ. **8** (1), 163–173 (2007)

A Conceptual Framework for a Communication and Collaboration Platform Within a European Transnational Logistics Knowledge Cluster of Universities and Companies

Mariela Castro Kohler and Tobias Hagen(✉)

Offenburg University of Applied Sciences, Klosterstraße 14, 77723 Gengenbach, Germany
{mariela.castro,tobias.hagen}@hs-offenburg.de

Abstract. The need for the logistics sector to timely respond to the increasing requirements of a globalised and digitalised world relies greatly on the competences and skills of its labour force. It becomes therefore essential to reinforce the cooperation between universities and business partners in the logistics and supply chain management fields across the European region and to build a logistics knowledge cluster supported by a communication and collaboration platform to foster continuous learning, skill acquisition and experience sharing anytime anywhere. In this paper we focus on designing the conceptual and technical framework for a communication and collaboration platform with the aim to establish the communication pipelines between the partner institutions, facilitating user interactions and exchange, leading to the creation of new knowledge and innovation in the logistics field. This framework is based on the requirements of the three main stakeholders: students, lecturers and companies, and consists of four functional areas defined according to the platform operational requirements. A working prototype of the platform was developed using the Moodle learning management system and its core tools to determine its applicability and possible enhancement requirements. In the next stages of the project some additional tools like a knowledge base and the integration of the partners' learning management systems to form the logistics knowledge cluster will be implemented.

Keywords: Communication · Collaboration · Elearning · Online platform · Knowledge base · Moodle · Erasmus+

1 Introduction

Globalisation, digitalisation and mobility are trends which currently have a significant impact on our society, transforming not only the workplace and traditional business models but also the skills and competencies required by the labour force [1, 2].

Logistics and supply chain management as one of the world's fast-growing sectors follows this trend, requiring flexible and attractive working environments and a skilled

Á. Herrero et al. (Eds.): ICEUTE 2020, AISC 1266, pp. 84–93, 2021.
https://doi.org/10.1007/978-3-030-57799-5_9

labour force, with internationalisation, interculturality, mobility and involvement in technological innovation playing an important role for the sustainability of the business [2].

1.1 Midplus Logistics Knowledge Cluster

The Midplus logistics knowledge cluster is an Erasmus+ funded project which aims at responding to this evolving scenario by developing an approach to promote internationalisation, innovation and exchange of good practices, reinforcing the cooperation between universities and strategic business partners in the logistics field. The project is seeking to enhance the intercultural and European mindsets of students, lecturers and companies connected to the demands of the regional, national and European/International economy [3]. Consequently, this project builds strongly on the mobility concept, considering not only student and lecturer mobility but also knowledge mobility. Based on the premise that each participating institution owns an internal knowledge cluster, the project integrates these to build a transnational knowledge chain, enriching it through the establishment of communication pipelines facilitating the active participation and interaction between partners and thus serving as a catalyst for innovation, learning, creation and transfer of new knowledge [4, 5].

Currently Midplus mainly involves three cooperating European universities of applied sciences: VIA University College (Denmark), Offenburg University (Germany) and Fontys University (Netherlands), which complement each other's competencies in the field of logistics and supply chain management. Based on their long-standing experience in the education of future logistics experts, the three institutions identified problem areas that affect the three main stakeholders - students, lecturers and companies:

- *Students* need to experience different cultures and teaching methods. They have sometimes trouble with getting merit at home university for their international learning activities; networking amongst alumni students happens rarely.
- *Lecturers* keep their expert knowledge mainly local; collaboration outside academia happens mainly with local companies.
- *Companies*: There is little knowledge about logistics education programmes. Expertise of universities is often not very well known.

To address these problem areas the following objectives for Midplus were defined:

- Promote lecturer and student mobility and experience exchange through the organisation of a series of "Topic weeks", involving all universities and offering elective courses and collaborative projects in an international and cross-cultural setting.
- Promote knowledge transfer and experience exchange in the logistics field through "company projects", with all universities and companies working collaboratively on selected case studies.
- Establish an international alumni network to promote communication between alumni and companies, offering a way for companies to locate ideal employee candidates and for alumni to seek for career opportunities.

1.2 Communication and Collaboration Platforms

To support the Midplus objectives, the implementation of an online platform as a digital service to facilitate the interactions and the management of data sources between the distinct but interdependent user groups of the project is required. Modern technologies have enabled the development and availability of numerous platforms and tools to work, communicate, learn and exchange knowledge online [6, 7], many of them published under open source licenses with source code that can be extended and modified [8]. In the logistics field there are currently online platforms available which act as transnational hubs for learning[1] or for communication and collaboration[2] or for knowledge management[3] but they do not integrate all three elements and include companies as a stakeholder.

This paper focuses mainly on the conceptual and technical design of a framework for the communication and collaboration platform as an online hub to support and integrate the objectives and activities of the Midplus logistics knowledge cluster, and the subsequent implementation of a functional prototype to be able to test usability and functionality throughout the project phases.

2 Conceptual Framework for the Communication and Collaboration Platform

We follow a software engineering methodology with requirements engineering based on user stories and agile development principles such as early user tests and continuous delivery [9]. A starting point for the conceptual design was to analyse and define use cases based on user stories of the stakeholders, the project objectives and the problem areas. As a result, a set of functional requirements was collected as shown below (Table 1):

Table 1. The three user groups and their main requirements

Students	Lecturers	Companies
- Access courses and materials - Work collaboratively in teams/projects - Participate in videoconferences - Create and search knowledge - Exchange experiences, contact experts - Communicate with participants - Search for career opportunities - Join the Alumni network	- Publish courses and materials - Publish call for projects/lectures - Run videoconferences - Create and search knowledge - Support group work - Communicate with participants - Exchange experiences - Manage calendar entries - Publish news	- Participate as guest lecturer - Participate in joint projects - Run videoconferences - Create and search knowledge - Communicate with participants - Publish career opportunities - Access student profiles for job offers

[1] e.g. https://www.mdgacademy.com/en.

[2] e.g. https://www.clustercollaboration.eu/cluster-organisations/bayerisches-it-logistikcluster.

[3] e.g. https://www.kennisdclogistiek.nl/kennisboxen.

Based on the requirements, four functional areas for the communication and collaboration platform were identified: service & support, learning, communicating & collaborating and managing (see Fig. 1):

- *Service & Support* for the administration of the platform. This includes user management, authentication and support, system updates, tool integration and further development, system security and data privacy requirements and site and learning analytics.
- *Learning* for the creation and administration of course categories, courses and projects to share materials, resources and activities. Enables learning and knowledge exchange interactions between participants and supports the required blended learning approach.
- *Communicating & Collaborating* enable both synchronous and asynchronous interactions between users. Capability to send messages and chat with individuals or groups, enable collaboration spaces to support group work, contact topic experts, and create and manage webinars and videoconferences.
- *Managing*: Users may maintain their profiles in the platform including their professional profile for career opportunities. They can share and search for knowledge in the platform's knowledge base, and receive news and announcements. This may also include alumni activities.

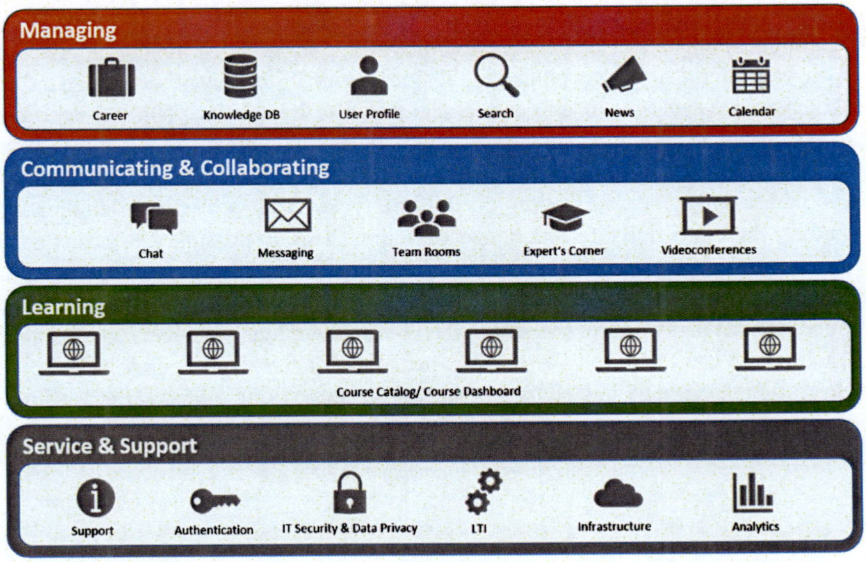

Fig. 1. Functional areas of the communication & collaboration platform

There is a significant degree of integration between the functional areas and their components. For example, communication and collaboration tools can be used inside and outside of courses. The knowledge base is built upon course content, outcomes from the collaboration activities including videoconferences or from user profiles as knowledge creators and topic experts. Service & support sustains to all other functional areas to ensure a reliable cross-operation performance between the different components. The following user roles were identified and defined to fit this structure:

- *Lecturers*: professors from the participating universities and guest lecturers.
- *Students:* students from the participating higher education institutions.
- *Alumni*: former students who were enrolled in the platform courses and work now in companies in the logistics sector.
- *Companies*: partners who work together with the universities in joint projects or as guest lecturers and who may offer career opportunities for the students and alumni. Companies provide content to the platform as experts but also will use the community functions which will be the main value add for them.
- *Platform Administrator(s)* for technical, operative and user support.

3 Design Approach and Implementation of a Prototype for the Communication and Collaboration Platform

A Learning Management System (LMS) provides a structure to include all elements and aspects of the learning processes, define learning objectives, competences and paths, carry out assessments and collect and organise knowledge [10]. Asynchronous and synchronous communication tools help to promote both formal and informal learning in a constructivist and social environment as well [11]. Current LMS platforms allow additionally the integration of third-party services, objects or applications, extending their functionality and scope of action. This is possible by means of the Learning Tools Interoperability standard (LTI), a single framework for integrating any LMS with any application through the use of a Uniform Resource Locator (URL), a key and a secret password for the required LTI-compliant tool[4] which is supplied by the LTI tool provider. These may be used as metadata for the resource link to be able to connect to the externally hosted application and launch it with a browser [12] (Fig. 2).

[4] List of LTI-compliant tools: https://site.imsglobal.org/certifications.

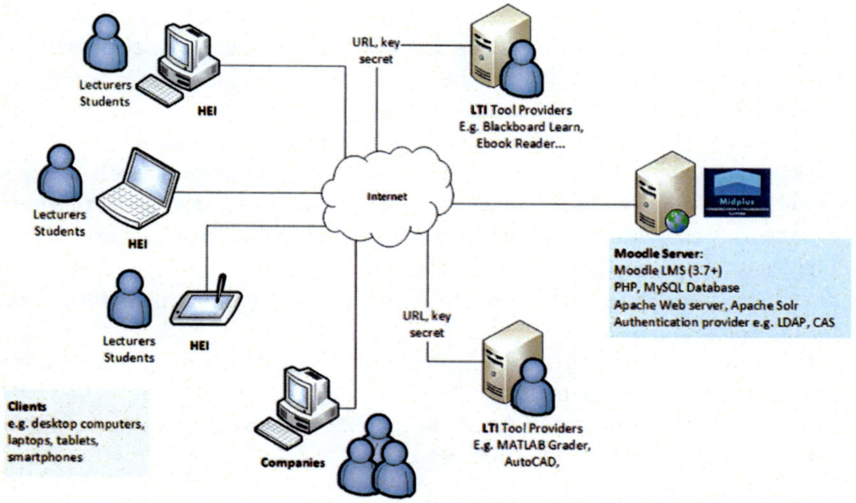

Fig. 2. Moodle topology for the communication and collaboration platform

Another approach for integration combine the local Learning Management Systems of the partner institutions into a cluster. The platform will then act as a community server which will authenticate and integrate LMS or community clients through the use of a Single-Sign-On protocol and the implementation of a middleware interface. This way, a course of one institution can be used in a course of another, or study programs can be run cooperatively [13, 14].

3.1 A Moodle Framework

The open source LMS which currently meets Midplus' requirements best is Moodle[5], a scalable and flexible LMS which offers a wide range of tools, including LTI integration, and provides a platform to experience both blended and fully online learning environments. Blended Learning, also known as Hybrid Learning, refers to the combination of virtual and non-virtual learning settings [15], an approach which perfectly supports the aspect of mobility and borderless communication and collaboration to be achieved through the platform.

Moodle is used all over the world in the higher education sector as a collaborative learning and content management system. The core Moodle system and extensions are continually being enhanced by the developers in the Moodle headquarters, who also work closely with the Moodle user community to further develop the system [16].

Similar platform projects have also considered Moodle as a suitable front-end system, for example, the European project "Open Virtual Mobility" [17] or the "Delta 3" project from three Austrian universities [18], to mention a few. Moodle was also our choice as a basis for the platform, particularly because most of the tools needed for the platform are already provided by the core system. This includes:

[5] Moodle (Modular Object-Oriented Dynamic Learning Environment): https://moodle.org/.

- A comprehensive user management with customizable user profiles to define different user groups/cohorts, including a detailed management of access rights.
- Integration with various authentication systems to be able to provide access to the platform across institutions, for example, LDAP[6] or CAS[7].
- Management of courses, course categories, materials and resources.
- Synchronous and asynchronous communication and collaboration tools to support group work, communication between participants, discussions and exchange of experiences, and also publish platform news and manage calendar entries.
- Plugins to integrate videoconference systems.
- Search functionality for courses, materials and people within the platform as a basis for the knowledge base functionality.
- A wide collection of freely-available plugins to enhance the core system and which are supported by the community.
- Responsive and customizable themes to provide consistent access to the platform and its content anywhere and from any device.
- Multi language support.
- Site and learning analytics.

The following diagram shows a Moodle framework for the communication and collaboration platform, abbreviated as MCC-platform (Fig. 3):

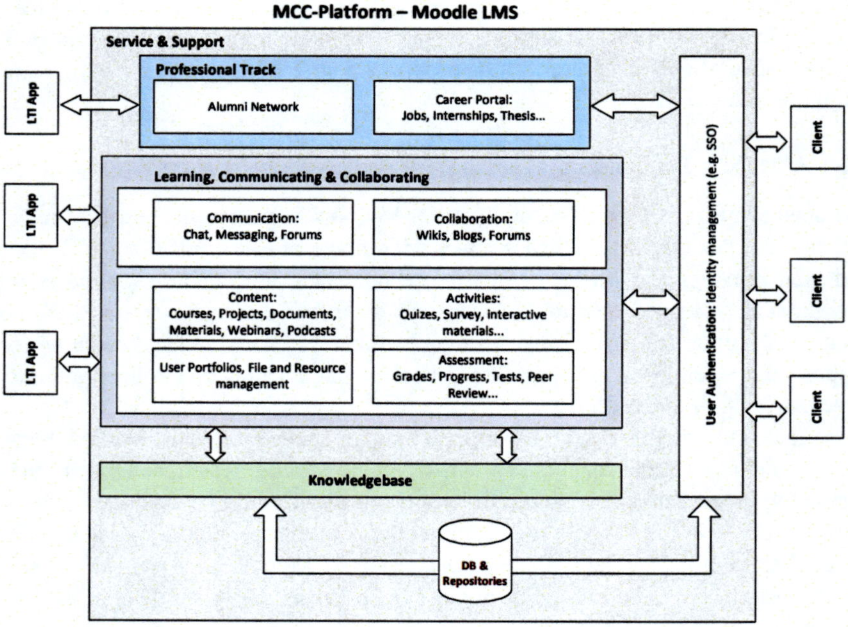

Fig. 3. Moodle framework model for the MCC-platform

[6] LDAP Server: Lightweight Directory Access Protocol e.g. MS Active Directory.

[7] CAS Server: Central Authentication Service, a single sign-on (SSO) authentication protocol.

3.2 Platform Mockups

For the platform mockups, research on different learning hubs and course aggregators[8] was carried out that included the analysis of design, structure and navigation.

Based on the platform's functional areas we developed the mockups of the prototype directly in a Moodle 3.7.2+ test server installation to be able to simultaneously evaluate different themes. The reviewed Moodle Themes were the native Boost, the popular Adaptable, EGuru and Moove[9], having chosen the last one. For the prototype, only core plugins were used, including BigBlueButton[10] as videoconferencing tool and the Moodle Global Search including the Solr extension, an enterprise-search platform[11], as basis for the knowledge base.

3.3 Current Implementation Status

The functional prototype of the communication and collaboration platform was implemented according to the designed structure and with the standard functionality provided by the core plugins from Moodle, meaning that it serves in its current state as a typical LMS, where each user has an account and authentication data and can publish/access courses and materials and can communicate and collaborate. The

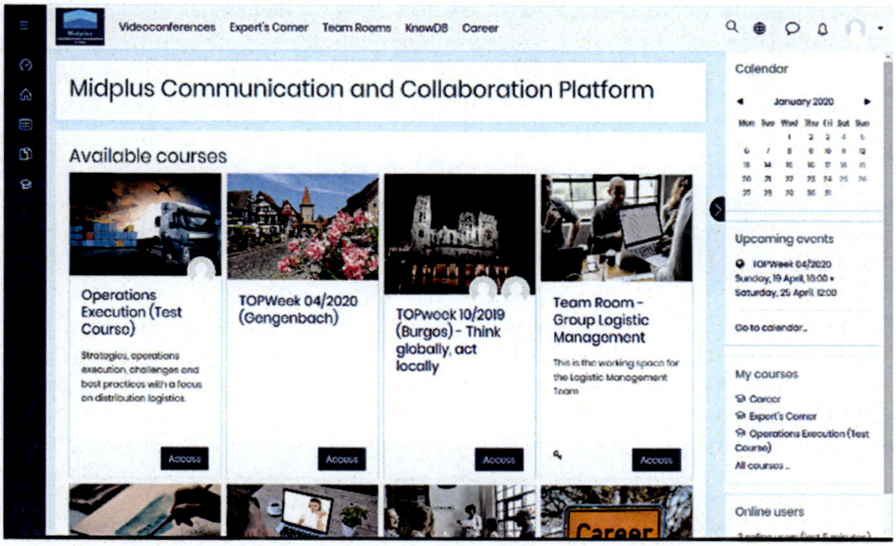

Fig. 4. The homepage of the MCC-platform prototype

[8] Course buffet: https://www.coursebuffet.com/, Oncampus Website: https://www.oncampus.de/kursangebot, Open Course World: https://www.opencourseworld.de.

[9] Moodle Theme Plugins: https://moodle.org/plugins/browse.php?list=category&id=3.

[10] BigBlueButton: https://moodle.com/certified-integrations/bigbluebutton/.

[11] Apache Lucene Solr search platform: https://lucene.apache.org/solr/.

platform was first used in this state and tested informally during a TOPWeek[12] mobility program in June 2019 serving its purpose to upload and share content and communicate with the participants (Fig. 4).

Currently, the platform is enhanced with additional functions and plugins such as competence matrix, student portfolios and a custom knowledge base plugin with a semantic search for retrieving knowledge within the platform. Furthermore, the appearance of the platform is improved by incorporating User Experience (UX) concepts [19] into the selected Moodle theme.

4 Conclusions and Future Outlook

We developed a conceptional framework for a communication and collaboration platform that will be used by a transnational logistics knowledge cluster of universities. The platform is being used by active and former students, lecturers and companies. A working prototype was implemented according to functional areas based on use case requirements of the stakeholders and the objectives of the logistic knowledge cluster. Limited by the functionality of a core Moodle LMS, the prototype is already capable of supporting the activities and interactions of a mobility program like a TOPWeek or collaboration activities within the company projects.

Ongoing and future activities will amongst others - focus on a semantic search engine and adoption of the user interface of the platform.

All additions and improvements will be continuously evaluated within the next Miplus activities, for example a first productive test will be conducted during a TOPWeek program, foreseeably in 2020. The aim is to achieve a fully operational framework by the end of the project in 2021, to be able to include other European universities and business partners in future, broadening the network and increasing the opportunities for transnational learning, collaboration and exchange of knowledge and expertise.

Acknowledgements. The authors would like to express acknowledgements and thanks to all partners in the project, especially M. Beyer-Pedersen, I. Dittrich, and S. Gruner for the initial use cases and requirements. This work is co-financed by the European Commission under the ERASMUS+ program (Project reference: 2018-1-DK01-KA203-047073).

References

1. PwC Global Homepage. https://www.pwc.com/gx/en/services/people-organisation/workforce-of-the-future/workforce-of-the-future-the-competing-forces-shaping-2030-pwc.pdf. Accessed 21 Nov 2019
2. McKinnon, A., Flöthamm, C., Hoberg, K., Busch, C.: Logistics Competencies, Skills and Training: A Global Overview. World Bank Studies, Washington (2017)

[12] The TOPWeek is another project part of the Midplus logistics knowledge cluster aimed at encouraging lecturer and student mobility.

3. European Commission Erasmus + Project's Site. https://ec.europa.eu/programmes/erasmus-plus/projects/eplus-project-details/#project/2018-1-DK01-KA203-047073. Accessed 21 Nov 2019
4. Bathelt, H., Malmberg, A., Maskell, P.: Clusters and know ledge: local buzz, global pipelines and the process of knowledge creation. Prog. Hum. Geogr. **28**(1), 31–56 (2002)
5. Maskell, P., Bathelt, H., Malmberg, A.: Building global knowledge pipelines: the role of temporary clusters. Eur. Plan. Stud. **14**(8), 997–1013 (2006)
6. OECD: An Introduction to Online Platforms and Their Role in the Digital Transformation. OECD Publishing, Paris (2019)
7. Amaro-Mellado, J.L., Antón, D., Pérez-Suárez, M., Martínez-Álvarez, F.: Game-based student response system applied to a multidisciplinary teaching context. In: Martínez Álvarez, F., Troncoso Lora, A., Sáez Muñoz, J., Quintián, H., Corchado, E. (eds.) International Joint Conference: 12th International Conference on Computational Intelligence in Security for Information Systems (CISIS 2019) and 10th International Conference on EUropean Transnational Education (ICEUTE 2019), CISIS 2019. Advances in Intelligent Systems and Computing, vol. 951. Springer, Cham (2020)
8. Opensource.com Homepage. https://opensource.com/resources/what-open-source. Accessed 09 Jan 2020
9. Kleuker, S.: Grundkurs Software Engineering mit UML: Der pragmatische Weg zu erfolgreichen Softwareprojekten. Springer Fachmedien Wiesbaden GmbH, Wiesbaden (2018)
10. Watson, W., Watson, S.L.: An argument for clarity: what are learning management systems, what are they not, and what should they become. TechTrends **51**, 28–34 (2007)
11. Jowers, I., Gaved, M., Dallison, D., Elliott-Cirigottis, G., Rochead, A., Craig, M.: A case study in online formal/informal learning: was it collaborative or cooperative learning? Des. Technol. Educ.: Int. J. **22**(1), 1–24 (2017)
12. IMS Global Learning Consortium Homepage. http://www.imsglobal.org/basic-overview-how-lti-works. Accessed 12 Jan 2020
13. Heiko Bernlöhr Homepage: ECSA – Elearning Community Service Architecture. https://www.freeit.de/documents/ecsa/ecsa.pdf. Accessed 10 Jan 2020
14. Hijar Miranda, J.A., Vázquez Sánchez, D., Vázquez Ceballos, D.E., Hernández Rubio, E., Meneses Viveros, A., Ruiz Ledezma, E.F.: Middleware to integrate heterogeneous learning management systems and initial results. Int. J. Adv. Comput. Sci. Appl. (IJACSA) **5**(10), 134–139 (2014)
15. e-teaching.org Homepage. https://www.e-teaching.org/lehrszenarien/blended_learning. Accessed 10 Jan 2020
16. Moodle Homepage. https://docs.moodle.org/38/en/About_Moodle. Accessed 08 Apr 2019
17. Buchem, I.F., Konert, J., Carlino, C., Casanova, G., Rajagopal, K., Firssova, O., Andone, D.: Designing a collaborative learning hub for virtual mobility skills – insights from the European project open virtual mobility. In: Zaphiris, P., Ioannou, A. (eds.) 5th International Conference LCT 2018, LNCS, vol. 10924, pp. 350–375. Springer, Cham (2018)
18. Csanyi, G.S., Herbst, I., Hruska, A., Kirisits, S., Michalek, C.R., Seidel, T., Weißenböck, J.: Delta 3 – moodle als gemeinsame lernplattform in einem interuniversitären e-learning strategieprojekt. In: Moodle-Konferenz-Info, pp. 1–11, Internationale Moodle-Konferenz, FH Campus Hagenberg (2006)
19. Mockplus Homepage. https://usabilitygeek.com/the-difference-between-usability-and-user-experience/. Accessed 16 Jan 2020

Variables Influencing University Dropout: A Machine Learning-Based Study

Irene Díaz[1], Ana B. Bernardo[2], María Esteban[2], and Luis J. Rodríguez-Muñiz[3](✉)

[1] Department of Computer Science, Universidad de Oviedo, 33005 Oviedo (Asturias), Spain
[2] Department of Psychology, Universidad de Oviedo, 33002 Oviedo (Asturias), Spain
[3] Department of Statistics and O.R. and Mathematics Education, Universidad de Oviedo, Asturias, 33005 Oviedo, Spain
luisj@uniovi.es

Abstract. University dropout is a growing problem having considerable academic, social and economic consequences. It may depend on several factors, such as for example the knowledge area. In previous works we studied dropout and transfer paths using machine learning, obtaining several key factors that are predictive for analyzing drop out and transfer paths patterns. In this work we delve into this topic, making a more exhaustive study using again machine learning. Results show that Polynomial SVM is the method that obtains the highest performance for predicting university dropout. On the other hand, it is possible to identify the key factors affecting university dropout, showing in addition different factors depending on the field of study.

Keywords: Dropout · Educational Data Analysis · Higher Education · Machine learning

1 Introduction

Governments had promoted widening policies for accessing higher education systems. These policies increased the interest in dropout phenomena, which was much more reduced when access was more elitist. Research about dropout grew up in parallel with the development of statistical techniques and computing capacities. Many references can be found in the literature analyzing dropout and establishing models for detecting, predicting and/or preventing it. Nevertheless, different definitions of dropout had been introduced, thus, through this paper, we assume the dropout definition used within the Spanish context, established by the National Agency for Quality in Higher Education, considering students as withdrawers when, after being registered in a program, they do not enroll again for the next two successive years. In Spain, the implementation of European Higher Education Area has led to a major concern in universities regarding student affairs, guiding different policies in most faculties, as implementing tutorial action plans, developing soft skills and cross-curricular issues, as well as other complementary

Á. Herrero et al. (Eds.): ICEUTE 2020, AISC 1266, pp. 94–103, 2021.
https://doi.org/10.1007/978-3-030-57799-5_10

activities, and, in general, a much greater monitoring of students' academic paths [1] in order to act before dropout becomes unavoidable.

In the literature it is possible to find different studies about the high cost of dropout not only for the students and their families, but also for universities and governments [2, 3]. Consequently, tools to reduce dropout rates and to mitigate the dangerous consequences of this educational problem become necessary [4].

These tools are provided by using different types of statistical data analysis [5, 6]. To overcome the difficulties of applying statistics data mining and machine learning (ML, in the following) methods [7] are considered a good tool, as well as they soften the required conditions on data. Therefore, the so-called Educational Data Analysis (EDA, in the following) includes different techniques based on ML, for making predictions or establishing classifications [8]. Focusing on ML, there are many different paradigms: lazy methods, such as KNN, methods based on tree construction as, for instance, C4.5, classification and regression trees or Neural and Bayesian networks. In predicting university dropout, the majority of the works using ML techniques try to predict students' grades or their persistence in the courses (mostly e-courses) [9, 10], but massive data from all the degrees of one university are no found in literature, except in [8] and [11], where authors used the current sample.

The aim of the present work is to undertake the analysis of dropout of a full cohort of students in the University of Oviedo (Spain) by using ML-based methods for extracting significant influent variables, so that we are able to identify dropout and attrition paths, depending on the five different study fields in which Spanish bachelor's degrees are classified.

2 Population and Sample

The population is the cohort of freshmen in academic year 2010/11 in a medium-size university (University of Oviedo) located in the Spanish region of Asturias. This university holds 54 bachelor's degrees. The studied cohort is particularly interesting since it was the first one within the Bologna process. That year, 5215 students entered the institution, two years later, 4149 students continued studying the initial degree, 363 transferred to another degree in the same university, and 658 left this university. Further details can be found in [11]. It is necessary to underline that this university is the unique one (either public or private) in the region of Asturias (Spain), therefore, for those students living in the region and dropping out, that also means either to abandon higher education or to move away from home to another university.

As the first stage of data collection, the following socio-academic information was extracted from the institutional data warehouse: identification data, sex, birthplace, nationality, disability, large family, parents' level of studies and current occupation, average score in previous education, score in admission exam to university, age when admitted, date of their first enrolment, priorities stated on the programs admission application, knowledge area corresponding to the student′s program, number of enrolled credits, passed credits and average score, scholarship holding, current academic situation and, in case of program or university transfer, destination of it.

To provide more information about students' context and situation other non-registered data were necessary. Hence, a random stratified sampling was selected to perform an ad hoc questionnaire. Lately, since questionnaire response rate is generally low when polling dropped out students, a non-probabilistic intentional sampling was performed until acceptable rates were reached in each of the strata and blocks explained below.

Spanish university programs are classified into five knowledge areas that guided a random stratified sampling, to obtain a sample of N = 1055 students, divided into two sub-samples, those who stay in the same program (N = 626) and those who dropout it (N = 429). Additionally, quotas were established in the dropout sub-sample, to distinguish those who transferred to another degree in the same institution (276) and those who left the institution (153). Thus, Table 1 shows the distribution of the sample.

Table 1. Sample distribution by knowledge area.

Knowledge area	Stay		Change		Dropout		Total	
Arts & Humanities	62	5.88%	14	1.33%	43	4.08%	119	11.28%
Engineering & Architecture	171	16.21%	51	4.83%	84	7.96%	306	29.00%
Health Sciences	83	7.87%	7	.66%	17	1.61%	107	10.14%
Law & Social Sciences	258	24.45%	69	6.54%	111	10.52%	438	41.52%
Sciences	52	4.93%	12	1.14%	21	1.99%	85	8.06%
Total	626	59.34%	153	14.50%	276	26.16%	1055	100%

This study adopted an ex-post facto design, under a holistic approach to analyze the dropout phenomenon, acquiring the university warehouse data, completed by an ad hoc design questionnaire performed either by telephone or by email to all the individuals in the sample (N = 1055). Items were related to marital status, income level, type of housing during the classes, motivations for choosing that program and this university, participation in welcome activities for freshmen and its assessment, dedication to studies, job, and housework, assessment of the demand of the program and of the satisfaction with scores, the value associated with personal relationships (professors and peers), dropout intentions and reasons, if proceed, degree of satisfaction with the University of Oviedo. The questionnaire was composed of closed questions with several alternatives varying from 3 to 8, closed questions with Likert-type scales, and few open questions. We included not only academic variables, but also others related to the student educational and socioeconomic background and current context, motivation, engagement, personal relationships within the institution, and indicators of his/her satisfaction with the institution. Variables considered where social, demographic, academic and economic ones (as age, civil status, place of residence, gender, number of family members, nationality, being handicapped or not, non-academic work and housework hours per week, parents' studies, etc.).

3 Methods

Once the information was gathered, next step was to model withdrawal as a ML problem. Thus, starting with the afore defined dataset, we try to predict Dropout, Change and Continue behaviors using ML. At the same time, we use this technique to identify factors characterizing the different withdrawal profiles.

It is well known that there is no ML method performing better than the others independently of the problem characteristics. Thus, the baseline selected method was identified after comparing different performances obtained by several techniques. We have tested different kinds of machine learners for this problem: C4.5, CART, Bayes Nets, Multi-Layer Perceptron, K-Nearest Neighbors and Support Vector Machines (SVM) [12]. All the machine learners afore mentioned were evaluated using cross-validation, with ten bins. The cross validation with ten bins is repeated 30 times. Each of the 30 times the training, validation and test sets are randomly selected. The performance of the classifiers is measured through a test set in terms of the well-known F-measure [13].

F-measure, defined as the harmonic mean between Precision (the fraction of relevant examples among the retrieved instances) and Recall (the fraction of relevant instances that have been retrieved over the total amount of relevant instances), is a quite common measure for weighting the well-known existing trade-off between Precision and Recall. F-measure, together with Accuracy are the used tools for evaluating the classifiers. Then, the method with the best F-measure across the 30 runs is selected and for this method the more relevant variables for predicting dropout are retrieved.

The experiments were developed using R version 3.5.3. and caret package version 6.0-82. The variable importance is computed using the specific caret VarImp method, a model-based approach closely tied to the model performance and that it may be able to incorporate the correlation structure between the predictors into the importance calculation. All measures of importance are scaled to have a maximum value of 100 [12].

4 Results and Discussion

In the following we present main results. As it is revealed, important and significant differences can be found regarding students' behavior about change, continuity or dropout, depending on their field of studies. Some of the differences can be explained by connecting the results of the analysis with the characteristics of the bachelor's degrees, and the academic policies and norms.

Table 2 shows the F1 averaged across the 30 runs and the standard deviation. As it can be seen, Polynomial SVM is the methods showing the best performance for predicting dropout independently on the field of study. Only in case of Health Sciences the performance of tree-based methods (C4.5 and CART) is close to the one obtained by Polynomial SVM.

Table 2. Performance of each method for the total sample. Averaged F1 plus standard deviation across the 30 experiments is shown.

Study field	C4.5	CART	Naive Bayes	Multilayer Perceptron	KNN	SVM-Poly
Arts & Humanities	.86 ±.07	.84 ± .06	.75 ± .07	.68 ± .06	.85 ± .08	**.92 ± .05**
Engineering & Architecture	.89 ± .04	.89 ± .03	.73 ± .07	.74 ± .09	.89 ± .03	**.94 ± .02**
Health Sciences	.99 ± .02	.99 ± 0.02	.66 ± .09		.98 ± .02	**.99 ± .01**
Sciences	.95 ± .06	.92 ± .04	.75 ± .10	.63 ± .03	.91 ± .03	**.97 ± .03**
Law & Social Sciences	.88 ± .03	.87 ± .03	.54 ± .03	.61 ± .14	.88 ± .02	**.91 ± .02**

The next step is to extract the most relevant variables for predicting dropout with Polynomial SVM. Figures 1, 2, 3, 4 and 5 show the relevant variables associated to each field of study and dropout situation.

As Fig. 1 shows, students enrolled in Arts and Humanities degrees are strongly influenced by the number of passed, attended and enrolled credits, as it is obvious, and this behaviour will stand in the rest of fields, so we will discuss this issue at the end of the Section. But if we look for singularities, we observe that the variable 'mother's studies' becomes important to determine whether a student changes or drops-out. Also 'relation with teachers' it is relevant to predict dropout and continuity. And, finally, the reason why they chose the degree becomes very relevant in all three classes.

Arts and Humanities degrees have, in general, a low demand. Additionally, they use to be quite connected to intrinsic motivations [14, 15]. This explains the importance of some intrinsic variables as 'studies choice reason' or intrinsically perceived, as it is 'relationship with teachers' that do not appear in other study fields. There is also a social perception about the difficulty of these studies, endorsed by the released national

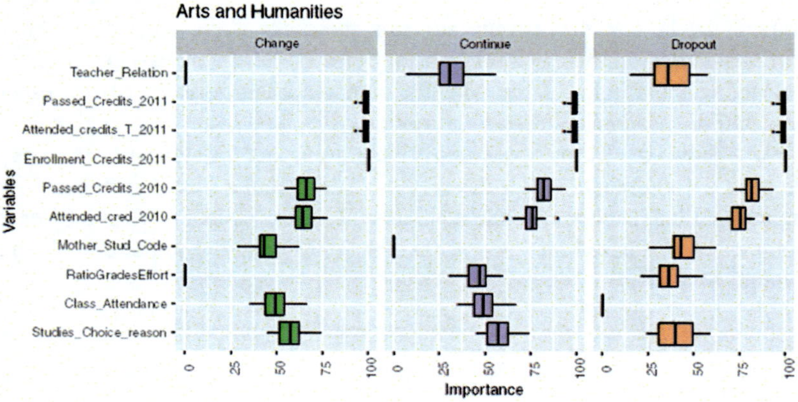

Fig. 1. Ten most important variables to predict dropout for Arts & Humanities students.

results about students' performance, which, in our opinion, justifies the importance of the balance scores/effort, that does not appear in other fields. Much more difficult to interpret is the surprising effect of mother's studies that arises only in this field. Despite several studies point out the importance of parents' studies to predict students' performance [16] we would need a deeper study in order to find a reasonable explanation about the significance of this variable in the case of Arts and Humanities students.

In Fig. 2 we can see how when considering students enrolled in Engineering bachelors the situation is significantly different. Mother's studies and relation with teachers do not have relevance now. But other variables appear: gender becomes relevant in all three classes (change, continuity and dropout), and age is important to predict continuity and dropout. Additionally, class attendance, which was relevant in Arts and Humanities for change and continuity, is now also important for predicting dropout.

Gender gap in Engineering and other STEM studies has been underlined in the last years [17], analysed data endorse the importance of this variable specifically in the case of Engineering students. It is also interesting to remark the importance of variable age, which only appears within this field. We think it is related to the average lower performance of Engineering students. For predicting change class, the admission grade and the way to get admitted to the university are also relevant (which can be High School, Vocational education or even other ways: older than 25 years, older than 45 years, etc.), being less important in the other two classes.

In Fig. 3 we also can find some singularities among Health Sciences students. For instance, the ration grades/effort is now more relevant for predicting continuity and dropout, but not so much for predicting change, which is exactly the opposite situation to that in Arts and Humanities. Class attendance is now very important for predicting continuity and change. There are also differences depending on the study. And the admission way (from High School or from Vocational Training) becomes now a relevant factor for changing.

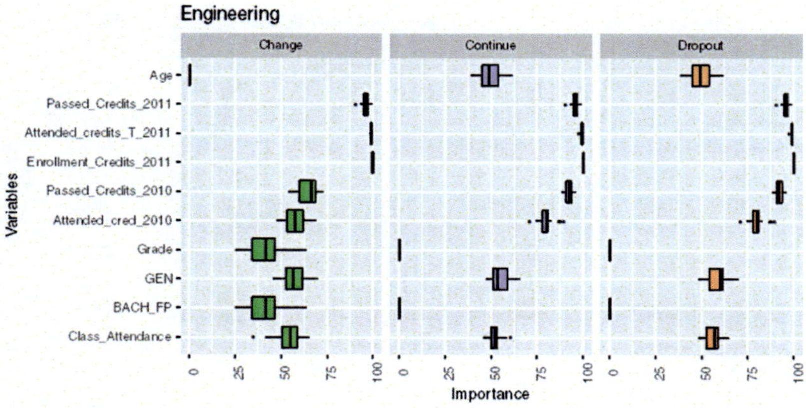

Fig. 2. Ten most important variables to predict dropout for Engineering students.

The importance of variables 'grade' and 'admission modality' (Bach/FP) is now given by a different situation to that in Engineering. In the case of Health Sciences, these are very demanded studies, in fact, the top demanded, and it becomes very difficult to get admitted (it depends only on the previous records and the scores in the university entrance exams). This situation makes that a relatively important part of students (quality analyses estimated around 5–8%) does not enter in the Medicine degree, which is the most demanded, and they enrolled into Physiotherapy or Nursery degrees, for one year, trying to improve their scores in the university entrance exam the next year and, so, they can be admitted in Medicine. Additionally, other students, instead of getting enrolled in a degree prefer to study a Vocational cycle (1 or 2 years) related to Health, which is also valid to be admitted. Thus, we can explain the relevance of the two variables by considering this situation. Gender is another relevant variable in this field that is strongly feminized, with percentages in Nursery and Physiotherapy over 85% of women.

Figure 4 shows the most important variables for predicting Sciences students' behavior. We can see how the lecture attendances increases its relevance with respect to the previous fields, particularly to predict dropout, but also in the other two classes. It also appears now the participation in the integration activities for freshmen. And priority in the election of bachelor's degree becomes significant to predict the change.

It is interesting to see the difference regarding integration activities between Sciences (that is, experimental Sciences) and Health Sciences. While in the former what is important is the participation in those activities, especially for change and dropout classes, in Health Science, it was the utility assessment that students gave to such activities what was important to predict change and dropout. So, it seems that Health Sciences students are more demanding than those in Sciences regarding the utility than they perceive in the activities, or maybe the difference is caused by a better profiling in the design of the activities in the case of Sciences.

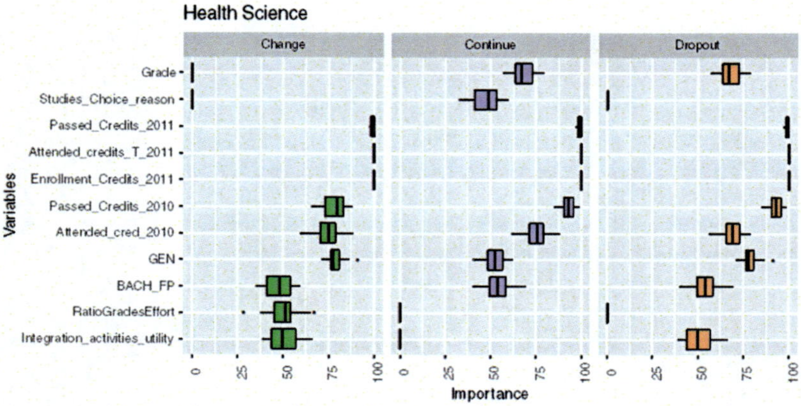

Fig. 3. Ten most important variables to predict dropout for Health Sciences students.

There is also another singularity in the case of Sciences that we think it is derived, again, from the high demand levels in the Health Sciences degrees, since an important amount of students that cannot get the admission to Health Sciences get enrolled in a Sciences degree waiting one year and trying to improve they scores. This is particularly relevant within the Chemistry degree, and, in our opinion, this is the reason of the important of the 'grade' variable in all three classes. This reasoning is also valid for explaining the importance of 'priority' variable, which is directly linked with the previous phenomenon. We also note the importance of gender (in continue class) that we think that this is linked to the STEAM gender gap. And we also notice the importance of the variable admission modality both in continue and dropout classes, which would need a deeper knowledge for providing a likely explanation.

Figure 5 illustrates the important variables for Law and Social Sciences students. This is the only field for which holding a grant (covering tuition fees and/or some expenses) becomes important, especially for continuing. Gender issues again raise in importance (as in Engineering), as well as the admission way (as in Health Science) and class attendance.

This is, followed with Engineering, the field including the greatest number of students in our university, and, subsequently, the largest quota in the sample. These studies offer many positions and the demand is not quite high, so that, it is not difficult to get admitted, making many students balance their expectations, and choosing this study field, which has quite good rates of incorporation into workplaces. Therefore, we think holding a grant becomes more important now, since the continuity can depend on whether the student keeps the grant or not. We do not have a supported explanation for the importance of gender, and it would need to be deeper investigated. It seems more natural to explain the importance of class attendance, in all three classes, not as relevant as in the case of Sciences, but still very important, which is consistent with what literature has underlined from the beginning of dropout studies [6].

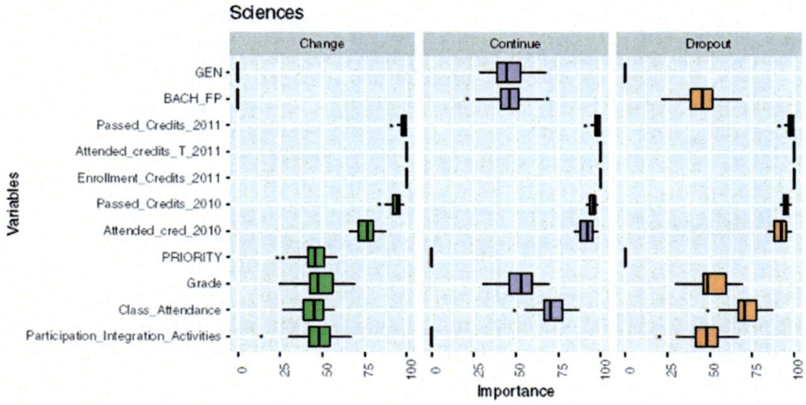

Fig. 4. Ten most important variables to predict dropout for Sciences students.

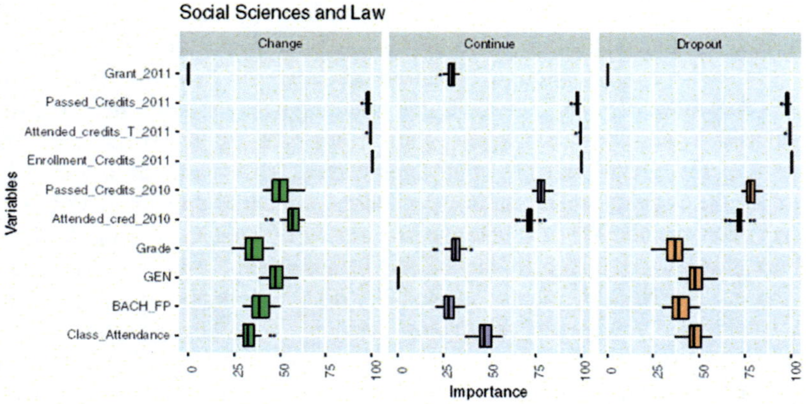

Fig. 5. Ten most important variables to predict dropout for Law and Social Sciences students.

Finally, a comment about the importance of the academic variables, which has been remarked in the literature [3]. We have also to consider that there is an academic policy about a minimum number of credits obtained every year that strongly influence on students' decision, even forcing to dropout the degree. Nevertheless, what is interesting in this study is checking how, being the same academic policy for all the fields, some academic variables have different importance in the five fields. While passed, attended and enrollment credits in 2011 have a very similar behavior all along the five fields and the three classes, the number of passed and attended credits in 2010 have different behaviors. The number of passed credits in 2010 is extremely important to predict all three classes in Sciences students. It is very important in Health Sciences and Engineering to predict continuity and dropout. And even being quite important, the relevance is reduced in the rest of the combinations of classes and fields, being the lowest (every around 50%) in the case of change for Law and Social Sciences. Particularly, the pair change-Law and Social Sciences is the only case in which the importance of the passed credits in 2010 is lower than that of attended credits in the same year. In the rest of the cases, holding an obvious correlation, attended credits have always lower importance than passed, being the highest difference between them in the case dropout-Health Sciences.

5 Conclusions

We have shown that by using ML methods, particularly Polynomial SVM, we can identify the most relevant variables influencing dropout in a cohort of undergraduate students, classified by the field of study. The relevance of the results yields on identifying different variables for different fields of study, this fact indicates that in order to analyse dropout educational measures must be necessarily different depending on the field of study. In a future research we are going to identify risky profiles so that we can provide stakeholders tools to locate them and plan customized interventions to prevent and reduce dropout.

References

1. Arce, M.E., Crespo, B., Míguez-Álvarez, C.: Higher education drop-out in Spain - particular case of universities in galicia. IES **8**(5), 247–264 (2015)
2. Braxton, J.M., Milem, J.F., Sullivan, A.S.: The influence of active learning on the college student departure process: toward a revision of tinto's theory. J. High Educ. **71**(5), 569–590 (2000)
3. Tinto, V.: Theories of student departure revisited. In: Smart, J.C. (ed.) Higher Education: Handbook of Theory and Research, vol. 2, pp. 359–384. Agathon Press, New York (1986)
4. Chen, R.: Institutional characteristics and college student dropout risks: a multilevel event history analysis. Res. High. Educ. **53**(5), 487–505 (2012)
5. Bernardo, A.B., Cerezo, R., Rodríguez-Muñiz, L.J., Núñez, J.C., Tuero, E., Esteban, M.: Predicción del abandono universitario: variables explicativas y medidas de prevención. Fuentes **16**, 63–84 (2015)
6. Esteban, M., Bernardo, A.B., Rodríguez-Muñiz, L.J.: Permanencia en la universidad: la importancia de un buen comienzo. Aula Abierta **44**, 1–6 (2016)
7. Cambruzzi, W.L., Rigo, S.J., Barbosa, J.L.: Dropout prediction and reduction in distance education courses with the learning analytics multitrail approach. J. UCS **21**(1), 23–47 (2015)
8. Delen, D.: A comparative analysis of machine learning techniques for student retention management. Decis. Support Syst. **49**, 498–506 (2010)
9. Rovira, S., Puertas, E., Igual, L.: Data-driven system to predict academic grades and dropout. PLoS ONE **12**(2), e0171207 (2017)
10. Tan, M., Shao, P.: Prediction of student dropout in E-learning program through the use of machine learning method. iJET **10**(1), 11–17 (2015)
11. Rodríguez-Muñiz, L.J., Bernardo, A.B., Esteban, M., Díaz, I.: Dropout and transfer paths: what are the risky profiles when analyzing university persistence with machine learning techniques? PLoS ONE **14**(6), e0218796 (2019)
12. Kuhn, M., Johnson, K.: Applied Predictive Modeling. Springer, Heidelberg (2013)
13. Powers, D.M.W.: Evaluation: from precision, recall and *F*-Measure to ROC, informedness, markedness & correlation. J. Mach. Learn. Technol. **2**(1), 37–63 (2011)
14. Areces, D., Rodríguez-Muñiz, L.J., Suárez-Álvarez, J., de la Roca, Y., Cueli, M.: Information sources used by high school students in the college degree choice. Psicothema **28**(3), 253–259 (2016)
15. Rodríguez-Muñiz, L.J., Areces, D., Suárez-Álvarez, J., Cueli, M., Muñiz, J.: ¿Qué motivos tienen los estudiantes de Bachillerato para elegir una carrera universitaria? Rev. Psicol. Educ. **14**(1), 1–15 (2019)
16. Stephens, N.M., Fryberg, S.A., Markus, H.R., Johnson, C.S., Covarrubias, R.: Unseen disadvantage: how American universities' focus on independence undermines the academic performance of first-generation college students. J. Pers. Soc. Psychol. **102**(6), 1178–1197 (2012)
17. Legewie, J., DiPrete, T.A.: The high school environment and the gender gap in science and engineering. Sociol. Educ. **87**(4), 259–280 (2014)

Prospective Teachers Creating and Solving a Probability Problem: An Exploratory Study

Marlén Alonso-Castaño[1](✉), Pedro Alonso[3], Maria Mellone[2], and Luis J. Rodríguez-Muñiz[1]

[1] Department of Statistics and Operational Research and Mathematics Education, University of Oviedo, Oviedo, Spain
alonsomarlen@uniovi.es
[2] Dipartimento di Matematica e Applicazioni "Renato Caccioppoli", Università degli Studi di Napoli "Federico II", Naples, Italy
[3] Department of Mathematics, University of Oviedo, Oviedo, Spain

Abstract. Probability teaching is conditioned by the teacher's knowledge, not only being related to the mathematical knowledge but also the pedagogical knowledge. In this paper, a content analysis has been used in an exploratory study to determine how prospective teachers manage to pose and solve a probability task in a specific educational context. Through this content analysis we have obtained a classification of prospective teachers' productions, according to how they pose and solve the problem and how they succeed or fail in adapting the problem to a certain level. Results indicate that working with prospective teachers in creating their own problems help them to implement their skills for an effective teaching of the subject of mathematics. We could also observe that many prospective teachers still show lack of mathematical or pedagogical knowledge. This fact underlines the need for working with complex mathematical tasks, so that the prospective teachers engage into different mathematical processes.

Keywords: Probability · Prospective teachers · Task design

1 Introduction and State of the Art

Probability has been incorporated into primary education curriculum in Spain quite recently [1]. Current curriculum incorporates simple activities where the student can identify what is a random experiment, determine whether an event is more or less likely to occur, or make predictions about the possibilities of obtaining different results in simple random experiments, such as games. But probability instruction cannot be constrained to a limited variety of built-for-school or game situations, it has also to address to problematic situations in daily life [2].

The teaching of probability is conditioned by the knowledge of the teacher [3]. According to Shulman [4], the knowledge that a mathematics teacher must have is divided into two main domains: mathematical knowledge and pedagogical knowledge. For effective teaching of mathematics, it is necessary for teachers to have a deep knowledge in both domains.

Á. Herrero et al. (Eds.): ICEUTE 2020, AISC 1266, pp. 104–113, 2021.
https://doi.org/10.1007/978-3-030-57799-5_11

Although many studies have been carried out in terms of teachers' mathematical knowledge, in the field of probability, we can find much less studies [5–12], so it still needs further research, especially with primary teachers.

In the aforementioned studies, prospective teachers' (henceforth, PTs or PT if singular) difficulties in probabilistic reasoning were addressed, even after having received training in this field. And not only prospective, but also in-service teachers have difficulty understanding and working with probabilistic concepts [13, 14]. Studies also detected deficiencies in combinatorial reasoning [15].

In [7] authors detected a lack of evidence of mathematical knowledge about probabilistic concepts and contents, whereas in [11] PTs showed they knew and were able to use the contents of a certain educational level to solve mathematical problems. However, researchers agree that the analysed PTs generally present a lack of mathematical and didactical knowledge, which endorses the need to reinforce initial teacher training.

Considering what type of tasks can help PTs to become skilled in their future professional activity, Llinares [3] understands that: "some of the professional tasks of the mathematics teacher are to choose, design and sequence the mathematically relevant tasks for the learning of their students" [p. 33]. Hence, asking them to design mathematical tasks allows us to analyse how their pedagogical and mathematical knowledge is. Based on these previous studies, we consider it is extremely important to work with PTs to reduce these difficulties from their initial training.

In this paper, we address the following research question: *Are prospective teachers able to pose and solve a probability problem in a specific school context using a focused statement?* By using a content analysis, we will check if the problem posing and also the problem solving are mathematically correct and if PTs' answers are adapted or not to a specific educational level, showing PTs' pedagogical knowledge.

2 Theoretical Framework

The concept of probability was introduced in primary education in most of the countries between the end of 20th century and the beginning of 21st century. In [16] it is indicated that, at this educational level, we can find several meanings of probability. In this paper we are intentionally restricting ourselves to the so called *classical* or *Laplacian* meaning of probability. When considering a random experiment with different possible outcomes, Laplace defined the probability of an event *A* as the ratio between the number of favourable outcomes to *A* and the number of possible outcomes. To use this formula, the sample space must consist of a finite number of possible outcomes, and we should have no reason to assume that any of them can occur more likely than the others, that is, that the events are equiprobable.

This meaning of probability is the most frequent in Spanish primary education textbooks [17], sometimes exclusive. Children have to apply their combinatorial reasoning for calculating probabilities, and this is often a complicated task for many of them [9]. But combinatorial reasoning is also complicated for PTs [18]. This implies that prospective teachers could have difficulties in teaching probability and its applications, especially when considering classical meaning, by using combinatorial skills.

When PTs pose and solve probability problems, they show their mathematical knowledge about probability. But, to teach this kind of content, it is also necessary to mobilize their pedagogical knowledge. In [19] two big domains are distinguished: mathematical content knowledge and pedagogical content knowledge. Developing these two domains is necessary for effective teaching. In our case, PTs should use their knowledge about probability relating it to their knowledge about how to teach probability. Being capable of adapting a content to the diverse educational levels distinguished a teacher from an expert in the field [4, 20].

Previous studies show that PTs have limited knowledge, both mathematical and pedagogical, such as [7]. Sometimes, even showing a deeper mastery of their mathematical skills, PTs have difficulties in relating this content knowledge with their pedagogical knowledge [9]. Llinares [3] indicates that providing PTs with tasks in which they must propose activities for their future students, encourages PTs to make sense of the mathematical concepts they already know. The creation of tasks will, therefore, help them to reason and reflect on their own teaching practice.

The proposed task for our group of PTs is aimed at detecting whether PTs manage to show their mathematical and pedagogical knowledge simultaneously. We would also want to observe how are these knowledges and we intend that PTs reflect on their own task. In order to determine how the answers given by the PTs are, a content analysis will be carried out [21]. Our interest is to determine whether students pose and solve problems in a way that is appropriate for the 6th grade of primary education under certain conditions and if they can reflect about their work.

3 Method

3.1 Population and Sample

During the academic year 2017/2018, the experiment was carried out with PTs from the Bachelor's Degree in primary education at the University of Oviedo. This study involved a group of 168 PTs who were enrolled in the course "Mathematics and its Didactics III", including topics of statistics, probability and problem-solving.

The experiment was carried out through the Moodle online campus of the University of Oviedo. The task, which was mandatory, was proposed to all the enrolled PTs, and 111 answers were collected (which is a response rate of 66.07%). Two empty answers were discarded. Therefore, the data consisted of n = 109 valid PTs' answers.

3.2 Instruments

PTs were asked to design and solve a probability problem, being adapted to a specific educational level in primary school. They had to create and solve their problem, and also to justify their creative process, and how their proposals were adapted to the required educational level.

The task was contextualized at the 6th grade of primary education. Therefore, our PTs have to use their mathematical knowledge for teaching to pose a problem that can be solved by using Laplace's formula and to solve it. This puts their knowledge of probability and combinatory into action. Not only their mathematical knowledge is activated with this task, but also their pedagogical knowledge, which helps them to decide which problem would be appropriate for the specified educational level.

The activity was divided into two phases, with a duration of two weeks for each. In the first one, PTs were asked to create and solve a probability problem by following a series of instructions given by the researchers (who were also part of the teacher educators team). Once the deadline for submissions had expired, the phase was closed, and the evaluation started. In the second one, PTs had to perform an anonymous co-assessment by revising the tasks performed by two of their classmates.

For this work we will focus only on the answers given by the PTs during the first phase, by analysing their creations to provide an answer to the problematic situation presented.

The statement proposed by the researchers for the study was as follows:

Within Euro currency, coins can be: 1, 2, 5, 10, 20, or 50 cents, or 1 or 2 euros.
Starting with this situation: "In my pocket I have 3 coins...", add constraints so that you can formulate a problem in which the question is: "What is the probability of having more than 2 euros in my pocket?" and the answer must be one number between .6 and .75 (both included).

Although it may seem to be a simple task, it presents a great complexity, both pedagogically and mathematically, when we ask for a mathematically correct problem to be posed, solved and adapted to a specific educational level.

Regarding the mathematical concepts, it is important to stress that, without any limitation, if we solve this problem considering that in our pocket we can have any combination of the coins (1, 2, 5, 10, 20, and 50 cents, and 1 and 2 euros) and that these coins can be repeated, then we are facing a combinatorial problem in which we have combinations with repetitive elements of 8 elements using 3 every time, that makes 120 possible cases. So, it becomes obvious that trying to solve this task without any kind of restrictions, would lead to a problem that can be mathematically solved, but with great combinatorial difficulty.

The estimation of the favourable outcomes to have strictly more than 2 euros in the pocket is even more difficult to calculate, being necessary to find strategies that allow us to determine how many and which are these cases. PTs should be aware that, in order to have more than 2 euros in their pocket, they must have at least two 1 euro coins or one 2 euro coin, considering all cases in which this occurs. By noticing this fact, the counting of the number of favourable cases get considerably reduced. After a complete counting, there are 43 favourable outcomes having more than 2 euros in the pocket. But doing these counts is still a difficult process.

The adaptation of this task to the 6th grade of primary education will show if the PTs can combine their mathematical and pedagogical knowledge, since such a problem with 120 possible cases and 43 favourable cases seems to be oversized for students in primary education. PTs' knowledge in probability, combined with their pedagogical skills, that allow them to adapt the task to the requested level, will determine the effectiveness of the task for the selected level with the specified mathematical content.

3.3 Analysis

After analysing the answers, we were able to classify them according to how the proposed problem was mathematically solved by the PTs. We have also considered whether the problems were adapted to 6th grade of primary education or not. So, in our classification of the content analysis, we found correct, partially correct and incorrect answers, with the specifications given below:

- **Correct:** the problem posing, and the solving process are correct (pedagogically and mathematically).
- **Partially correct:**
  - **Regarding the statement:** the problem posing is not adequate to the level of 6th grade of primary education because the sample space is too large (restrictions would be needed), the wording makes a difficult problem, or it is ambiguous. In these answers, the problem-solving is mathematically correct.
  - **Regarding the problem-solving process:** the sample space is poorly defined because it is incomplete (but if the PTs would have considered all the possible outcomes it could be adequate to the level) or the outcomes in which the sum equals 2 are considered to be favourable. That is, pedagogically seems to be correct but mathematically it is not.
- **Incorrect:** The problem posing and the problem solving do not fit with the 6th grade of primary education and are mathematically incorrect, they do not make sense, the solving does not correspond to the statement or there is no solving process.

4 Results

Once the 111 answers have been collected, we have proceeded to their revision, discarding those that could not be considered as valid (two empty answers). So, being $n = 109$ the sample size, and after conducting the content analysis, we have obtained the following results:

- **59 answers were correct:**
  - 34 of them were mathematically correct and adapted to the educational level and completely followed the statement proposed by the researchers.
  - 25 of them were mathematically correct and adapted to the educational level, but they did not follow the suggested model, starting with *"In my pocket I have 3 coins…"* and ending with *"What is the probability of having more than 2 euros in my pocket?"*. We can distinguish two situations in these answers:
    - The statements do not begin with the sentence *"In my pocket I have 3 coins…"* nor end with the question *"What is the probability of having more than 2 euros in my pocket?"*, but PTs themselves try to find a real-life situation in contexts close to primary education students. They pose problems maintaining the use of 3 coins and asking for the probability of having more than 2 euros in their pocket, but in an unlimited manner in the wording of the statement. They are correct and appropriate to the requested educational level. For example: *"Marcos went to a bookstore to buy a new comic.*

His mother gave him 3 euros, but the comic costs between 5 and 6 euros. In his pocket he still has three coins that can be 2 euros, 50 cents or 20 cents What is the probability of Marcos having more than two euros in his pocket?".

- The statements begin with the sentence *"In my pocket I have 3 coins..."* but they continue by adding extra coins. This is because they consider that *"3 coins"* correspond to 3 coins of 1 euro each (this is a contextual problem for using the Spanish expression "monedas de euro" as a synonym of "monedas de 1 euro"). Assuming this, in order to create a situation in which the probability of having more than 2 euros in the pocket could be calculated, they need more coins of other different values (not just 1 euro) so that the probability is not always 1, and they can adapt the result to the requested interval. In most cases they maintain the initial structure *"In my pocket I have 3 coins..."*. Other PTs make a free interpretation of the statement, as in the previous section, but they also add more coins.

- **13 answers were partially correct**
  - **Regarding the statement:** There were 6 answers in which the sample space was too large (for example, some of the problems would exceed 64 possible outcomes) for 6th graders. Thus, this implies that there is no adaptation to the level requested. When trying to solve the problem, these PTs involuntarily consider fewer possible outcomes, which results in a simpler resolution. If their sample space were correct, then the problem-solving would be mathematically correct. We can also find some ambiguities in the statement in some of the problems, but they can be easily solved by restructuring the wording.
  - **Regarding the problem-solving process**
    - There were 4 answers where the PTs considered as a favourable case to having more than 2 euros, that the sum of the coins that we have in our pocket equals 2 euros. This increases the number of favourable outcomes, so that it is easier to adapt the solution to the requested interval. However, this error does not seem to come from adapting the solution of the problem to the interval, but it looks more like a mathematical mistake, by confusing *"more than"* and "*greater or equal to*".
    - There were 3 answers having an incomplete sample space but, if we complete the sample space, the proposal would be adequate for the level.
- **37 answers were incorrect:**
  - 9 of them neither followed the suggested structure for the wording (starting and ending sentences) and the problem-solving process was also mathematically incorrect.
  - 4 of them used too large sample spaces, so they were not adapted to the 6th grade of primary education, and the problem-solving process did not make sense.
  - In 3 of them there was no solution to the problem, only the statement was provided.
  - 21 of them present one or more than one of the sections considered in the incorrect classification which were not mentioned in the previous points.

It is important to remark that, in most of the cases, the lack of adaptation to the level is due to the lack of restrictions to the problem. Adding restrictions would considerably reduce the sample space, but in many PTs' answers too many coins were considered without providing further restrictions.

A very frequent tendency is that PTs fixed the value of one of the coins (usually 1 euro coin, but also 50 cents, or the case in which the value of the first coin can be 1 or 2 euros) so that the sample space was considerably reduced, and its construction became much simpler. Even there were some cases, the value of the second coin or the possible values of all the three coins were fixed. In total, we have found 21 answers in which the value of at least one of the coins is fixed. Only 3 of these answers were incorrect, being 8 correct and 10 partially correct.

Another tendency in the answers was to work with a sample space without the smaller coins, using only 20 and 50 cents, and 1 and 2 euro coins; or, additionally, working with only one 10 or 20 cents coin. It seems that PTs assumed that the use of small coins hardened to obtain probability results within the requested interval [.6, .75].

5 Discussion and Conclusions

For this work, a task has been created for PTs in the third year of the Bachelor's Degree in primary education in which, under pre-established conditions, PTs had to create and solve a probability problem adapted to the 6th grade of primary education.

The aim of this task was to detect whether PTs manage to show their mathematical and pedagogical knowledge simultaneously when they design, solve and justify their election of a probability task, strongly related with combinatorial thinking to determine what kind of mathematical knowledge do they use in this specific school context.

After collecting data and, subsequently, analysing them by using a content analysis we created a classification of types of answers, depending on the mathematical and pedagogical knowledge showed by the PTs: mathematically correct, partially correct or incorrect and their adaptation to the level in each section. This classification can provide a basis for analysing similar data in the same educational context, and new categories can be added if new data samples are collected.

We have been able to observe that more than half of the PTs participating in the study (59 of them) managed to create a statement and a result appropriate to the 6th grade of primary education. This is a large group of PTs who showed evidence of deep mathematical and pedagogical knowledge at the same time which is not usual in large-scale studies like this one.

But there is also a large group of PTs who had difficulties in developing the task so, this is consistent with other previous studies [7, 13–15], and this fact endorses the importance of emphasising both the pedagogical and the mathematical instruction during initial teacher training. As noted in [10], it has been found that both the teaching of probability and the probability itself, are complicated for too many of our PTs, who showed a lack of combinatorial reasoning since, in many cases, PTs were not able to determine the different combinations of coins, both possible and favourable. At the same time, we were faced with situations in which PTs were not able to correctly determine the sample space, as in [10, 17]. We can conclude, therefore, that the treatment of chance and the use of combinatory greatly complicated the knowledge of the subject and its teaching. This produces PTs having a limited subject content and pedagogical content knowledge about probability, being this result consistent with the findings in [7].

In summary, despite having obtained good results in half of the analysed answers, there are still many PTs with difficulties, both in the development of their mathematical and pedagogical skills. This fact leads us to some conclusions. The first one regards the instructional process for future. In Spain in terms of teacher training, there are differences in the teaching of mathematics and its didactics among universities. In some of them (as in the University of Oviedo) the teaching of mathematics and its didactics are combined within the same subject while, in others, there are courses for mathematical contents separated from the didactics courses. Reinforcing their mathematical and pedagogical skills all together is essential, given that many PTs have insufficient pedagogical and mathematical knowledge, as it was also pointed out in [22]. We think that the obtained results endorse the structure of combined instruction in the mathematical and the didactical contents, since we can control the interactions between both domains.

A second conclusion regards the type of task, and we consider it is the main finding of this paper, especially relevant in the case of probability education. We have formulated a task that can be considered to be rich. When we consider it under the perspective of the NCTM [23] processes we are also clearly mobilizing PTs' abilities for problem solving (obviously, because they have to solve the problem they create), communication (since they have to correctly explain their mathematical thinking), connection (due to the need of computing and calculating different ways to sum more than 2 euros with 3 coins), and representation (because the problem was contextualized in a daily-life situation, so that they have to exchange the verbal and the probabilistic representations of the phenomenon). Moreover, they could also mobilize their ability for reasoning and proof, if they analyse different situations with the coins when trying to define the problem and to adapt it to the required educational level. Thus, we are convinced we have to keep on working in this type of tasks as the way to develop a proper mathematical knowledge for teaching probability in PTs.

The main limitation of this work yield on the non-random sampling, that prevent us from knowing the answers of one third of our PTs who simply dropped-out the course. Nevertheless, the sample size is quite large, especially since we are performing an exploratory analysis, so we did not look for statistical representativeness of the sample in terms of the population of PTs in Spanish universities. Nevertheless, results are mostly consistent with previous studies in other universities, therefore, this support their validity.

As a continuity of this research line, in a forthcoming work, we will analyse the data according to this classification to look for evidences of some of the sub-domains of the MKT model [19] (CCK or common content knowledge, SCK or specialized content knowledge and KC or knowledge of curriculum) to determine the relationships between the kind of answers and the different domains of the mathematical knowledge which are evident in the development of the task. This will help us to determine how is our PTs' mathematical knowledge for teaching.

Acknowledgements. This research has been partially supported by the Spanish MINECO project (TIN2017-87600-P).

References

1. Ministerio de Educación, Cultura y Deporte: Real Decreto 1513/2006, de 7 de diciembre, por el que se establecen las enseñanzas mínimas de la Educación primaria. Boletín Oficial del Estado, núm. 293 (8-XII-2006), pp. 43053–43102 (2007)
2. OECD: Marco De Evaluación y de Análisis de PISA para el Desarrollo: Lectura, matemáticas y ciencias, Versión preliminar, OECD Publishing, París (2017)
3. Llinares, S.: Experimentos de enseñanza e investigación. Una dualidad en la práctica del formador de profesores de matemáticas. Educación matemática **26**(Especial), 31–51 (2014)
4. Shulman, L.S.: Knowledge and teaching: foundations of the new reforms. Harv. Educ. Rev. **57**(1), 1–22 (1987)
5. Azcárate, P.: El conocimiento profesional de los profesores sobre las nociones de aleatoriedad y probabilidad. Su estudio en el caso de la educación primaria. Dissertation, Universidad de Cádiz (1995)
6. Cardeñoso, J.M.: Las creencias y conocimientos de los profesores de primaria andaluces sobre la matemática escolar. Modelización de concepciones sobre la aleatoriedad y probabilidad. Dissertation, Universidad de Cádiz (2001)
7. Contreras, J.M., Batanero, C., Díaz, C., Fernandes, J.A.: Prospective teachers' common and specialized knowledge in a probability task. In: RESME Seventh Congress of the European Society for Research in Mathematics Education, pp. 766–775. European Society for Research in Mathematics Education (RESME), University of Rzeszów, Poland (2011)
8. Estrella, S., Olfos, R., Mena-Lorca, A.: El conocimiento pedagógico del contenido de estadística en profesores de primaria. Educação e Pesquisa **41**(2), 477–493 (2015)
9. Gómez-Torres, E., Batanero, C., Díaz, C., Contreras, J.M.: Developing a questionnaire to assess the probability content knowledge of prospective primary school teachers. Stat. Educ. Res. J. **15**(2), 197–215 (2016)
10. Mohamed, N.: Evaluación del conocimiento de los futuros profesores de educación primaria sobre Probabilidad. Dissertation, Universidad de Granada (2013)
11. Ortiz, J.J., Batanero, C., Contreras, J.M.: Conocimiento de futuros profesores sobre la idea de juego equitativo. Revista latinoamericana de investigación en matemática educativa **15** (1), 63–91 (2012)
12. Ortiz, J.J., Mohamed, N., Batanero, C., Serrano, L., Rodríguez, J.D.: Comparación de probabilidades en maestros en formación. In: Bolea, M.P., Moreno, M., González, M. J. (eds.) Investigación en educación matemática: actas del X Simposio de la Sociedad Española de Investigación en Educación Matemática, pp. 267–276. Instituto de Estudios Altoaragoneses, Huesca (2006)
13. Batanero, C., Díaz, C.: Training school teachers to teach probability: reflections and challenges. Chil. J. Stat. **3**(1), 3–13 (2012)
14. Batanero, C., Godino, J.D., Roa, R.: Training teachers to teach probability. J. Stat. Educ. **12** (1) (2004)
15. Serrano, L., Batanero, C., Ortiz, J.J., Cañizares, M.J.: Heurísticas y sesgos en el razonamiento probabilístico de los estudiantes de secundaria. Educación Matemática **10** (01), 7–25 (1998)

16. Batanero, C., Henry, M., Parzysz, B.: The nature of chance and probability. In: Jones, G.A. (ed.) Exploring Probability in School: Challenges for Teaching and Learning, pp. 16–42. Springer, New York (2005)
17. Mohamed, N., Ortiz, J.J.: Evaluación de conocimientos de profesores en formación sobre el juego equitativo. Números. Revista de Didáctica de las Matemáticas **80**, 103–117 (2012)
18. Batanero, C., Godino, J.D., Navarro-Pelayo, V.: Combinatorial reasoning and its assessment. In: Gal, I., Garfield, J.B. (eds.) The Assessment Challenge in Statistics Education 2016, pp. 239–252. IOS Press International Statistical Institute, Amsterdam (1997)
19. Ball, D.L., Thames, M.H., Phelps, G.: Content knowledge for teaching. J. Teach. Educ. **59** (5), 389–407 (2008)
20. Cochran, K.F., DeRuiter, J.A., King, R.A.: Pedagogical content knowing: an integrative model for teacher preparation. J. Teach. Educ. **44**(4), 263–272 (1993)
21. Raigada, J.L.P.: Epistemología, metodología y técnicas del análisis de contenido. Socioling. Stud. **3**(1), 1–42 (2002)
22. Vásquez, C., Alsina, Á.: El conocimiento del profesorado para enseñar probabilidad: un análisis global desde el modelo del Conocimiento Didáctico-Matemático. Avances de Investigación en Educación Matemática **7**, 27–48 (2015)
23. National Council of Teachers of Mathematics (NCTM): Principles and standards for school mathematics, Reston, VA (2000)

Introducing Active Methodologies in Renewable Energy Engineering Bachelor in Mathematical and Numerical Analysis Subject

Oihana Aristondo[1], J. David Nuñez-Gonzalez[2(✉)], and Manuel Graña[3]

[1] Department of Applied Mathematics, University of the Basque Country (UPV/EHU), Leioa, Spain
[2] Department of Applied Mathematics, Computational Intelligence Group, University of the Basque Country (UPV/EHU), Leioa, Spain
jdnunez001@gmail.com
[3] Department of Computer Science and Artificial Intelligence, Computational Intelligence Group, University of the Basque Country (UPV/EHU), Leioa, Spain

Abstract. Renewable Energy Engineering Bachelor trains students as interdisciplinary professionals with solid knowledge in basic, common subjects of industrial engineering and specific renewable energy technologies. This bachelor considered that students should have a solid knowledge on MATLAB powerful calculation tool, graphical representation and simulation. The subject Mathematical and Numerical Analysis is a subject of the first course that need to be dedicated 37.5 h to laboratory lessons using MATLAB. For this purpose we will apply a PBL method that will involve different mathematical skills and competences necessary to acquire in the Renewable Energy Engineering Bachelor. In addition, the resolution of the PBL problem will include the implementation of other subjects of the first course of the bachelor.

Keywords: Renewable Energy Engineering · Active methodologies · Mathematical curricula

1 Introduction

The jump from Secondary Education to University is noticed by students once they have seen the bad obtained results after the first ordinary exam session they deal with. Many factors suchs as group integration problems, lack of motivation and/or requiered study methodology turn the jump a difficult wall to overcome. The aim of this work is start a discussion of some ideas to introduce active methodologies in Mathematics forRenewable Engineering Bachelor degree [1–6].

Renewable Energy Engineering Bachelor trains students as interdisciplinary professionals with solid knowledge in the basic and common subjects of industrial

Á. Herrero et al. (Eds.): ICEUTE 2020, AISC 1266, pp. 114–120, 2021.
https://doi.org/10.1007/978-3-030-57799-5_12

engineering and in specific renewable energy technologies. Students receive a solid theoretical and practical training in the fields of hydraulic, wind (terrestrial and marine), solar (photovoltaic, thermal, thermoelectric), geothermal, marine, as well as bioenergy, electric vehicles, etc.

In addition, students will know the electrical, electronic and mechanical devices that are associated with renewable energy technologies, and also the measures of saving and efficient use of energy and its applications in transport, building or industry in general. Students acquire the necessary technical knowledge to carry out projects related to the generation, transport, distribution and storage of energy.[1]

Among the main competences for which this bachelor qualifies, it is worth highlighting the following that justify the technological vertebration that it exposes throughout this article:

- Get scientific-technical capabilities for the exercise of the profession of Renewable Energy Engineer and knowledge of the functions of consulting, auditing, analysis, calculation, project, construction, maintenance and operation of facilities.
- Get knowledge in basic and specific subjects that allow the learning of new methods, theories and modern engineering tools, providing sufficient versatility so that the graduates are able to adapt to new situations in the exercise of their profession.
- Analyze and assess the social, economic and environmental impact of technical solutions.
- Apply the own strategies of the scientific methodology: to analyze the problematic situation qualitatively and quantitatively, to propose hypotheses and solutions using the own models of the engineering of Renewable Energies.
- Search and interpret relevant data to make judgments, which include a reflection on social, scientific or ethical issues, applying them to measurements, calculations, assessments, appraisals, appraisals, studies, reports, work plans and other similar work.

2 MATLAB as an Emerging Step Towards Integration of Active Methodologies

During the preparation of the Curriculum (in the 2009/10 academic year) it was considered that students should learn MATLAB from the first course, as it is a powerful calculation tool, graphical representation and simulation, which would help students solve problems including those that have no analytical solution or that it is very complicated to obtain it. Problem solving is a priority task in the science and engineering degree programs.

The use of MATLAB since the first year of the Degree in Renewable Energy Engineering could help establish and/or reinforce active methodologies in the teaching of the different subjects and their orientation towards problem solving.

[1] https://www.ehu.eus/es/grado-ingenieria-energias-renovables.

MATLAB, an acronym that stands for MAtrix LABoratory, is an extraordinarily powerful software for performing numerical calculations, graphical representations and data processing.

Since MATLAB integrates the calculation, visualization and programming in a simple environment to use and that the problems and their solutions can be expressed in a flexible way, understandable by the students without the strict requirements of the classical languages such as Java, C++, etc; The use of MATLAB has the following advantages:[2]

- Power. More complex problems can be solved in less time and with less effort.
- Ease of learning It takes much less time to learn the basics of MATLAB than with any other language.
- Emphasis on solving the problem rather than the difficulties of the programming language or the application development environment. Students have more time to reflect on the problem and analyze the solutions instead of using the effort to find or discover tricks to solve them.
- Many fewer mistakes are made, and it is not necessary in most cases to use debugging tools. Errors are shown and help is provided as the code is written.

The final objective is to be achieved in less time, so that the student can raise the problem, analyze it by dividing it into parts, translate each part into lines of code and organize the code in the form of functions, script or command calls.

3 New Horizons for Active Methodologies Integration in the Subject Mathematical and Numerical Analysis

The subject Mathematical and Numerical Analysis taught in the second semester of the first course of Renewable Energy Engineering This subject has 45 teaching hours, in the classroom, of which 30 lectures hours and another 15 attributed to group work. In addition, it has 15 h of computer laboratory sessions. However, students must devote another 67.5 h for classroom lessons and 22.5 h for laboratory lessons, see Fig. 1 and Fig. 2.

The 45 h of work in classes are devoted to study the basic concepts of the subject of Mathematical and Numerical Analysis. In the 30 h devoted to the master classes, different types of active methodologies are used, such as the inverted classroom (active learning and cooperative learning methodology) that will be subsequently used to work in groups in the other 15 h attributed to group work.

However, in this work we will focus on the 15 h corresponding to the computer laboratory session. These sessions will focus on applying the topics studied during class time of Mathematical and Numerical Analysis and other previous subjects. For this purpose we focus on Problem Based Learning methodology, hereafter PBL. In addition, as a global objective of the Renewable Energy engineering Bachelor, these classes will also help students to obtain a solid training in the MATLAB software package.

[2] http://www.sc.ehu.es/sbweb/energias-renovables/MATLAB/intro.html.

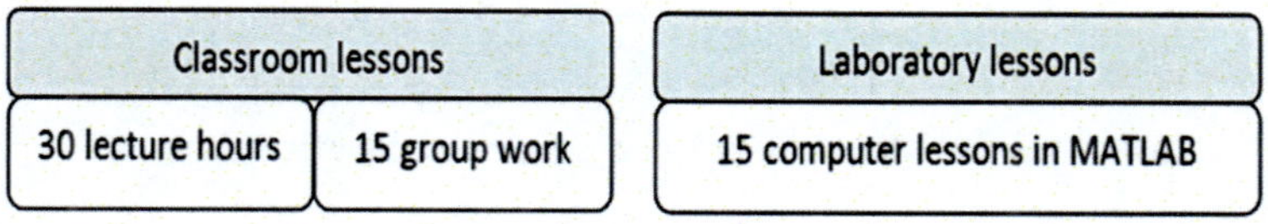

Fig. 1. Workload distribution in the school

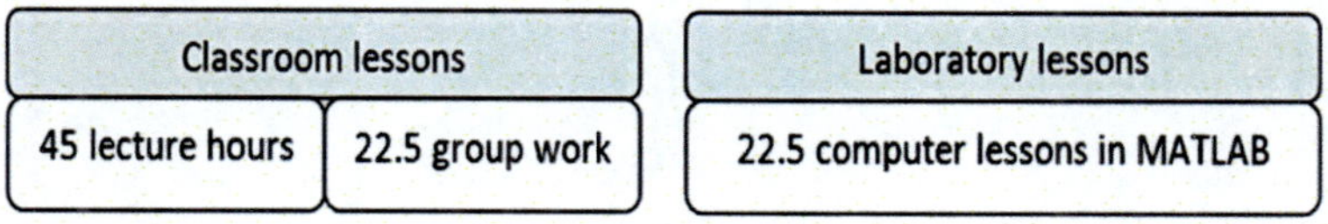

Fig. 2. Workload distribution at home

The engineering school of Gipuzkoa considered that the students of engineering in renewable energies should have a solid formation in a computer package and it bets on the MATLAB program. It should be noted, although this package is not free, there is another free analogue package, named OCTAVE, and all the results obtained with the MATLAB could be copied to the OCTAVE.

Following with the PBL method applied in the subject of Mathematical and Numerical Analysis, in this work we will build on the mathematical competence to be acquired by the Danes Niss (2003).

As Niss cite, to possess a competence in some domain of personal, professional or social life is to master essential aspects of life in that domain. Mathematical competence then means the ability to understand, judge, do, and use mathematics in a variety of intra- and extra-mathematical contexts and situations in which mathematics plays or could play a role. Niss, 1999, identify these skills in eight competencies : Thinking mathematically (1), Posing and solving mathematical problems (2), Modeling mathematically (3), Reasoning mathematically (4), Representing mathematical entities (5), Handling mathematical symbols and formalisms (6), Communicating in, with, and about mathematics (7) and Making use of aids and tools (8).

As we have commented previously in this work, we propose a proposal for the laboratory hours assigned to the subject of Mathematical and Numerical Analysis. We have 15 h of class and 22.5 h of work at home. Hence, the PBL proposed must restrict to these class and home hours. The problem could be the following shown in Fig. 3.

This resolution of this PBL problem will include the implementation other subjects of the first course of the Renewable Energy engineering Bachelor.

Firstly, students need you identify different the mathematical functions a wind blade could follow. In this case, students need the knowledge of the subject Calculus and Algebra, Informatics. Once students have identified the mathematical equations of the wind blades the will reduce all this equations using MATLAB tools that have been obtained in the Informatics subject. The distribution of the indicated subjects can be seen in Fig. 5.

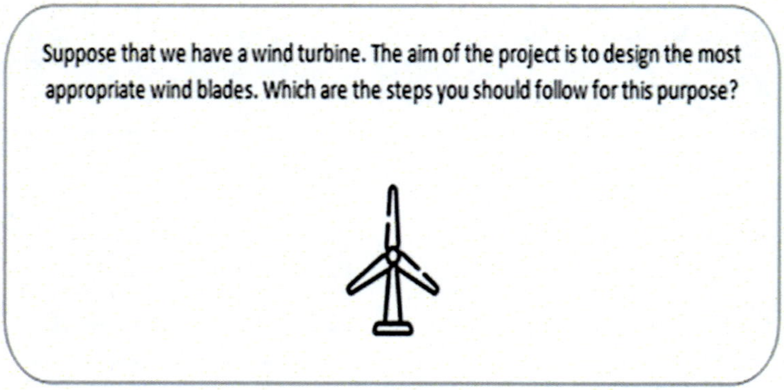

Fig. 3. PBL problem for Mathematical and Numerical Analysis subject

Clarify, that all this way will always be supervised by the teacher giving them clues for example of the mathematical equations that could have a wind blade, such as circumferences, ellipses and so on.

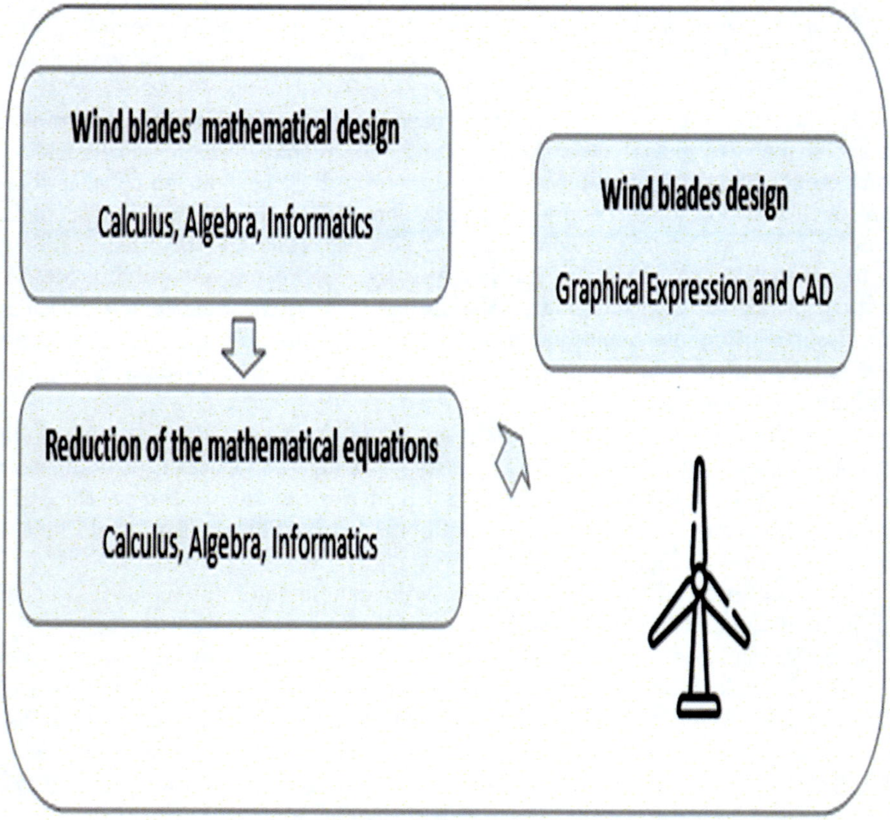

Fig. 4. Competences involved in the PBL

Subsequently, they will have to draw the wind blade to computer for later 3D printing and testing in the wind tunnel. However, in our Mathematical and Numerical Analysis subject we will focus on the design of different wind blades with different geometric shapes.

In addition, we would like to say that these PBL could be taken to the Wind Energy subject of the third year and implement the results obtained in a wind tunnel.

To finish, we want to add the competences worked in the PBL presented in the paper. I fact, this PBL will be able to provide the students with the eight competences proposed by Niss (2003), see Fig. 4.

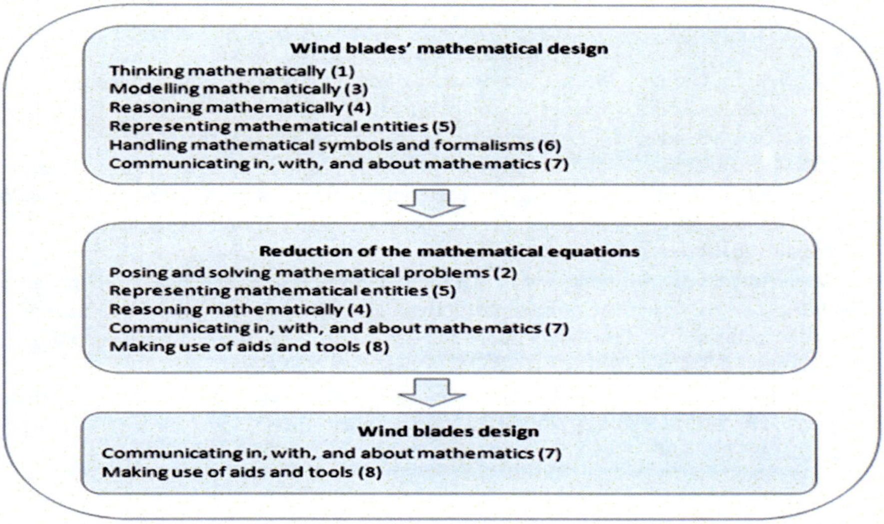

Fig. 5. Subjects involved in the PBL

4 Conclusion

Renewable Energy Engineering Bachelor trains students in different subjects in other to provide students with acknowledgment on basic, common subjects of industrial engineering and specific renewable energy technologies. This Bachelor also provide students a solid knowledge on MATLAB.

For this purpose we have applied a PBL method in the subject Mathematical and Numerical Analysis that will also provide students with different competencies as thinking, reasoning and modeling mathematically, solving mathematical problems, representing mathematical entities and communicating in, with, and about mathematics making use of aids and tools.

Acknowledgments. Work supported by FEDER funds for the MINECO projects TIN2017-85827-P and ECO2015-67519, by Departamento de Educación, Política Lingüistica y Cultura del Gobierno Vasco under the projects IT1367-19 and IT1284-19, and by UPV/EHU funding UFI 11/46 BETS.

References

1. Ayalon, M., Hershkowitz, R.: Mathematics teachers' attention to potential classroom situations of argumentation. J. Math. Behav. **49**, 163 – 173 (2018). http://www.sciencedirect.com/science/article/pii/S0732312317301141
2. Gonzalez-Marcos, A., Olarte-Valentin, R., Sainz-Garcia, E., Mugica-Vidal, R., Castejon-Limas, M.: A virtual learning environment to support project management teaching. In: Alfonso-Cendon, P., Corchado, S.Q. (eds.) International Joint Conference SOCO17-CISIS17-ICEUTE17, pp. 751–760 (2017)
3. Musgrave, S., Carlson, M.P.: Understanding and advancing graduate teaching assistants mathematical knowledge for teaching. J. Math. Behav. **45**, 137 – 149 (2017). http://www.sciencedirect.com/science/article/pii/S0732312316302012
4. Nunez-Gonzalez, J., Grana, M., Lopez-Guede, J.: Prelude of an educational innovation project: discussing a redesign of the continuous assessment in mathematics for chemistry and geology bachelor degree students. Adv. Intell. Syst. Comput. **771**, 536–543 (2019)
5. Olarte-Valentin, R., Mugica-Vidal, R., Sainz-Garcia, E., Alba-Elias, F., Fernandez-Robles, L.: Analysis of the online interactions of students in the project management learning process. In: Alfonso-Cendon, P., Corchado, S.Q. (eds.) International Joint Conference SOCO17-CISIS17-ICEUTE17, pp. 743–750 (2017)
6. Wasserman, N.H.: Knowledge of nonlocal mathematics for teaching. J. Math. Behav. **49**, 116 – 128 (2018). http://www.sciencedirect.com/science/article/pii/S0732312317300597

Estimating Expected Student Academic Performance

Walter Orozco[1,2](✉), Miguel Ángel Rodríguez-García[1], and Alberto Fernández[1]

[1] Universidad Rey Juan Carlos, Madrid, Spain
wa.orozco@alumnos.urjc.es, {miguel.rodriguez, alberto.fernandez}@urjc.es
[2] Universidad Península de Santa Elena, La Libertad, Ecuador
worozco@upse.edu.ec

Abstract. In recent decades, society has a primary need for improving education systems. Predicting the performance of students has become a reference topic very analyzed by the research community. Currently, there are several cutting-edge technologies that make very easy to collect educational data in institutional systems due to the new information management systems where almost everything is digitalized. The analysis of this information offers unique opportunities that have a direct impact on students, instructors and academic institutions programs. Concretely, we propose a modular system to evaluate teaching performance by considering several primary factors related to students learning process. In this work, we present the first module, a statistical model that aims at obtaining the expected student's achievements in a particular course. We first analyzed students' performance primary factors on Higher Education Systems. To identify these factors, we have conducted a literature review. Then, we examine three different techniques to build the prediction model: Multiple Linear Regression (MLR), Support Vector Machine (SVM) and Artificial Neural Networks (ANN). To select the suitable technique, we carried out a variety of experiments by using a real dataset from University of Santa Elena (Ecuador). The achieved experiments revealed promising results.

Keywords: Machine learning · Prediction algorithms · Academic performance

1 Introduction

Evaluating teaching activity is a fundamental task in current university systems. Professors are essential elements of the education process and student performance is highly correlated with the performance of their instructors. Thus, higher education institutions are interested in evaluating the efficiency of their employees so as to take actions to improve the overall system.

Several models have been defined to evaluate teaching performance, which usually combine different types of data that are processed by innovative technologies. Traditionally, the basis for the evaluation of teaching activity has focused on student achievements and opinion polls [1]. More elaborated models such as Value Added Model (VAM) [2] and Linear Hierarchical Model (HLM) [3] are based on statistical

Á. Herrero et al. (Eds.): ICEUTE 2020, AISC 1266, pp. 121–131, 2021.
https://doi.org/10.1007/978-3-030-57799-5_13

techniques that use information generated from students achievements in academic processes. However, to evaluate the teaching effectiveness, it is primary to not only considering information of student grades, due to there are other factors such as background, group size, resources, etc. which have usually a direct influence on students' performance. With the goal of evaluating professors' performance accounting for multiple criteria, several accreditation agencies have proposed complex methods (e.g. DOCENTIA [4], NBPTS [5]).

Currently, available technologies allow professors to generate information that can be further analyzed for the improvement of teaching and learning processes [6]. In some cases, data collection instruments such as surveys are used, in others, data are obtained from institutional databases [7].

Our research aims at developing a framework to evaluate teaching performance that uses objective information available in institutional information systems. In particular, our approach combines four main components, namely, Student Achievements, Professor Profile, Academic Process and Support Activities [8].

This paper focuses on the student achievements component. We present a method for predicting student grades in a particular subject. The motivation is that professors' performance can affect positively or negatively on students' results. Thus, analyzing the deviation with regards to the expected achievements is an indicator of the instructor's performance. We analyze the three algorithm Multiple Linear Regression (MLR), Support Vector Machine (SVM) and Artificial Neural Networks (ANN) to predict the academic performance of the students by using data obtained from the Nursing program at University of Santa Elena (Ecuador).

The rest of the manuscript is organized as follows: in Sect. 2, relevant works related to predicting students' performance are reviewed. In Sect. 3, our framework for evaluating professors' efficiency is summarized so as to set the context of this paper. Section 4 details the proposed predicting model, which is evaluated in Sect. 5. Finally, we conclude the paper and point out to future research lines.

2 Related Works

There are several studies associated with the identification of primary factors related to the academic performance of university students. Each investigation develops its own methodology and validation strategies. Here, we present some of the variables usually included in those performance studies: Performance self-efficacy, Grade goal, Frequency of class attendance, High school Grade Point Average (GPA), Admission test results, Effort regulation, Strategic approach to learning, Achievement motivation, Academic self-efficacy, Intelligence, Conscientiousness, Financial support, Concentration, Academic goals, Emotional intelligence, Need for cognition, among others [9, 18]. In this line, Schneider and Preckel [10] presented a systematic review that includes 105 variables associated with students' academic performance. They classified and ranked them according to their effect on student achievement. In a similar study, Richardson et al. [11] review the research on factors associated with academic performance. In this case, each factor is considered as an independent variable that must be analyzed in relation to the output variable represented by the student's final average. They only take into account relevant variables those with a Pearson's correlation index higher than 0.6.

From the point of view of applied technologies, there is a variety of works to estimate the students' academic performance. Shahiri et al. [7] provide an overview of the data mining techniques that have been used for this purpose. Their work also indicates how to use the prediction algorithms to identify the most important attributes associated with the student, including Artificial Neural Networks (ANN), Decision Trees (DT), Support Vector Machines (SVM) and Multiple Lineal Regression (MLR), among others.

Ibrahim and Rusli [12] describe a methodology that evaluates the results generated by applying ANN, DT and Lineal Regression (LR). The used variables describe the student's demographic profile (Information technology application knowledge, previous school, programming knowledge and family financial status) and Cumulative Grade Point Average (CGPA).

Yang et al. [6] use Multiple Linear Regression (MLR) to predict students' success. Inputs are associated with activities such as video review, work at home, exercises and quizzes, applied during an academic period. Variables are selected applying Principal Component Analysis (PCA).

Huang and N. Fang [13] focuses the analysis on the prediction of student performance in the course "Dynamics" in an Engineering degree. The inputs of the model are the average grades of the previous subjects Statics, Calculus I, Calculus II, and Physics. The validation strategy includes experimentation with four types of mathematical modeling techniques: MLR, MultiLayer Perceptron (MLP), Radial Basis Function (RBF) network, and SVM. They analyze six combinations of predictor variables with the idea of identifying the best combination.

Iyanda et al. [14] use student achievement as a single predictor variable and select the best algorithm among MLP and Generalized Regression Neural Network (GRNN) for Computer Science students.

After this intensive literature review, we construct the pipeline to develop the proposed statistical model. First, we identified the principal variables utilized to train it. We implemented several models by using different techniques, and finally, we tested the model performance by using statistical methods used by other works too.

3 A Model for Analyzing Professor Performance

In this section we summarize a model we are developing for analyzing teaching activities by professors at University. More details can be found in [8].

Our proposed assessment method, which is depicted in Fig. 1, comprises four components that analyze different aspects of professors' teaching activities. The results of each element are combined into a final number that represents the professor's performance according to the model.

The **Student Achievements** component uses as inputs students results obtained in their evaluation process. The basic motivation of this module is that good professor's performance causes better students' results. Student performance is associated with many factors both internal and external to the institution [11]. The objective of this component is to estimate, for a given subject, the grades students are expected to obtain according to their profile and the historic results of past students. Data existing in institutional databases are used and, after some statistical analysis, are processed using AI technologies. This paper focuses on this module, thus next sections detail the process, prediction models and results obtained.

The **Professor Profile** component aims at measuring professors' academic and professional experience in the subjects they teach. Furthermore, it identifies the teaching experience in the subject and value the relationship between the knowledge requirements and training. In that context, the goal is to assess the relationship between the demands of the subjects and the characteristics of the professor, supported by knowledge representation and reasoning techniques.

The **Academic Process** component contemplates activities that complement students' academic process, for instance, to support learning and therefore a better performance reflected in their achievements. In the development of the class, it is necessary to assess the contribution of the professor in the student's performance. Metrics such as academic planning, attention to students, compliance with responsibilities, learning environment, among others, can be evaluated using machine learning techniques.

The **Support Activities** component evaluates additional activities that can affect student achievements, e.g. thesis supervision, participation in committees, postgraduate classes.

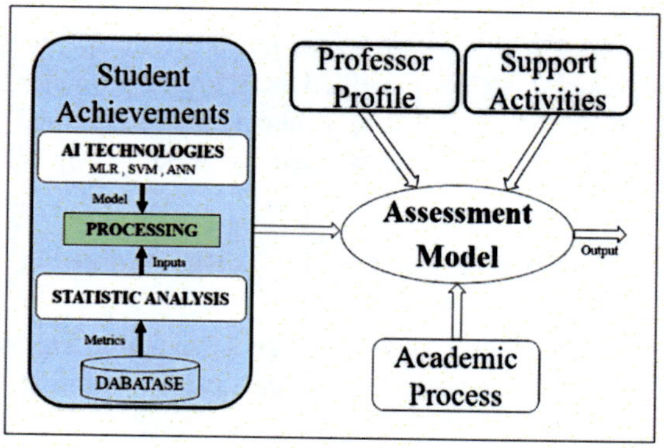

Fig. 1. Professor effectiveness assessment architecture.

4 Student Achievements Module

In this section, we analyze the Student Achievements module, which aims at obtaining information about how professor performance influenced on students' achievements. It mainly consists of a statistical model capable of predicting the academy students' performance. As inputs, the model uses real data obtained from a University information system. In the following, we describe the process pursued to develop such a model.

4.1 Primary Factors Analysis

According to the study developed by Schneider and Preckel [10], there are 105 different indicators classified into 11 categories which can be directly linked to high student achievement. Besides, they demonstrated that these correlations could be quantified by using the effect size, a statistical measure obtained through Cohen's index "d". From this list and following the authors' suggestions, we only considered those indicators with a medium ($0.35 \leq |d| < 0.66$) and large ($|d| \geq 0.66$) effect to build our features list. Then, we filtered out the resulting list finding only those which was available in the database. Here, we provide a list of features that we have used for our study:

- *Previous average (prevAvg)*: It represents the average mark of student's previous academic period.
- *Accumulated average (cumAvg)*: It stands for the cumulative student average mark before the start of a new period.
- *Subject Average (subAvg)*: It depicts the general average of the subject.
- *Period Average (perAvg)*: It defines the student's average mark in all subjects in the current period.
- *Attendance (atten)*: Is the percentage of hours the student attended to class.
- *Gender (gender)*: To represent if the student is a man or woman.
- *Scholarship (sch)*: To identify if the student has financial aid from the university.
- *Mark (mark)*: It represents the student's achievement in each subject. This is the variable to be predicted by the model.

4.2 Data Analysis

To accomplish this experiment, we have utilized a dataset that was created from the *Nursing* Program of the University of Santa Elena (Ecuador). It has a total of 2488 records, corresponding to the last five years (ten semesters), distributed among 33 subjects. Marks were normalized in the range [0, 10]. Before applying the predicting models, we analyzed which of the variables selected in Sect. 4.2 were meaningful. Thus, we applied Principal Component Analysis (PCA) [17] to analyze its correlation. Only those variables that obtained a correlation index greater than 0.5 were considered for building the prediction model [15]. Table 1 shows the results obtained from this statistical analysis.

If we analyze in detail the results (Table 1), we realize only *cumAvg*, *subAvg* and *perAvg* reached an index greater than 0.5, which means only those variables have a significant contribution. Therefore, we discarded the others.

Table 1. Nursing program variable correlation index

	prevAvg	*cumAvg*	*subAvg*	*perAvg*	*atten*	*gender*	*sch*	*mark*
prevAvg	1							
cumAvg	0.67	1						
subAvg	0.10	0.13	1					
perAvg	0.53	0.81	0.28	1				
atten	0.01	0.00	−0.01	0.01	1			
gender	0.07	0.06	0.02	0.08	−0.01	1		
sch	0.24	0.34	0.04	0.28	0.00	0.01	1	
mark	0.34	**0.52**	**0.58**	**0.64**	0.01	0.05	0.18	1

4.3 Predicting Models

To build our prediction model, we follow an iterative process which consisted of selecting a set of algorithms, establishing different settings, analyzing their behavior with different sets of chosen variables to optimize their precision. After a literature review, we decided to choose Support Vector Machine (SVM), Artificial Neural Networks (ANN) and Multiple Linear Regression (MLR). i) SVM is a supervised learning technique that utilizes hyperplanes to carry out linear or non-linear classification depending on the data distribution [19]. Concretely, in this work we used SVM with radial kernel given the dimensionality of the features; ii) ANNs are being used on a wide variety of research areas [20, 21]. In particular, in this work we used it as a supervised learning method to predict student's performance supported by a multilayer neural network; finally, iii) MLR is a supervised learning technique is a multivariate statistical technique to analyze the linear correlation between a single dependent variable and a set of independent variables and predict a dependent variable from a set of independent variables [22]. In particular, we utilized this statistical tool to analyze the influence of the primary factors of student performance on their marks.

5 Experiments

To evaluate the model accuracy, we reproduced the statistical criteria described by Piñeiro et al. [16]. In this work, the authors used Root Mean Square Error (RMSE) and Coefficient of determination (R^2) as metrics to assess model predictions. In particular, they used RMSE to analyze the model performance. Concretely, it estimates the observed values upon the regression line of observed vs predicted values. Furthermore, they include in the analysis the Coefficient of determination to measure the dimension of the total variance computed by the regression model.

As the basis for assessment, we have defined two simple predictor models based on baselines: BASELINE#1 and BASELINE#2. The former utilizes to predict the students' previous average (prevAvg) and the latter employs the students' accumulated average (cumAvg). The target of these baselines is to provide an evaluation framework that enables us to compare the more complex models against the simple ones and evidence on their improvement. For the implementation of our models we used RStudio v1.2. The parameters that we have utilized to conduct the experiments on each algorithm are: i) SVM: we have tested three types of kernel (linear, radial and polynomial) obtaining the best performance with *radial* kernel and parameters *cost* = 1, $\gamma = 0.33$ and $\epsilon = 0.1$. ii) ANN: we tested configurations with several hidden layers and neurons and chose the one that fit better with our data, resulting in an architecture with one hidden layer with 3 neurons, threshold = 0.05 and maximum number of iterations = 10000. iii) MLR: we used the default parameters.

Table 2 depicts the RMSE and R^2 resulting of applying the two baseline models and the three algorithms to a sample of 8 subjects (one per semester of the Nursing program). The last row shows the average of the whole set (33 subjects).

Table 2. Index prediction by subject and average

Subject	BASELINE#1		BASELINE#2		MLR		SVM		ANN	
	RMSE	R^2	RMSE	R^2	RMSE	R^2	RMSE	R^2	RMSE	R^2
Introduction to Nursing (S1)	0.92	0.02	0.53	0.46	0.48	0.73	0.44	0.78	0.41	0.82
Microbiology (S2)	0.75	0.26	0.67	0.48	0.45	0.83	0.43	0.84	0.43	0.85
Semiology (S3)	0.72	0.19	0.54	0.41	0.41	0.82	0.37	0.86	0.40	0.83
Physiopathology (S4)	0.70	0.17	0.64	0.26	0.48	0.74	0.47	0.75	0.48	0.73
Nursing of Mother and Newborns (S5)	0.54	0.25	0.48	0.39	0.36	0.81	0.36	0.81	0.37	0.82
Alternative Medicine (S6)	0.63	0.23	0.56	0.27	0.36	0.78	0.34	0.80	0.36	0.78
Clinical Nursing (S7)	0.78	0.03	0.59	0.38	0.26	0.91	0.27	0.91	0.31	0.91
Nursing of Public Health (S8)	0.76	0.10	0.59	0.27	0.30	0.84	0.31	0.82	0.30	0.82
...										
Average	**0.71**	**0.16**	**0.60**	**0.30**	**0.40**	**0.79**	**0.39**	**0.80**	**0.40**	**0.80**

The results show that the complex models overperform than the basic ones. For instance, taking into account the subject "Introduction Nursing (S1)", the BASELINE#1 estimator reaches an RMSE of 0.92, the higher error reported by a model in all experiments, if we compare to 0.48, 0.44 and 0.41, the errors reached by MLR, SVM or ANN respectively. Conversely, the minimum error made by the baselines models is reported in the subject "Nursing of Mother and Newborns (S5)" where the error made by BASELINE#1 was 0.52, and the BASELINE#2 was 0.42. On the contrary, if we analyze the stats reported by the more complex estimators, we will find lower errors than the ones analyzed. For instance, the best results were reported in the subject "Nursing of Public Health (S8)", where MLR, SVM and ANN reduced the error to 0.30, 0.31 and 0.30, respectively. In the case of the lower results, the experiments carried out over the subject "Physiopathology (S4)" have reported the highest RMSE value obtaining MLR 0.48, SVM 0.47 and ANN 0.48. On the other hand, if we take into account the correlation coefficient, the lowest results obtained from both baselines reveal a weak correlation between variables, what it confirms the need of a more complex estimator to be developed.

On the other hand, the results presented in Table 3 shows that the three algorithms obtain similar prediction results. The three models have a good fit (R^2 = 0.8) and low error (RMSE = 0.4, i.e. 4%). The prediction of each subject is not very different to each other. Nevertheless, we observed that, in general, predictions for courses in the first semester are worse than the others, which may be due to the fact that, in those cases, *cumAvg* is obtained from the access grade to University studies. Furthermore, we have tried and compared our proposal to other approaches to predicting student's marks. For this purpose, we have considered approaches that use the variable "student achievement-mark" in their experiments.

It is noteworthy that it is not an easy task to compare our results to other models. The reasons are due to i) their datasets are not available; ii) there are some variables which are not clearly defined; iii) they usually considered variables not available in our database. For instance, Iyanda et al. [14] use ANN to predict the average mark a student will get in the next semester based on the results obtained in subjects of the previous semester. Whereas we have that type of information in our database, we cannot compare to it since we are interested in average grades in a specific subject rather than the whole semester. Ibrahim and Rusli [12] use *cumAvg* and other variables that are not normally easy to obtain (e.g. IT knowledge, previous school, family financial status, etc.).

The only work we found to carry out some kind of comparison is Huang and Fang [13]. That work analyzes for a particular subject which other subjects are relevant, i.e. may impact on the expected results. Then, they tested different combinations of variables including *cumAvg*, grades of key subjects as well as mid results of the predicted subject. The best prediction is obtained when all variables are used.

Table 3 shows two prediction results reported by work [13]: their best model and the one only using *cumAvg* ([13] bis). In addition, we include the results of our proposal (Model1), which were already presented in Table 2, as well as set of experiments (Model2) in which we only used *cumAvg*. Therefore, [13] bis and Model2 use the same inputs and outputs of the model, but each of them tested with different datasets.

In Table 3 we show three prediction quality metrics: R^2, RMSE and APA (Average Prediction Accuracy). APA is obtained from the average relative error [13] as: 100 * (1 – Average Relative Error).

Table 3. Prediction indices for compared models.

Approach	MLR			SVM			ANN		
	R^2	APA	RMSE	R^2	APA	RMSE	R^2	APA	RMSE
[13]	0.48	89.70	–	–	90.10	–	–	89.60	–
[13] bis	0.20	88.60	–	–	86.50	–	–	88.50	–
Model1	0.79	96.10	0.40	0.80	96.30	0.39	0.79	96.10	0.40
Model2	0.54	94.60	0.55	0.53	94.60	0.53	0.52	94.50	0.55

The results show that [13] bis and Model2 differ in the MLR approach, which is because they were tested with different data. In fact, our approach is more general, while [13] requires choosing each subject carefully and requires mid-exams. Furthermore, if we analyze the best models in each paper, we can observe that a similar accuracy increase is obtained when all variables are used. Taking that observation as a reference, we can see that although SVM and ANN report similar results, the SVM has achieved slightly better results. Therefore, we finally determine that the SVM will be the estimator selected to comprise the Student Achievement module in the architecture.

6 Conclusions

Evaluating the teaching activity is a great challenge of Education Institutions. To guarantee high-quality education systems, improve the students learning process or reduce the distance between students and professors are some of the primary targets to achieve. However, the evaluation is not an easy task because we can find many factors which may affect the ternary relation between student, teacher and learning process. Therefore, in this work, we introduce an initial design of a framework for assessing professor effectiveness, and we describe one of the main base modules, the Student Achievements. In the framework, this module is responsible for providing relevant information regarding student performance. In particular, we proposed a method for estimating the performance students were expected to achieve in a given course, so a comparison with their actual results could give hints about their professor's performance. We have carried out a literature review to identify the primary factors related to university academic performance. Then, considering these factors, we have created three statistical models based on Support Vector Machine (SVM), Artificial Neural Networks (ANN) and Multiple Linear Regression (MLR) techniques. To train the models, we have utilized a real dataset obtained from the University of Santa Elena (Ecuador). Then, we carried out experiments obtaining good results in terms of accuracy and error. Compared to other works, our proposal uses variables that are usually available in academic information systems.

As future work, we expect to keep on working on developing the remaining modules, and we would like to extend this study, including new machine learning techniques to analyze the impact of new indicators in the student learning process.

Acknowledgements. Work partially supported by the Community of Madrid, through the Talent Attraction Program (grant 2017-T2/TIC-5664) and the Young Researchers R+D Project. Ref. M2173 – SGTRS (co-funded by Rey Juan Carlos University); and by the Spanish Ministry of Science, Innovation and Universities, through grant RTI2018-095390-B-C33 (MCI/AEI/FEDER, UE).

References

1. Alenoush, S., Cheryl, A.: Evaluating university teaching: time to take stock. Assess. Eval. High. Educ. **26**, 341–353 (2010)
2. Heneman, H., Kimball, S., Milanowski, A.: The teacher sense of efficacy scale: validation evidence and behavioral prediction. WCER (2006)
3. Ita, G., Kreft, G.: Using Multilevel Analysis to Assess School Effectiveness: A Study of Dutch Secondary Schools, pp. 104–129. American Sociological Association, Washington, D.C. (1993)
4. Galán, A., Amilburu, M.G., Muñoz, I.: The effects of implementing the ESHE on the assessment of teaching. Revista catalana de dret públic **44**, 349–370 (2012)
5. Yeh, S.: The Cost Effectiveness of NBPTS Teacher Certification. SAGE, Thousand Oaks (2010)
6. Yang, S., Lu, O., Huang, A., Huang, J., Ogata, H., Lin, A.: Predicting students' academic performance using multiple linear regression and principal component analysis. J. Inf. Process. **26**, 170–176 (2018)
7. Shahiri, A., Husain, W., Rashid, N.: a review on predicting student's performance using data mining techniques. In: The Third Information Systems International Conference, pp. 414–422 (2015)
8. Orozco, W., Rodríguez-García, M.A., Fernández, A.: Teaching effectiveness: an innovative evaluation model. Learning Technology for Education Challenges (LTEC). Communications in Computer and Information, vol. 1011, pp. 450–461 (2019)
9. McKenzie, K., Schweitzer, R.: Who succeeds at university? Factors predicting academic performance in first year Dutch Australian university students. High. Educ. Res. Dev. **20**(1), 21–33 (2001)
10. Schneider, M., Preckel, F.: Variables associated with achievement in higher education: a systematic review of meta-analyses. Am. Psychol. Assoc.: Psychol. Bull. **143**(6), 565 (2017)
11. Richardson, M., Abraham, C., Bond, R.: Psychological correlates of university students' academic performance: a systematic review and meta-analysis. Am. Psychol. Assoc. **138**(2), 353–387 (2012)
12. Ibrahim, Z., Rusli, D.: Predicting students' academic performance: comparing artificial neural network, decision tree and linear regression. In: Annual SAS Malaysia Forum (2007)
13. Huang, S., Fang, N.: Predicting student academic performance in an engineering dynamics course: a comparison of four types of predictive mathematical models. Comput. Educ. **61**, 133–145 (2013)
14. Iyanda, A.R., Ninan, O.D., Ajayi, A.O., Anyabolu, O.G.: Predicting student academic performance in computer science courses: a comparison of neural network models. Int. J. Mod. Educ. Comput. Sci. **10**(6), 1–9 (2018)
15. Miller, R.G.: Beyond Anova: Basics of Applied Statistics. Taylor & Francis Group, United States of America (1998)
16. Piñeiro, G., Perelman, S., Guerschman, J.P., Paruelo, J.M.: How to evaluate models: observed vs. predicted or predicted vs. observed? Ecol. Model. **216**(3–4), 316–322 (2008)

17. Wold, S., Esbensen, K., Geladi, P.: Principal component analysis. Chemom. Intell. Lab. Syst. **2**(1–3), 37–52 (1987)
18. VanDinther, M., Dochy, F., Segers, M.: Factors affecting students' self-efficacy in higher education. Educ. Res. Rev. **6**(2), 95–108 (2011)
19. Raghavan, V., Gudivada, V., Govindaraju, V., Rao, C.R.: Cognitive Computing: Theory and Applications. Elsevier, Amsterdam (2016)
20. Patel, J.L., Goyal, R.K.: Applications of artificial neural networks in medical science. Curr. Clin. pharmacol. **2**(3), 217–226 (2007)
21. Ebesu, T., Fang, Y.: Neural semantic personalized ranking for item cold-start recommendation. Inf. Retr. J. **20**(2), 109–131 (2017)
22. Field, A.P.: Discovering Statistics Using SPSS, 3rd edn., pp. 197–263. Sage Publications, London (2009)

International PhD Quality Assessment: Growing Awareness of Transnational Projection

Bruno Baruque-Zanón[1(✉)] and Ana María Lara-Palma[2]

[1] Computer Science Department, University of Burgos, Avd. Cantabria s/n., 09006 Burgos, Spain
bbaruque@ubu.es

[2] Management Engineering Department, University of Burgos, Avd. Cantabria s/n., 09006 Burgos, Spain
amlara@ubu.es

Abstract. The University of Burgos (UBU) offers eleven Programs of Doctorate and, the Doctorate School, as its managing office, has been evolving towards a space where knowledge, reflection, learning and innovation coexist from its creation in 2012, until this point in time. Derived benefits from it are multiple, such that students reach their maximum academic level with the Doctorate Title, and their research and impact publications contribute to improve the nearby socio-economic surroundings. Consequently, the relation of the institution with its social environment become stronger due to the impact attained by the scientific contributions of its community. The purpose of this paper is to evaluate the quality of PhD studies at UBU, specifically, the European Doctorate Mention (RD 778/1998) and International Mention (RD 99/2011) in such a way that is it possible to detect the key parameters that enhance international doctorate. EHEA (European Higher Education Area) statements run in these objectives, among others, and UBU needs to refer an excellence quality reference in PhD International Programs, as a real example of transnational education. The methodological approach consists in a qualitative evaluation (12 items in rubric format) that gather interest data in International PhD studies, such as the approach of learning, competences, level of participation, evaluation, transdisciplinary, impact and social projection, professional field, resources and multiculturalism. The concluding results allow to establish actions of progress and future work lines and, fundamentally, to open a dialogue on metrics of doctorate studies in the context of transnational education at UBU.

Keywords: International PhD studies · Quality Assessment · Transnational education

1 Doctoral Programs

1.1 Doctoral Programs in Europe Universities. A Global Approach

PhD studies are a stimulus for all those researchers that, once finished their studies of Degree and Master, decide to keep on contributing knowledge, innovation and evolution in the society.

Á. Herrero et al. (Eds.): ICEUTE 2020, AISC 1266, pp. 132–141, 2021.
https://doi.org/10.1007/978-3-030-57799-5_14

Quoting the Salzburg Principles (conclusions and recommendations of the Bologna Seminar on Doctoral Programs for the European Knowledge Society) related in the Report for European Universities and Ministers of Higher Education of the European University Association (EUA) in 2003 [1], Ministers "added an Action Line to the Bologna process entitled European Higher Education Area and European Research Area, two pillars of the knowledge based society that underlines the key role of Doctoral Programs and research training in this context".

According to this, to suit the guidelines of the EHEA, and summarizing the basic mandates of all the action lines regarding PhD studies (Salzburg Principles in 2005, Leadership seminars in London 2005, Brussels Workshops in 2006, Research Conference in Vienna 2006 and Bologna Seminar in Nice 2006), EUA pointed out that, it is a responsibility of the universities to stay awake because "these studies are innovative and original, incorporate lines that favor the professional career development, be aware about the importance of the inclusion and the diversity, serve for social use, get ready of sufficient resources and, thus, the consequent academic support to reach the excellence". This means, that one of the most ambitious objectives stablished in the Doctorate Studies has to do with the transversal competences of the students.

These have served to highlight not only the creativity or self-discipline [2], passion for long-term goals [3], but the importance of motives and aspirations such as improving prospects career or personal development [4].

As far as the current regulations are concerned, universities have to draw up evaluation, academic monitoring, academic projection and academic support, relevance and visibility.

Some authors go more in detail, focusing predominantly in the performance of doctoral students (scores) [5], assessing the program's design, development (educational period and research period) results and final product (doctoral thesis) [6].

Finally, the core pillar is the promotion of the internationalization, which allows for national and foreign students to reach the doctorate (European Mention) (RD 778/1998) or International Mention (RD 99/2011); it means, "to promote the novelty of the creation of doctoral schools, specialization of research training through new doctoral programs, training in research and knowledge transfer competences" [7].

In the following subsection, we tackle the concrete aspects of the internationalization in a particular Spanish example, the University of Burgos.

1.2 Doctoral Program in Spain. A Particular Case of Study (UBU)

With the entry in the XXI Century, the European Space of Higher Education regulates formation with the RD 1125/2003 and 1393/2007, and from this moment on, universities are provided with a specific unit dedicated to the doctorate studies, it is, the Doctorate School.

It means doctorate studies are regulated under a set of requirements and rules such as structure (management, specializing training, assessment, cooperation between disciplines, quality supervision, regulations, financing), coordination with other schools, professorship, students (selection, evaluation, internationalization), lecturer program, labor world (acquisition of transferable workmanship, agreements), infrastructure (resources, international visibility) [8].

In the last report elaborated by CRUE Spanish Universities, the number of international students of doctorate represents a 1.7% with regard to the nationals; more specially, international doctorate in Engineering and Architecture Programs represent 7.3% out of the whole of the registered PhD enrolled [9].

At regional level, in the community of Castilla y León, the information indicates that there is a lack in the evolution of students enrolled in International Doctorate Studies from the academic year 13–14.

In addition, the registration ratio in national doctorates has been fluctuating, staying stable in the last academic year [9].

Figure 1 gives a summary of the number of International and national PhD students enrolled in Castilla y León.

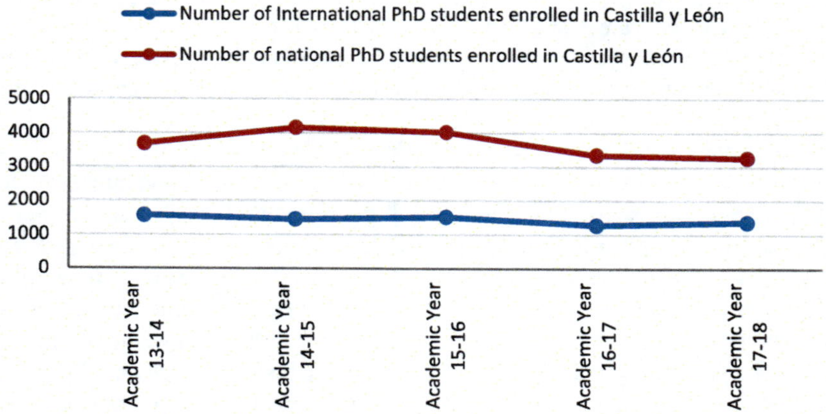

Fig. 1. Number of students enrolled in international and National PhD in Castilla y León

Currently, the University of Burgos occupies the TOP300 of the best young universities of the world (less than 50 years of antiquity), and in the TOP200 ranking of teaching excellence of all the European Universities; thus, it is visible in the Young University Ranking and Europe Teaching Rankings, by Times Higher Education. According research status (international doctorate), this institution occupies the 12th position out of 49 public universities evaluated [10]. More specifically, the Civil Engineering Program Doctorate data reflects a 6.7% of international theses.

At present, this Engineering Program supervises four international PhD in co-direction (collaboration Agreement) with a German University (TH Köln) by a

Collaboration Agreement. Transnational Doctorate is a target to keep latent, as "they are the base of the academic formation and the investigation" [11]. The following Table 1 gives a summary of all sources analyzed.

Table 1. Research Ranking by contribution at UBU

Source	Research Ranking	Contribution
U-Ranking 2019	10th position	Number of high-level impact publication (Q1) and join publication with foreign researchers
Fundación CYD (Conocimiento y Desarrollo)	4th position	Number of high-level impact publications (Q1)
	11 out of 70	Students mobility
	27 out of 70	Join publication with foreign researchers
U-MULTIRANK	2 out of 77	Interdisciplinary publications
	14 out of 49	Students mobility
	12 out of 49	**international PhD**
RANKING CYD (Conocimiento y Desarrollo)		3 indicators: interdisciplinary publications, licence profits and spin-offs

As far as PhD studies are concerned, transnational education refers that the academic training has to be alive in a window of knowledge transference and wisdom sharing. Its impact on society, students and institutions will be, for sure, highly positive and enriching [12]. The following Fig. 2 shows the evolution in time of PhD status at the Civil Engineering Doctorate Program at UBU.

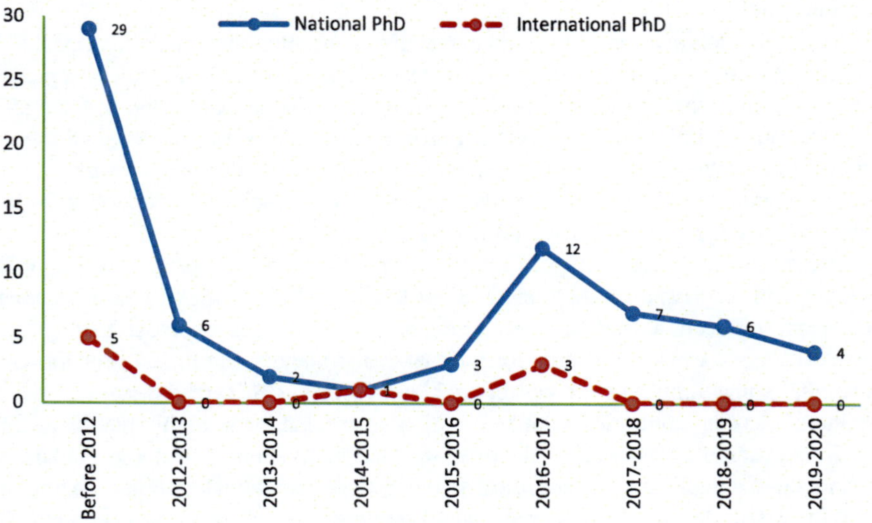

Fig. 2. PhD Status at the Civil Engineering Doctorate Program at University of Burgos (before 2012–2019)

2 Enhancing International PhD Quality at UBU

2.1 Structural Areas for Measuring Quality at PhD Studies

The crucial discussion about how to enhance international PhD quality Programs at UBU focuses on the supervision of the parameters related to them. Our contribution highlights a rubric for self-assessment valuing 12 items codified in learning approach, competences, level of involvement, evaluation, academic support, resources, relevance and visibility. It is an exercise that fits with the Bologna commitment and a necessary purpose for consolidate the relevance of the international PhD at UBU.

The Rubric aims to be a scoring guide and it is based on [13] and [14] and own elaboration linked to the objectives of the study. To clarify the conceptual taxonomy, next we introduce a brief context of all items:

Learning approach corresponds with the rich potential for transformative learning, which falls within the scope of being creators of their reality, say, transculturation [15].

Success in doctorate studies demand strength capacities and abilities. These **competences and skills** support in each stage the key stones of the research periods and inspire candidates with creativity, autonomy, personal initiative and sensitivity [16].

Student's participation ensures and reinforces their outcomes; thus, quoting [17], the process of feedback and feedforward respectively.

Assessment and evaluation of the student's research work is a mechanism of sustainability. The extensive work period needs continuous support and feedback; not only to develop the student's line of work, but also to motivate and keep alive their creativity and innovation [18].

As far as **academic monitoring** item, [19] points out that, "quality of cooperation between the participating organizations as well as other relevant stakeholders run a cohesive consortium with active involvement, common goals, […], cooperation and commitment".

Transdisciplinarity is a trend and demanding concept in the transition of past PhD Programs to the current ones, named as co-determination by some authors; it is a practical scenario to crosslink different fields of knowledge and complement between them [20].

PhD studies influence the **transformation** of all the actors involved. Students, supervisors, institutions, industry and society enrich with doctoral research. Some authors emphasize "postgraduate knowledge is seen as crucial in theorized and practical contributions to social development" [21].

Regarding **professional field**, [22] this aspect shows the importance of providing students with suggestions and guidelines for the PhD journey thinking in the strategy and social implications and its potential relationship with their expertise field.

Visibility is a major resource for international student's enrollment. For this purpose, disseminate and exploit the research results means to spread wisdom, and, in addition, a strong connection with recognition as specialist and highly qualified [23].

To contribute in fostering the involvement and motivation of doctorate students to reach their accomplishments, specifically within the international mention, institutions need to contribute with **academic support resources**. These sometimes penalize "the invisible others" (international PhD students [24]). Overall, the description of a well-ranked university must provide "barrier-free and inclusive doctoral experience".

Drawing on the Rubric, next section shows findings of the methodological approach (empirical study).

3 Empirical Study

In order to complete the study of the PhD program at the University of Burgos, with special emphasis in its international dimension, a survey has been designed and sent to PhD students who have recently completed or are completing at the moment the international program. The students completed this survey in an anonymous way in order to prevent unintended data recollection and to encourage the students to answer in the most honest way possible.

3.1 Survey Description

The survey, which was conducted in an on-line form, consisted in a few initial questions designed to help segment the students answering the question into different programs and academic years and a rubric form. This rubric was adapted from the one described in [13] to refer to PhD studies rather than graduate level studies. The rubric consisted in 12 questions, which matched the structural areas of analysis described in Sect. 2.1. The complete questionnaire can be found on-line[1], but a sample of the rubric used is included in Table 2. All are qualitative questions can be transformed to numerical values quite straightforwardly; by assigning a score number ranging from 1 to 4, as they can be interpreted as a scale from less desirable (leftmost option) to most desirable (rightmost option). The exception are questions "Competences" and "Social Impact and Projection", where the student can select more than one answer, as the options do not conform a well-defined scale.

Table 2. Sample of questionnaire used to assess student's perception of the University of Burgos' PhD program with international mention

1. Learning Approach	PhD studies are based on institutional requirements with no possible attitude changes	PhD studies develop learning that aims to make students change their way of seeing the world, be creators of their reality and encourage metacognition	PhD studies develop learning that aims to make students change their way of seeing the world, be creators of their reality. In addition, there is a specific space for reflection on PhD studies	The PhD studies also propose learning strategies that are based on students'; interest in developing a PhD and use them to maximize understanding and satisfy their curiosity. There are specific spaces for this

(*continued*)

[1] https://forms.gle/evpBxa2wRngfdgD96.

Table 2. (*continued*)

2. Competences	PhD studies promote the learning of the transversal competences of: collaborative work, communication skills and empathy	PhD studies focus on the above transversal skills and on the autonomy, creativity, critical thinking, personal initiative and sensitivity of the students	PhD studies encourage the learning of professional skills specific to the student's studies	PhD studies enhance research and investigation skills of the students
3. Level of involvement	The student is not involved in the preparation or decisions on the content and development of their annual PhD curriculum. Student participation is simple	The student can give his or her opinion on the annual academic plan. Student participation is purely consultative	The student participates in the definition of his annual academic plan, in determining its meaning and objectives. It also participates in the design, planning, implementation and evaluation. Participation is projective	The student requests, demands or generates new spaces and mechanisms of participation in the PhD studies. Equates to a maximum level of participation: meta-participation
...	...	...	...	...

3.2 Analysis of Results

Due to time constraints, in the current state of our study it has been impossible to reach a good number of students, so the results shown here are obtained from a relatively low quantity of students (8 students agreed to complete it). Nevertheless, the sample size corresponds to the majority of students having completed the PhD program with International Mention, as shown in Fig. 2. This study is an ongoing research that will try to reach a much greater number of students and from a wider variety of disciplines in order to obtain more statistically significant conclusions.

Our current sample can be described as follows: The majority of the international students reached are German citizens (75%) completing their Third Cycle studies at the University of Burgos, while a 25% of them were Spanish students completing part of their formative period abroad (at least 3 months, although some were abroad for 12 months). In the sample, it coincides that none of the non-Spanish students were living in the country while they completed those studies. In addition, 87% of them completed the program of Technological Industries and Civil Engineering, while 12% completed the program of Law, Economic and Social Sciences.

Regarding the answers to the quantitative questions proposed the results obtained are summarized in Figs. 3 and 4.

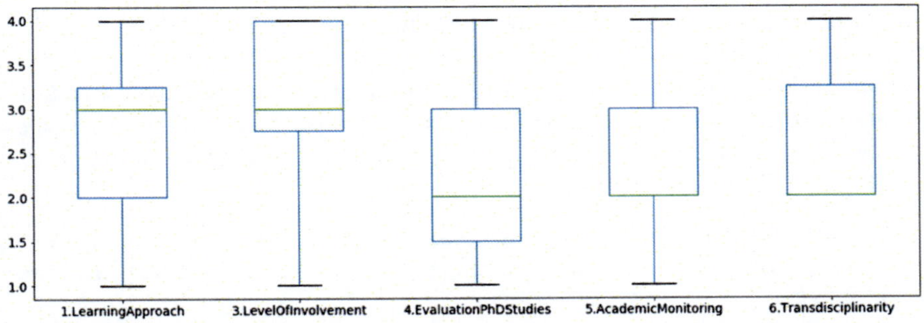

Fig. 3. Boxplot of the distribution of answers of the students to the first half of quantitative questions presented to them, corresponding to the quality areas described in Sect. 2.1. The line in the middle of each plot represents the median of all answers, while the borders of the box represent the 1st (bottom) and 3rd (top) quartiles of the data samples.

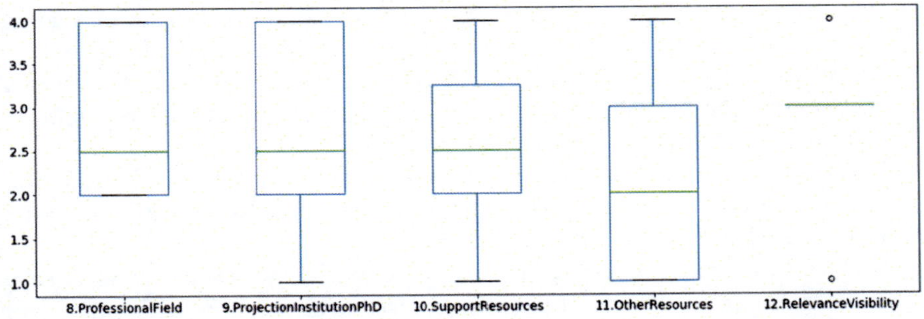

Fig. 4. Boxplot of the distribution of answers of the students to the second half of quantitative questions presented to them, corresponding to the quality areas described in Sect. 2.1.

Among the items worth of highlighting as good results, students consider the learning approach and the level of involvement above the average, specially this second one. Among the least valued items, we can find the evaluation of their studies and specially the additional resources at their disposal.

Regarding the qualitative questions (corresponding to the subjects of "Competences" and "Special Impact"), the combination of the two assertions "PhD studies encourage the learning of professional skills specific to the student's studies." and "PhD studies enhance research and investigation skills of the students." were the most repeatedly selected for the question related to Competences, while the single assertion "PhD studies work on real and proximate needs and influence society." was selected with a difference of four times more any other assertion in the question related to Social Impact.

4 Conclusions and Recommendations

As conclusions, taking into account the results obtained from the survey among the students and the experiences of the authors as PhD supervisors, a list of recommendations aimed to improve satisfaction among participants, attending to each of the areas of study could be the following:

- Learning Approach: it would be interesting to generate a specific space for students where they can devise their own learning strategies based on their particular topics of interest.
- Competences: it would be beneficial to promote doctoral studies where the competences of collaborative work, communication skills, critical thinking and autonomy are considered.
- Evaluation: we recommend that the evaluation consider the contribution of the research to the community.
- Academic Monitoring: most of the student consider that the follow-up is punctual but sparse in time. It would be constructive to consider if, in case of students needing continuous evaluations, the doctoral program incorporates that aspect.
- Transdisciplinary: it is proposed that stronger connections between the different doctoral programs would be identified and promoted so that students from different fields work together on complementary lines of research.
- Projection: it is proposed to revise the website and information offered to non-Spanish speaking students, as well as the virtual secretariat to make communication with the students enrolled closer and more fluid.
- Support Resources: the work of the staff involved in third cycle studies to some extent makes up for the lack of English resources and it would be advisable to open a debate on the needs of both the student and the tutor in this regard.
- Other Resources: it would be appealing to standardize the resources used by doctoral students, since, the disposition they currently have of them is not interpreted in the same way.
- Relevance and Visibility: it would be very positive to revise the regulatory norms; some nuances would improve the visibility, publicity, and ranking of our studies given that international doctorate students are affected (scholarships, grants, travel, etc.).

References

1. Doctoral Programmes in Europe's Universities: Achievements and challenges. Report prepared for European Universities and Ministers of Higher Education. European University Association (EUA) Bélgica, European University Association Publications (2007)
2. Montero, F.A., Heráldez, A.M., López, C.A.: Criteria and instruments for doctoral program admission. Electron. J. Res. Educ. Psychol. **12**(3), 853–886 (2014)
3. Duckworth, A.L., Peterson, C., Matthews, M.D., Kelly, D.R.: Grit: perseverance and passion for long-term goals. J. Pers. Soc. Psychol. **92**(6), 1087–1101 (2007)

4. Brailsford, I.: Motives and aspirations for doctoral study: career, personal, and inter-personal factors in the decision to embark on a history PhD. Int. J. Doct. Stud. **5**, 15–27 (2010)
5. Kuncel, N.R., Wee, S., Serafin, L., Hezlett, S.A.: The validity of the graduate record examination for master's and doctoral programs: a meta-analytic investigation. Educ. Psychol. Meas. **70**(2), 340–352 (2010)
6. de Miguel Díaz, M.: PhD assessment and improvement. Revista de Educación **352**, 569–581 (2010)
7. Jiménez-Ramírez, M.: New doctoral studies in Spain: progress and challenges for convergence with Europe. Revista Iberoamericana de Educación Superior **8**(1), 123–137 (2017)
8. Castro, A., Guillén-Riquelme, A., Quevedo-Blasco, R., Ramiro, M.P., Bermúdez, M.T., Buela-Casal, G.: Las escuelas doctorales: evolución histórica, características y aspectos relevantes para su consolidación en España. Aula Abierta **38**(2), 17–28 (2010)
9. Hernández, J., Pérez, J.A.: La Universidad Española en cifras 2013/2014, 2014/2015, 2015/2016, 2016/2017, 2017/2018. CRUE Universidades Españolas
10. UBU. https://www.ubu.es/portal-de-transparencia/resultados/posicionamiento-en-rankings-de-universidades
11. Bermúdez, M.P., Castro, A., Sierra, J.C., Buela-Casal, G.: A descriptive transnational analysis of the doctoral degree programmes in the EHEA. Revista de Psicodidáctica **14**(2), 193–210 (2009)
12. Lara-Palma, A.M., Cámara Nebreda, J.M., Vicente Domingo, E.M.: Impact level within the English Friendly Program in Virtual Degrees at the University of Burgos. A real case scenario of transnational education. In: Herrero, A., Baruque, B., Sedano, J., Quintián, H., Corchado, E. (eds.) CONFERENCE 2015. Advances in Intelligent Systems and Computing, vol. 369, pp. 525–532. Springer, Heidelberg (2015)
13. Campo, L.: Una rúbrica para evaluar y mejorar los proyectos de aprendizaje servicio en la Universidad. RIDAS, Revista Iberoamericana de Aprendizaje y Servicio **1**, 91–111 (2015)
14. European Commission: Guide for Experts on Quality Assessment. Actions Managed by National Agencies, Luxemburgo (2019)
15. Gjotterund, S.M., Ahmad, A.K.: Tranformative power of cross-cultural PhD supervision. Int. J. Doct. Stud. **13**, 441–456 (2018)
16. Milos, D.: Embedding transferrable skills into the higher degree by research candidature. Stud. Grad. Postdoct. Educ. **10**(3), 173–179 (2019)
17. Fernando, G.S., Eduardo, G.J.: Assessment of academic literacy. RELIEVE-Revista Electrónica de Investigación y Evaluación Educativa **21**(1), 1–25 (2015)
18. Acker, S.: Reflections on supervision and culture: what difference does difference make? Innov. Educ. Teach. Int. **48**(4), 413–420 (2011)
19. European Commission: Guide for Experts on Quality Assessment. Actions Managed by National Agencies, Luxemburgo (2020)
20. Conrad, A., Oberc, H., Wannöffel, M., Kuhlenkötter, B.: Co-determination – an interdisciplinary concept to train PhD students from different disciplines. Proc. Manuf. **31**, 129–135 (2019)
21. Wisker, G., Robinson, G.D., Leibowitz, B.: The purpose and impact of postgraduate knowledge. Stud. Grad. Postdoct. Educ. **10**(3), 160–172 (2019)
22. Syed Mohamed, A.T.F., Amir, A.F., Abd Rahman, N.K., Abd Rahman, E., Abdul Nasir, A. Q.: Preparing for PhD: exploring doctoral student's preparation strategy. Stud. Grad. Postdoct. Educ. (2020, article in press)
23. Caparros-Ruiz, A.: Doctorate holder´s careers in Spain: does international mobility matter? Eur. J. Educ. **54**(1), 117–136 (2019)
24. Laufer, M., Gorup, M.: The invisible others: stories of international doctoral student dropout. High. Educ. **78**(1), 165–181 (2018)

Trends and Patterns of International Student Mobility: The Case of Bachelor's Degrees in Computer Science at the University of Burgos

Ángel Arroyo[1(✉)], Secil Bayraktar[2], Carlos Cambra[1], Daniel Urda[1], and Álvaro Herrero[1]

[1] Grupo de Inteligencia Computacional Aplicada (GICAP), Departamento de Ingeniería Informática, Escuela Politécnica Superior, Universidad de Burgos, Av. Cantabria s/n, 09006 Burgos, Spain
{aarroyop, ccbaseca, durda, ahcosio}@ubu.es
[2] Department of Human Resources Management and Business Law, TBS Business School, 1 Place Alphonse Jourdain, 31068 Toulouse, France
s.bayraktar@tbs-education.fr

Abstract. As in many other fields, mobility of people is a fact in higher education. Although attention has been devoted to this topic, new challenges emerge at the present time. Globalization is an increasing reality and investigating the recent trends associated with it is required also in the educational context. In keeping with this idea, the present paper analyzes student mobility in both European and Non-European contexts under the frame of the Computer Science Bachelor at the University of Burgos. The findings are useful for students and lecturers together with university managers and policy makers in order to improve student mobility programs.

Keywords: International studies · Student mobility · Erasmus program · University of Burgos

1 Introduction

The international mobility of talent has been a rising trend as a direct consequence of globalization and internationalization of labor markets. In this context, where the workforce around the world has become larger and increasingly diverse and educated [1], attracting and retaining talent globally has been a major competition faced by the organizations [2, 3]. Also, given the changing requirements of the international arena, organizations have started searching and competing for people who can effectively manage through the ambiguous, challenging and changing parameters of today's business context [4].

The competition for global talent and the changing requirements from the workforce that require international competencies have also reflected on the higher education system around the world. Globalization, combined with the conditions of the knowledge economy, created a context for a more strategic approach to internationalization in higher

Á. Herrero et al. (Eds.): ICEUTE 2020, AISC 1266, pp. 142–153, 2021.
https://doi.org/10.1007/978-3-030-57799-5_15

education, placing it at the top of the agenda [5, 6]. Nowadays, it is vital that the students are equipped with the necessary set of skills to succeed in an international environment to be able to compete in the labor market. As a result, academic institutions have become more internationally oriented than ever [7].

One of the most effective ways to prepare students for the international arena is student mobility, which enable students to spend one or more semesters abroad. It has now become essential to benefit from international mobility programs to prepare students with the skills needed for their future employment. Therefore, this has translated into higher education institutions to incorporate international mobility programs as a part of their internationalization efforts. Student mobility is defined as a limited period of study varying generally between 6 and 12 months in an international environment [8]. There is no doubt on the numerous benefits of student mobility programs for the students, home and host institutions, and communities in general [9]. Especially for the students, going to a different country enhances students' personal development, expands their mental horizons, develops intercultural competencies, new language skills, and ability of students to address international aspects of work better, and increases their employability [10–13].

Following these benefits, and many others, it is not surprising that there is vast evidence of the increasing interest on this topic, such as the enrolment into international Massive Online Open Courses (in 2018 exceeded 100 million students for the first time ever according to ClassCentral [14]) and the "European Universities" initiative of the European Commission [15] (first call in 2019). Statistics show that in the period 2010–2020, the number of international students doubled to 5 million [6]. According to the Global Student Mobility 2025 report [16], international student mobility is expected to reach a student enrollment of 7.2 million by 2025. European policies involving the Erasmus+ program or other mobility strategies also provide good evidence of support for both student and staff mobility [17].

Like the rest of the world and Europe, Spanish universities are also undergoing a process of continuous internationalization. The Spanish legislation further supports internationalization in higher education by emphasizing the necessity of fostering student mobility in Europe and other parts of the world as a part of the degrees granted [18]. However, studies that focus on internationalization of higher education are mainly conducted in the United States or some European countries, pointing to the need for more research in this field in Spain [18].

This study, therefore, intends to provide some insights on the patterns and trends regarding student mobility as a part of internationalization processes in universities, focusing on bachelor students from the Computer Science (CS) Department in the University of Burgos (UBU), Spain. Given the acceleration of student mobility and the continuous pace of change in international education, it is crucial to study this phenomenon and identify some important trends. The main contribution of present paper is the trend analysis of CS student mobility from/to the UBU. The associated findings aim to identify main patterns, strengths, and areas to improve with the intention of defining future actions which can improve the internationalization of Spanish universities in general and the UBU in particular. The implications of the results also strive to enhance the experience of students in their mobility programs.

To conduct this study, data have been gathered by the International Office of the University of Burgos. Once preprocessed, these data are analyzed through a descriptive study. To the best of the authors' knowledge, this is the first study regarding the international mobility of CS students from the University of Burgos.

The rest of the paper is organized as follows: a historical perspective about the Computer Science studies at the University of Burgos is presented in Sect. 2. The performed analysis, together with the main results are compiled in Sect. 3 while the main conclusions are described in Sect. 4.

2 Computer Science Studies at the University of Burgos

The very first degree in the CS area at the University of Burgos started in the year 1995. According to Spanish law, at that time, universities were allowed to teach (upon authorization) the following Computer Science degrees:

- Ingeniería Técnica en Informática de Gestión - Bachelor.
- Ingeniería Técnica en Informática de Sistemas - Bachelor.
- Ingeniería en Informática - Master.

The only degree that started in 1995 was the Technical Engineering in Computer Management (Ingeniería Técnica en Informática de Gestión - Bachelor). After 6 years (in 2001), the master-level degree (Ingeniería en Informática) started too.

Due to the so-called "Bologna process" the above-mentioned degrees were modified in order to comply with the European Higher-Education Area (EHEA). As a result, two new "official" degrees were defined in Spain:

- Grado en Ingeniería Informática - Bachelor.
- Máster Universitario en Ingeniería en Informática - Master.

Accordingly, the new degrees were deployed at the University of Burgos; the new Bachelor started in 2010 and the new Master in 2012. Although degrees at both levels (Bachelor and Master) have been taught at the UBU, due to the reduced number of students in the Master degree, only the Bachelor-level degrees are analyzed in present study. Figure 1 shows the temporal evolution of enrolled student for these two degrees, namely Ingeniería Técnica en Informática de Gestión (ITIG) and Grado en Ingeniería Informática (GII).

Since the starting of international agreements, some initiatives have been run to promote mobility of students. It is worth mentioning the European Computer Science course [19], organized by some of the Erasmus+ partners.

In the present paper, the international flow (both incoming and outgoing) of students involved in the ITIG and GII degrees is studied. International mobility of students started in 2008 and since then it has been carried out under the frame of Erasmus+ agreements for EU countries and bilateral agreements for non-EU countries (UBU-GLOBAL program at the UBU). These agreements have been signed with different higher-education organizations in the case of CS studies. They are shown, together with the starting and ending years in Table 1. Those agreements in which no students have participated have been omitted.

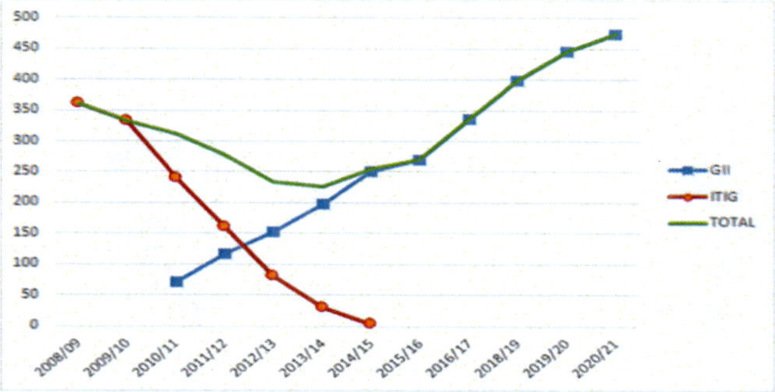

Fig. 1. Enrolled students in CS Bachelor's degrees at the UBU.

Table 1. List of Erasmus+ and UBUGLOBAL agreements for CS studies at UBU, comprising starting and ending year of the agreement. In the "End" column the symbol '–' means that the agreement is in force at present time.

COUNTRY/University	Start	End
CZECH REPUBLIC		
Czech Technical University in Prague	2013/14	–
FINLAND		
Turku University of Applied Sciences	2004/05	–
FRANCE		
Bordeaux INP	2014/15	–
University of Lorraine	2013/14	–
GERMANY		
Hamburg University of Applied Sciences	2005/06	–
Köln University of Applied Sciences	2010/11	–
ITALY		
University Ca′ Foscari Venezia	2013/14	–
University of Genoa	2004/05	–
POLAND		
Gliwice Silesian University of Technology	2014/15	–
Warsaw University of Technology	2010/11	–
Wroclaw University of Technology	2008/09	–
PORTUGAL		
Polytechnic Institute of Coimbra	2013/14	–
Polytechnic Institute of Castelo Branco	2016/17	–
ROMANIA		
West University of Timisoara	2010/11	–
UK		
Edinburgh Napier University	2009/10	2012/13
University of the West of Scotland	2012/13	2013/14

(*continued*)

Table 1. (*continued*)

COUNTRY/University	Start	End
TURKEY		
Altinbas University	2013/14	–
USA		
Dakota State University	2012/13	–
SOUTH KOREA		
University of Dankook	2014/15	–
MEXICO		
Technological Institute of Monterrey	1999/00	–

3 International Mobility Data Analysis

As previously stated, figures about the international students in the CS studies (ITIG and GII) at the UBU are analyzed. Figures have been gathered between 2008/2009 and 2018/2019 courses as some of the data are not available for the present year yet. Figure 2 shows the total number of outgoing students per course for the two degrees. Data for the incoming students in the ITIG degree are not available and hence, incoming students are not included.

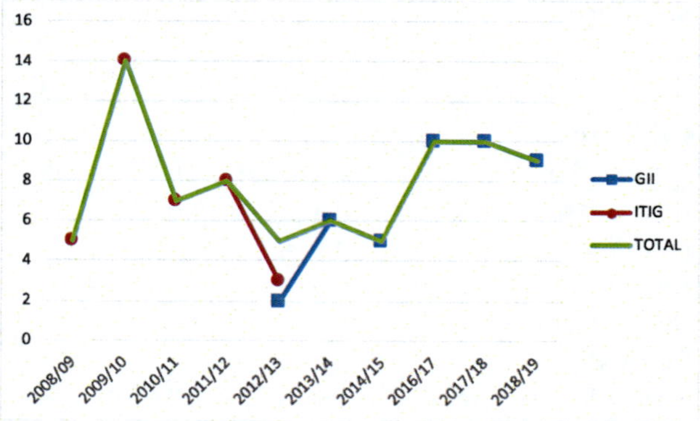

Fig. 2. Outgoing international students in CS Bachelor degrees at the UBU.

From this first graphical representation of the data, some ideas can be highlighted; the highest number of international students corresponds to year 2009/10 while the lowest one corresponds to 2012/13. This trend is somehow similar to that of the total number of enrolled students; it was reduced from 360 in 2008/09 to 227 in 2013/14 and increased up to 445 in 2018/19. Although the two highest rates of enrolled students were obtained in 2018/19 and 2017/18, the highest number of outgoing students was obtained in 2009/10. Authors consider that the economic situation may have influenced the mobility trend; there is a grant for international students and the associated financial aid has been reduced due to the crisis.

From the series comprising both degrees (TOTAL), it can be seen the important decrease of outgoing students between 2010 and 2014. It must be noted that during this years, the ITIG degree was coming to an end, while the GII studies were starting. There is no available data to study the impact of such circumstances in the general trend. According to official statistics, student mobility in Europe under the frame of the Erasmus+ program [20] has been constantly increasing from 2008 to 2020. A similar trend is observed in the case of students and trainees coming to Spain during the period under study. However, the number of outgoing students from Spain started to decrease in 2012/13 from 39,249 up to 31,955 (in 2014/15).

In the following subsections, data are analyzed considering the involved countries (Subsect. 3.1) and academic matters (Subsect. 3.2).

3.1 Geographical Analysis

In present section, data disaggregated by country are considered. As destination continents vary according to the agreement, Erasmus+ and UBUGLOBAL are independently analyzed. Figures 3 and 4 show the distribution by country of the outgoing and incoming students under the Erasmus+ program.

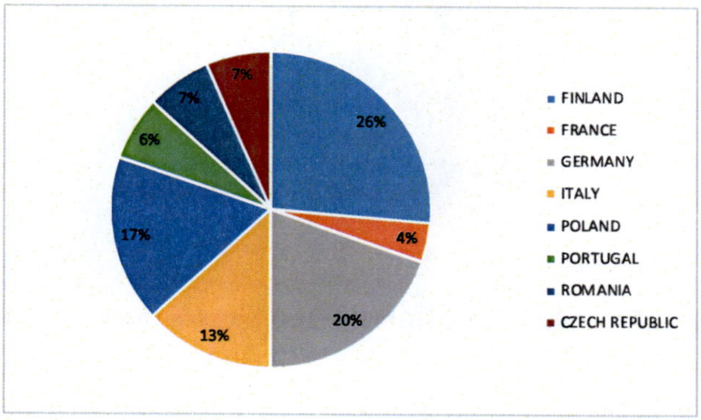

Fig. 3. Distribution of the Erasmus+ outgoing students by country.

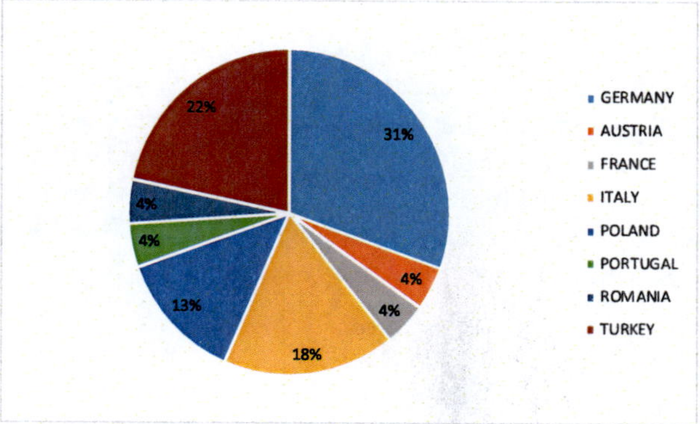

Fig. 4. Distribution of the Erasmus+ incoming students by country.

The most popular destinations chosen by the Erasmus+ CS UBU students are Finland (26%), Germany (20%), Poland (17%), and Italy (13%). All together, these four countries amount to 76% of the outgoing students. Concerning the incoming students, the most usual origins are Germany (31%), Turkey (22%), Italy (18%), and Poland (13%). All together, these four countries amount to 84% of the incoming students. It can be said that there is a reciprocal interest in students from UBU and those from Germany, Poland, and Italy. It can be noted that although Finland is the most attractive destination for UBU students, no Finnish students moved to Burgos. It is the other way round concerning Turkey; although it is the second most common country for incoming students, no UBU students moved there.

Both home and host countries are also studied in the case of non-EU agreements (UBUGLOBAL program). The distribution per country for outgoing and incoming students is shown in Figs. 5 and 6 respectively.

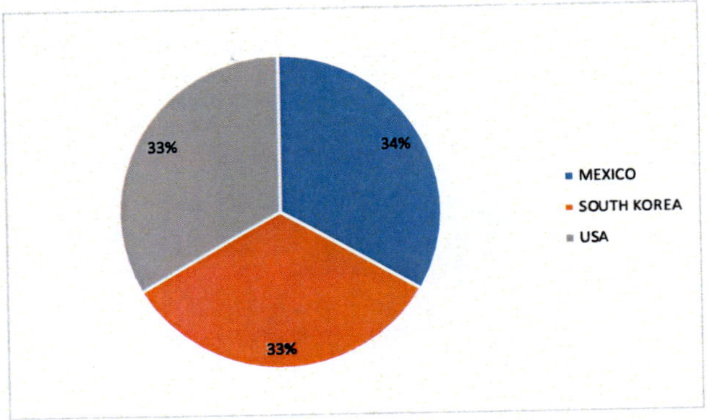

Fig. 5. Distribution of the UBUGLOBAL outgoing students by country.

It can be said that there is not a preferred destination for UBUGLOBAL CS outgoing students. The interest is equally distributed between the three destinations: Mexico, South Korea, and the USA. In the case of the incoming students, most of them (63%) come from Mexico, while there is a similar rate for the other three countries (Chile, Peru, and the USA).

In order to see whether the geographical trends vary along time, a joint study of Erasmus+ and UBUGLOBAL students is performed. Figures 7 and 8 show the total number of outgoing and incoming students by year and destination country.

From the figures above it can be said that the interest on Germany as a destination for outgoing students lasted for four years (2010/11–2013/14). In the case of Poland, it has been chosen by students mainly in recent years (2016/17–2018/19). In the case of incoming students, interests vary on a yearly basis.

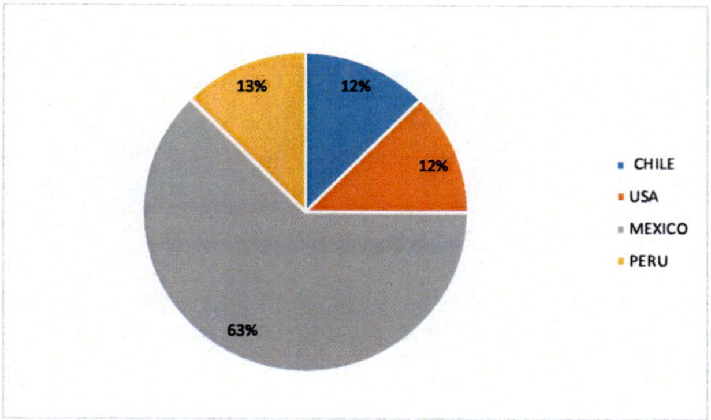

Fig. 6. Distribution of the UBUGLOBAL incoming students by country.

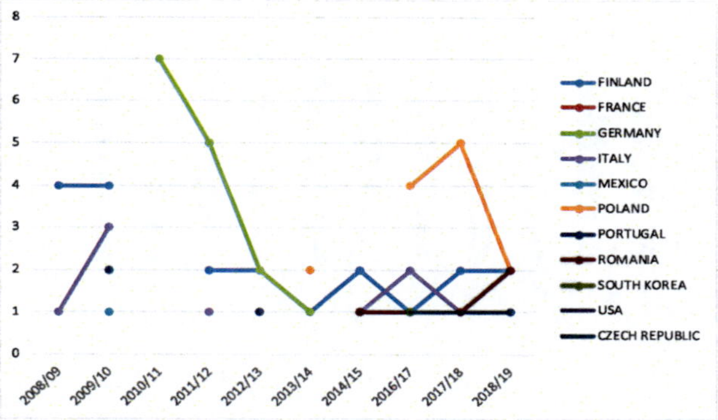

Fig. 7. Total number of outgoing (both Erasmus+ and UBUGLOBAL) students by year and destination country.

Fig. 8. Total number of incoming (both Erasmus+ and UBUGLOBAL) students by year and destination country. Data before 2011/12 are not available.

3.2 Academic Analysis

International mobility is also studied from a two-fold academic perspective for outgoing students (data are not available for incoming); both the effort and the results are considered. The former is studied by means of the average number of subjects (home institution) comprised in the Learning Agreement (LA) while the latter is studied by means of the average mark at UBU and the average percentage of subjects with the "No Show" mark.

From Fig. 9 it can be said that Poland, Czech Republic, and Portugal are is the Erasmus+ destination with the highest average number of subjects in the LAs. It is South Korea in the case of UBUGLOBAL destinations, with a rate similar to that of Portugal.

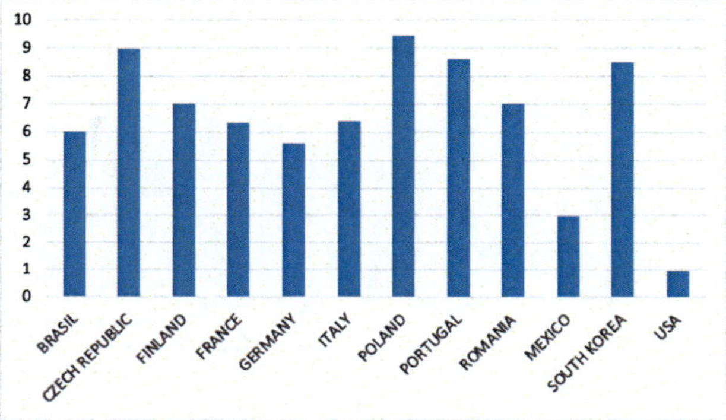

Fig. 9. Average number of subjects in the LA of outgoing students (both Erasmus+ and UBUGLOBAL) by country.

Figure 10 shows the average mark obtained by all outgoing students by country and year.

It is worth mentioning the good academic results obtained by the students that moved to Finland. In general terms, it can be said that the obtained academic results are positive. There is only one case (Portugal - 2014/15) with an average mark under 5 (fail).

Finally, Fig. 11 shows the percentage (out of the total number of subjects in the LA) of subjects with the "No show" mark for all outgoing students.

It can be seen in Fig. 11 that the percentage of subjects graded as "No Show" is generally below 50%. There are two cases (Portugal - 2009/10 and Italy 2011/12) with a high rate, that are associated to only one student each.

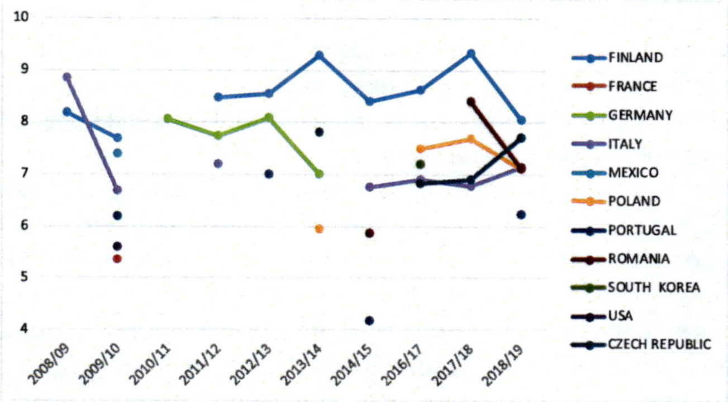

Fig. 10. Average mark of outgoing students (both Erasmus+ and UBUGLOBAL) by country and year.

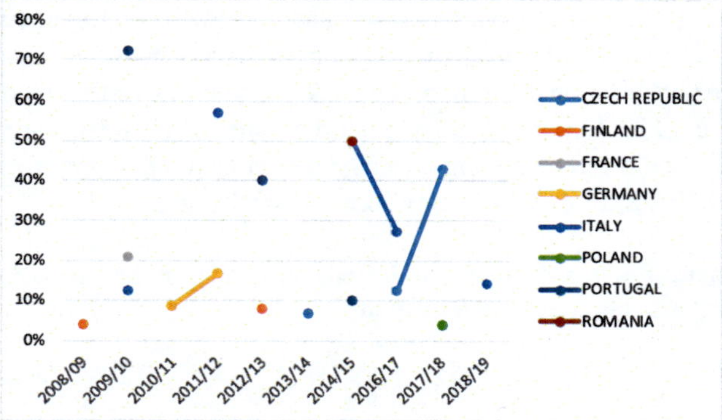

Fig. 11. Average percentage of "No show" mark of outgoing students (both Erasmus+ and UBUGLOBAL) by country and year.

4 Conclusions

The mobility of CS students at the UBU are studied in the present paper. The findings present the statistics regarding both incoming and outgoing students within and outside of Europe. It can be observed that the student preferences cannot be limited to a narrow pattern, and instead there are multiplicity of motivations behind the choice of location for mobility. For both incoming and outgoing students in Europe, while countries such as Italy and Portugal may reflect motivations due to cultural similarities or proximity in language, other choices (e.g. Finland and Germany) may show that students may consider other factors going beyond such similarities. Looking at the non-European countries, we see similar findings. While students prefer exchanges between countries that are similar in terms of culture and language, as would be expected, such as to and from Mexico, Chile or Peru, students display other motivations for distinct locations such as USA or South Korea. USA may be a preferred destination for popularity, future career opportunities or improving English language skills and Asian countries such as South Korea may be perceived as exotic places to meet people from distinct cultures and develop cultural awareness. The findings also point to the fact that Spain constitutes an attractive region for a wide range of countries. Future research may delve into the distinct motivations behind these choices.

The results also point to various locations losing or gaining attraction in student mobility preferences. While the changes in the trends over the years may be related to the overall economic or political situation concerning each country, the strength of certain programs available to apply for, or the trends following popularity of the destination, these dynamics and reasons behind the changing preferences in years need to be analyzed in line with each country's and the institutions' conjuncture and fluctuations across country level variables.

All in all, the results point to the fact that students are not necessarily accumulated in zones with similar culture and language aspects and motivations do not follow a single pattern. Therefore, higher education institutions, Bachelor's degrees in Computer Science at UBU in this specific case, have to be aware of the wide range of student choices based on various motivations and consider offering exchange programs in different countries. Future work may focus on studying the effect of some other variables on the international mobility of students, such as online studies, and similarities in language, culture, and other potential aspects related to students' choices. Additionally, it would be interesting to analyze data from some other universities and countries, if available, in order to study more general trends.

Acknowledgement. The authors thank the International Office at the University of Burgos for collecting the raw data analyzed in present study.

References

1. Briscoe, D., Schuler, R., Claus, E.: International Human Resource Management. Routledge, London (2009)
2. Beine, M., Docquier, F., Rapoport, H.: Brain drain and human capital formation in developing countries: winners and losers. Econ. J. **118**, 631–652 (2008)
3. Vaiman, V., Scullion, H., Collings, D.: Talent management decision making. Manag. Decis. **50**, 925–941 (2012)
4. Tarique, I., Schuler, R.S.: Global talent management: literature review, integrative framework, and suggestions for further research. J. World Bus. **45**, 122–133 (2010)
5. Altbach, P.G., Knight, J.: The internationalization of higher education: motivations and realities. J. Stud. Int. Educ. **11**, 290–305 (2007)
6. de Wit, H.: Internationalization of higher education. J. Int. Stud. **10**, i–iv (2020)
7. Teichler, U.: The changing debate on internationalisation of higher education. High. Educ. **48**, 5–26 (2004)
8. Kehm, B.M.: The contribution of international student mobility to human development and global understanding. US-China Rev. **2**, 18–24 (2005)
9. Agarwal, P., Said, M., Sehoole, M., Sirozi, M., de Wit, H.: The dynamics of international student circulation in a global context. In: Higher Education in the New Century: Global Challenges and Innovative Ideas, pp. 109–144. Brill Sense, Rotterdam (2007)

10. Bracht, O., Engel, C., Janson, K., Over, A., Schomburg, H., Teichler, U.: The professional value of ERASMUS mobility. Final report of the VALERA project. International Centre for Higher Education Research, University of Kassel (2006)
11. de Villé, P., Martou, F., Vandenberghe, V.: Cost-benefit analysis and regulatory issues of student mobility in the EU. Eur. J. Educ. **31**, 205–222 (1996)
12. Marcotte, C., Desroches, J., Poupart, I.: Preparing internationally minded business graduates: the role of international mobility programs. Int. J. Intercult. Relat. **31**, 655–668 (2007)
13. Wiers-Jenssen, J.: Background and employability of mobile vs. non-mobile students. Tert. Educ. Manag. **17**, 79–100 (2011)
14. Shah, D.: By The Numbers: MOOCs in 2018. ClassCentral: MOOC Report (2018)
15. European Universities Initiative. https://ec.europa.eu/education/events/european-universities-initiative_en. Accessed 07 Apr 2020
16. Bohm, A., Davis, D., Meares, D., Pearce, D.: The global student mobility 2025 report: Forecasts of the global demand for international education (2002)
17. Teichler, U.: Internationalisation of higher education: European experiences. Asia Pac. Educ. Rev. **10**, 93–106 (2009)
18. Sánchez-Sánchez, L.C., Salaberri, S., del Sánchez-Pérez, M.M.: Multiculturalism and internationalization in Spanish universities: ¿North-South Socio-cultural Differences? Proc. - Soc. Behav. Sci. **237**, 1125–1130 (2017)
19. Baruque, B., Herrero, A., Corchado, E., Sedano, J.: Implementation of the European computer science course under the Spanish university education system. In: International Conference on European Transnational Education, ICEUTE 2010, pp. 77–84 (2010)
20. Erasmus+ - Statistics. https://ec.europa.eu/programmes/erasmus-plus/about/statistics_en. Accessed 07 Apr 2020

Special Session: Role of English in Transnational Education and Teacher Training

The European Foreign Language Teacher Training Programme: A Comprehensive Proposal

María Amor Barros-del Río(✉)

University of Burgos, Burgos, Spain
abarros@ubu.es

Abstract. Mobility has been a key and central concept for the Bologna Process. Labour mobility of the teaching and learning community within the European borders is in continuous growth. But so far, despite this dynamic reality, there is no unified format for teacher training at a European level. In the light of this, a virtual solution for a comprehensive and coordinated training for foreign language teachers in Europe is needed. Within the frame of the VIR_TEACH Erasmus+ project, this article outlines the creation of a digital tool to improve the postgraduate certificates on Masters on Education (Foreign Languages), and to provide teachers, researchers, student teachers and policy-makers with open-source tools and resources.

Keywords: Teacher Training · Language Teacher · European education

1 Foreign Language Teacher Training in Europe: An Overview

Labour mobility within the teaching and learning collectives is in continuous growth within the European borders. This fact is clearly detected in the Eurydice report on teacher mobility which states that:

In half of the education systems surveyed, less than a third of all teachers appear to be transnationally mobile. In the EU, Executive Summary 13 12.4% of respondents reported that they had gone abroad solely when they were already practising teachers, whereas 5.9% had done so only during their Initial Teacher Training, and 3.6% only in both cases. In all countries surveyed except Iceland, modern foreign language teachers are the most transnationally mobile, compared to teachers of four other main subjects [1].

Many factors contribute to this situation, among which we can point to a constant increase in opportunities for university study abroad [2], which favours the

The contents of this article fall within the scope of the Erasmus+ funded VIR_TEACH project (A VIRtual Solution for a comprehensive and coordinated training for foreign language TEACHers in Europe, Project Reference: 2018-1-ES01-KA203-050045). The European Commission's support for the production of this publication does not constitute an endorsement of the contents, which reflect the views only of the authors, and the Commission cannot be held responsible for any use which may be made of the information contained therein.

Á. Herrero et al. (Eds.): ICEUTE 2020, AISC 1266, pp. 157–163, 2021.
https://doi.org/10.1007/978-3-030-57799-5_16

internationalisation of teacher training, and a static labour market resulting from the last years of economic crisis [3], which impels graduated students to seek jobs beyond their frontiers.

Despite these trends, there is no unified format for teacher training at a European level. Added to that, university trainers do not usually have a space where they can share their worries and advances in their education practice. This lack of common environments also affects teacher students who miss unified digital tools that cover their supervision and tutoring needs, especially during their internship phase. In terms of educational policy, national agencies tend to design and develop their lines of work on teacher training in each country separately.

The evident unbalance between nationally-oriented training and international practice demands urgent action towards the internationalization of the teaching profession as a whole. Within the European Higher Education Area, this may be facilitated thanks to the possibility of homologation of university degrees. However, so far there is no unified format for teacher training at the international level. At the level of educational policies, there is also a need to converge towards a more unified and coherent set of principles among the different European countries, particularly in terms of curricular content and experiences of internships. In general terms, Initial Teacher Education (hereinafter ITE), suffers from an excessively local projection that limits the future teaching exercise of teacher students beyond their borders. Hence, a greater degree of internationalization and convergence of the training of future language teachers is an urgent need within European borders.

2 Data Gathering: Consultation and Report

As Freeman et al. affirmed in the prologue of their famous volume *Teacher Learning in Language Teaching* [4], "in order to better understand language teaching, we need to know more about language teachers". Hence, before addressing the construction of a coherent proposal for foreign language teacher training, we deemed necessary to collect the opinions and experiences of the stakeholders involved in the matter. This, would help build consensus on the lacks and needs of current foreign language training programmes, and improve them coherently at a European scale.

For these purposes, in October–November 2019, a survey was passed among stakeholders to test the perception of language teacher training in Europe.

The main objectives of this survey were to detect current weak aspects regarding language teacher training and to implement positive measures for the advancement of foreign language teacher education in Europe. In particular, this survey sought to:

- Collect comparative data.
- Build consensus during the project.
- Fix potential issues.
- Improve methodological approaches, materials and resources.
- Prioritise intellectual outputs, contents and events' main topics.

The survey was launched in November 2019 and was open for 3 weeks. It reached a total of 508 participants from different gender, ages, countries of residence and profiles, which ensured the heterogeneity of the sample.

The survey was intended to offer a national and international picture of the current situation of foreign language teacher training in Europe. For this purpose, the survey was conducted in several European countries. The countries of residence of the participants were closely related to this fact, with 41,6% of respondents living in Portugal, 27,0% from Spain, 12,0% from Romania, 10,5% from Poland and 5,6% from Belgium. Other countries of residence such as Finland, Iceland, Germany and Turkey, also participated in the survey but with a much smaller percentage (less than 4% in total).

The procedure and results of this survey have been published in detail elsewhere [5] with the following conclusions:

1. There is an urgent need to move towards homogenization of teacher training.
2. Digital platforms are particularly useful tools for internationalisation of foreign language teacher training.
3. The contents of the training programmes need modernisation and upgrading, providing tools, skills and resources to help professionals adapt to the continuous transformations taking place in a changing and dynamic society.
4. Investment in ITE, particularly in the financing of student and teacher mobility, is essential.
5. Feminisation of the teaching practice is related with a low status of the profession and unappealing salaries. Governing bodies and universities should invest in attracting high quality people into the profession, irrespective of their sex/gender. In the medium term, these measures should impact on gender balance in teaching and learning.

3 Language Teacher Training Through a Virtual Learning Environment

In order to move towards language teacher training homogenization, virtual learning environments seem to be particularly useful tools for sharing and learning successful teaching formulas specific to this field of higher education. This way, cultural diversity and geographical distance could be overcome by a multi-national alliance of higher institutions. Moreover, this solution would facilitate cultural adaptation and favour international communication, two essential elements for teaching at an international level. However, a lack of coherence between teaching digital platforms that promote mobility and digital tools that guide the needs of students in their training and internship phases is problematic. As pointed out by Townsend and Bates [6], it is a set of tensions and pressures, inherent in an unstoppable process of globalization that makes a dent in the quality of teaching. This issue becomes more acute, if possible, in the training of language teachers, whose orientation towards communicative teaching requires the implementation of a series of strategies and predisposition to interaction and mutual understanding.

As Stanley [7] has affirmed, language learning with technology provides a wide range of interesting and useful activities, shows how any teacher can use technology, and offers a range of activities, which are suitable for all levels of proficiency to both novice and experienced teachers. Hence, in order to create a European Foreign Language Teacher Training Programme, it seems mandatory the development of a multinational dynamic and lasting over time environment that offers useful tools for the training of future teachers of languages. In the light of these reflections, a communicative approach seems a mandatory element to infuse in the training of 21st century language teachers. For this to happen, prospective teachers must acquire a reflective and critical training [8], oriented towards social commitment [9]. This implies the construction of a critical teaching identity among the candidates during their training process [10]. Added to that, the critical pedagogy received must be closely linked to a culturally receptive teaching practice [11, 12], especially in the multinational contexts in which future professionals of the teaching of languages will have to move. Finally, in a changing and global reality, future language teachers must know how to manage and apply to their teaching the digital technologies that best adapt to their teaching-learning objectives. In the same way, teachers that had been exposed to a learning process through MOOCs "expressed positive attitudes towards their learning experience and a desire to participate in more courses in the future" [13] showing not only gains in content knowledge but also in pedagogical knowledge, which is fundamental not only to the MOOC course design, but also to the overall future teacher's learning process.

4 Design and Methodological Guidelines for a European Language Teacher Training Programme

The creation of an open digital space to promote the exchange of methodologies and good practices that inspire a common, comprehensive and effective training of future professionals of language teaching at European level. This idea must be reflected in the design and development of a platform for online courses, the development of coherent levels of competence at the international level and the advancement of methodologies for better regulation in education policy. In addition, it must promote a greater degree of transparency of the training processes and the creation of a series of digital tools that favour the convergence of educational systems and the exchange of good practices.

Likewise, it must offer a consistent functionality in progressive, significant and cumulative learning sequences developed through a wide variety of learning resources. These must incorporate in turn the diversity that new technologies offer today. Therefore, these resources will have multiple formats, among which videos, readings, infographics, discussion forums, and so on, can be counted. Besides, the design of the exercises will seek to prioritize the interaction through formulas such as multiple choice, open questions with peer review, drag and drop, etc.

Added to that, the implementation of digital tools that facilitate coherence and follow-up between the phases of training and practice in the education of future language teachers should be considered. This measure should include a common protocol and interactive communication spaces. In addition, as indicated by the Finnish Institute for Educational Research, the training of future language teachers should contemplate competences related to pedagogical, communicative and reflection skills. Furthermore, attention to diversity and collaboration with colleagues and peers should be contemplated too. Therefore, the practical phase, or Practicum, must be oriented towards a reflection on the action that the future teachers carry out in the classroom [14]. For this purpose, it is appropriate to incorporate a digital tool that, based on a predetermined protocol, allows to monitor the internship period and to facilitate communication between the different agents involved: the teacher students, their respective mentors from the educational centre, and their academic tutors. As a result, the role of the mentors is expected to acquire more relevance and professionalism in the learning process, a concern addressed in numerous studies [15, 16]. Besides, this tool will allow the establishment of more intense interaction and support among the participants, an issue that brings positive outcomes according to Tomaš, Farrelly and Haslam [17]. Particularly, this virtuous communicative triangle will greatly facilitate the interaction between the aforementioned stakeholders, displaying a reflective, dynamic and enriching learning experience for all parties.

Also, the establishment of international networks that facilitate mobility and exchange among future teachers in training would be desirable. Although there are previous experiences of networks that have been pursuing the internationalization of educational practice, such as the Thematic Network on Teaching Education (TNTEE) and the Policy of Teaching Education in Europe (TEPE), one of the main challenges has been the lack of communication, cooperation and collaboration between the actors of the different countries. The VIRTEACH project tries to fill these gaps through the establishment of organized professional groups and networks of teacher educators.

At the same time, this question poses an enormous challenge that goes beyond the purely academic sphere, bringing into play factors such as intercultural communication or the development of the ability to adapt. As demonstrated by He, Lundgren and Pynes [18], the integration of teaching-learning experiences of languages abroad, included within a study program, offers future teachers a unique opportunity to develop empathy and innovative strategies for instruction. In order to collect the best results from this type of experience, it is advisable to follow up through open reflection journals, in which teaching beliefs and previous intercultural experiences are included, as well as the development of individual goals [19]. For these purposes, the European Portfolio for Student Teachers of Languages (EPOSTL) is an adequate tool for reflection and self-assessment of the knowledge and didactic skills necessary to teach languages [20]. It is an instrument that takes as reference the Common European Framework of Reference for Languages (CEFR) and that allows future teachers to connect their experiences with their teaching practices, so that these same reflections serve as evaluation and self-assessment tools. With permission already granted by the European Centre for Modern Languages of the Council of Europe, VIRTEACH will adapt EPOSTL to the requirements of the virtual environment to maximize its functionality.

The design of protocols and lines of action common to the European Higher Education Area in relation to the training of future language teachers. In this respect, it is necessary to point out the necessary collaboration of the political institutions with competences in educational matters. From the universities themselves, which are the immediate training entities, to the competent councils and governmental delegations, it is necessary to work on the convergence and coherence of curricula for the training of future language teachers with an international perspective, oriented to intercultural communication and that promote a collaborative, empathetic and reflective learning.

To finish, the national and international recognition of this training is essential and can be an interesting pilot project that ultimately would result in good teaching professionals who are dynamic and capable of adaptation and improvement.

5 Final Remarks and Conclusions

The previous sections have dissected the current state of Foreign Language Teacher Training in Europe and shed light upon a comprehensive proposal for convergence and quality enhancement in this field of education.

The suggested measures are expected to improve the quality of language teacher training programs and the coordination among the different stakeholders, through the design of comprehensive but coherent and converging curricula for ITE in the Member States of the European Union. The combination of open courses and teaching practices of a transnational nature must combine with a personalized orientation of the learning process. Coherence between theoretical and practical training, and the exchange of good practices, are the hallmarks of this proposal.

All in all, the final goal is to foster a more exigent training and selection of future educators of languages at an international level, to raise the quality of language teaching programs within the European Union, and to strengthen and unify European educational policies.

References

1. European Commission/EACEA/Eurydice: The Teaching Profession in Europe: Practices, Perceptions, and Policies. Eurydice Report. Publications Office of the European Union, Luxembourg (2015)
2. Martínez Rodríguez, J.B.: Movilidad/movilización de profesorado y estudiantes para la formación. Revista interuniversitaria de formación del profesorado **51**, 233–250 (2004)
3. ET2020 Working Group on Schools Policy. 2014/15: Shaping Career-long Perspectives on Teaching: a Guide on Policies to Improve Initial Teacher Education. European Commission - Directorate-General for Education and Culture, Brussels (2015)
4. Freeman, D., Freeman, D.J., Richards, J.C., Swan, M. (eds.): Teacher Learning in Language Teaching. Cambridge University Press, Cambridge (1996)
5. Barros-del Río, M.A., Mediavilla-Martínez, B.: The European perception of Foreign language teacher training. Findings on a survey in 13 countries. Ann. "Dimitrie Cantemir" Christ. Univ. Linguist. Lit. Methodol. Teach. **2**, 67–81 (2019)
6. Townsend, T., Bates, R.: Handbook of Teacher Education: Globalisation, Standards and Professionalism in Times of Change. Springer, Dordrecht (2007)
7. Stanley, G.: Language Learning with Technology: Ideas for Integrating Technology in the Classroom. Cambridge University Press, Cambridge (2013)

8. Barros-del Río, M.A.: Cómo fomentar el espíritu crítico en los futuros docentes de inglés desde la formación: fundamentos y propuestas. Íkala: Revista de Lenguaje y Cultura **24**(3), 607–618 (2019)
9. Yogev, E., Michaeli, N.: Teachers as society-involved "organic intellectuals": training teachers in a political context. J. Teach. Educ. **62**(3), 312–324 (2011)
10. Dvir, N., Avissar, I.: Constructing a critical professional identity among teacher candidates during service-learning. Prof. Dev. Educ. **40**(3), 398–415 (2014)
11. Abednia, A.: From teacher training to teacher education in second language education. AUC TESOL J. **1**(1), 4–12 (2011)
12. Jiménez, R.T., Rose, B.C.: Knowing how to know: building meaningful relationships through instruction that meets the needs of students learning English. J. Teach. Educ. **61**(5), 403–412 (2010)
13. Donitsa-Schmidt, S., Topaz, B.: Massive open online courses as a knowledge base for teachers. J. Educ. Teach. **44**(5), 608–620 (2018)
14. Manso Ayuso, J., Martín Ortega, E.: Valoración del Máster de Formación de Profesorado de Educación Secundaria: estudio de casos en dos universidades. Ministerio de Educación de España, Madrid (2013)
15. Andreucci, P.A.: La supervisión de prácticas docentes: una deuda pendiente de la formación inicial de profesores. Estudios pedagógicos (Valdivia) **39**(1), 7–26 (2013)
16. Sepúlveda Ruiz, M.: Las prácticas de enseñanza en el proceso de construcción del conocimiento profesional. Educar **36**, 71–93 (2005)
17. Tomaš, Z., Farrelly, R., Haslam, M.: Designing and implementing the TESOL teaching practicum abroad: focus on interaction. TESOL Q. **42**(4), 660–664 (2008)
18. He, Y., Lundgren, K., Pynes, P.: Impact of short-term study abroad program: inservice teachers' development of intercultural competence and pedagogical beliefs. Teach. Teach. Educ. **66**, 147–157 (2017)
19. He, Y., Levin, B.B., Li, Y.: Comparing the content and sources of the pedagogical beliefs of Chinese and American pre-service teachers. J. Educ. Teach. **37**(2), 155–171 (2011)
20. Newby, D., et al.: European Portfolio for Student Teachers of Languages EPOSTL. Council of Europe Publishing, Strasbourg (2007)

The European Portfolio for Student Teachers of Languages: A Reflection and Communication Tool in Teacher Education Programmes

Aránzazu-Lucía Cosido García(✉)

Facultad de Humanidades y Educación, Universidad de Burgos, Burgos, Spain
alcosido@ubu.es

Abstract. Language teacher education in Europe has become an international and transnational lifelong process and, as such, calls for tools to enable communication between the key players in the process. The European Portfolio for Student Teachers of Languages was developed to serve this purpose. This paper tries to shed some light on the usability of the document and to propose some possible future applications overcoming its possible shortcomings for its implementation in teacher education programmes.

Keywords: EPOSTL · Teacher education · Student teacher · Language teaching · European Portfolio

1 Student Teacher Education: State of the Art

To be a competent speaker of two foreign languages is nowadays regarded by the Modern Languages Section of the Council of Europe as an objective for all European citizens and consequently, they set themselves to promote, facilitate and improve language learning and teaching across Europe as an imperative.

As Jack Richards and Theodore Jones (2014) explain [1] language education has moved its focus from being teacher-centered to learner-centered, from being a mere transmission of knowledge to a conception of learning and teaching as interactive processes and to enable language learners to become competent users of a language, not to know a lot about it. This change in language learning and teaching meant that the languages studied were no longer Latin and Greek but modern languages which allowed people to communicate when travelling or doing business, especially in a globalized world with people travelling for pleasure, work or study purposes.

Learning autonomously is another key feature in present learning theories and practices. Independent learners of languages make their own decisions about their learning and take advantage of the opportunities they have to learn and use the target language outside the classroom environment. Language teachers should promote this and offer enough opportunities and strategies to learn independently.

Portfolios are more and more frequently used to foster learner autonomy as they strengthen the connection between theoretical instruction and practice, promote

Á. Herrero et al. (Eds.): ICEUTE 2020, AISC 1266, pp. 164–172, 2021.
https://doi.org/10.1007/978-3-030-57799-5_17

reflection [2] and support learner's autonomy as the learner is responsible for what is incorporated into the portfolio. Besides, a portfolio can be applied for external evaluation as it serves to showcase the language learners' work, and, moreover, for self-assessment in formative evaluation processes and a move toward another aim, promote learner's autonomy.

The European Union has meant an increase in the opportunities for teachers to study, train and work abroad and this new transnational environment demands some common ground in teacher training and with this aim, several documents have been released in the last decades.

1.1 European Documents

Since the 1960s, several policies have been designed and as a result several documents have been produced to try to standardize and provide transparency to the levels of linguistic competence. As a result, the Language Policy Department of the Council of Europe promoted the development of *Threshold Level*, "as part of a project to investigate the feasibility of a unit/credit system for adult language learning in Europe" [3].

Continuing with this idea, the Common European Framework of Reference [4] was presented in 2001 as a set of "I can ..." statements covering different linguistic and communicative competencies and organizing them in six levels of competence from A1 (Beginner) to C2 (Master). The CEFR has made the process of grading course contents, examinations and language learners easier as it provides a transparent framework to be applied for any language and type of learner or course.

In 2001, The *European Language Portfolio* (ELP) was launched to support the development of learner autonomy, plurilingualism, and intercultural awareness and competence; and to allow users to record their language learning achievements and their experience of learning and using languages by means of compiling samples of what they are able to do in the foreign language [5].

Language learner competencies were now clearly defined since both the CEFR and the ELP are transparent tools to be applied in different learning contexts, for different languages and learners. However, an instrument to determine and assess didactic and teaching competencies language teachers require is still to be formulated. At present, student and professional mobility have made this even more needed because pre-service language teachers take part in international schemes which have different structures and components to qualify as a language teacher and institutions from different countries need to work together on common grounds to assess the competence of the student teacher.

A project coordinated by Michael Grenfell analyzed language teacher education programmes in 25 countries and published *the European Profile for Language Teacher Education - A Frame of Reference* (EPLTE) [6]. Its aims were to provide a framework for transnational cooperation, facilitate movement and to share knowledge and possible future developments. It comprises a list of forty elements which should be present in language teacher education syllabuses, such as the combination of academic study and classroom practice, the course structure, skills and competencies to be acquired and values to be promoted.

1.2 The European Portfolio for Student Teachers of Languages

The CEFR and the ELP contributed to determine the competencies a language learner should acquire, then, the EPLTE provides a framework for course development but teacher educators and student teachers still require some guidance about the competencies all language teachers should gain not only during their training but also for lifelong professional development.

In 2007, experts from different countries and the European Centre for Modern Languages (ECML) published the *European Portfolio for Student Teachers of Languages* (EPOSTL): a document which defines didactic and pedagogical competencies involved in language teaching, is based on the student learners' self-assessment and reflection and includes a dossier where they can include samples of their teaching (a portfolio). Thereby, it encourages reflection, helps student teachers to become autonomous learners and teachers, who then are able to foster autonomous learning in their teaching practice. At the same time, it tries to be a tool to be used in different teaching education contexts as a common framework for reflection and practice used by the student teacher himself, teacher educators and school mentors during the teaching practice part of their training.

The EPOSTL defines as its aims:

1. "to encourage you to reflect on the competencies a teacher strives to attain and on the underlying knowledge which feeds these competencies;
2. to help prepare you for your future profession in a variety of teaching contexts;
3. to promote discussion between you and your peers and between you and your teacher educators and mentors;
4. to facilitate self-assessment of your developing competence;
5. to provide an instrument which helps chart progress."

(Newby et al. 2007: 5) [7]

Thus, it is clear its authors pursued critical reflection on the competencies a language teacher should have, their own level of attainment on the one hand, and to do so through self-assessment and dialogic interaction with other student teachers, teacher educators and mentors, on the other hand.

The document is structured in six parts: a Personal Statement (student teachers reflect on their experience as language learners, their expectations about the teacher education course they are doing, and some aspects of the teaching profession), a Self-assessment section where language teaching competencies are identified; a Dossier: the student teacher's portfolio (lessons taught, lessons observed, feedback received…); a Glossary, an Index of terms, and a User's guide explaining what the EPOSTL is and what information to introduce. The main part of the document (the self-assessment section) comprises 193 descriptors formulated as "can-do" statements and organized in seven categories, which cover areas such as understanding the teaching context, methodology aspects, lessons planning or assessment of the learning process. Each descriptor is presented with a progress bar where to register their level of achievement on different dates.

2 Applications of the EPOSTL

2.1 As a Reflection and Self-assessment Tool

Both the CEFR and the ELP were the result of a series of changes in language education from being teacher-centered to learner-centered, moving from a mere transmission of knowledge to a conception of learning and teaching as interactive processes [8]. Student teachers need to critically think about their previous experiences, the context they find themselves in, the competencies language teachers should have and their own attainment levels for these competencies. Besides, when creating a teaching portfolio, the selection of appropriate materials which demonstrate their competence requires a deep understanding of the competence descriptor and assessment of the connection between the activity or materials included and the level of competence they show.

Reflection is defined as careful thought over a period of time and as a rational act that is required to reach an aim. Reflective practice is an essential part of the teaching profession and should thus be encouraged and promoted in teacher education. The EPOSTL calls for the student teacher's reflection before teaching practice, in the Personal Statement and during and after it, following Schön's reflection in-action and on-action principles [9]. Critical thinking and reflection on our own abilities and areas for improvement are essential to become an autonomous learner. As Ekbatani [2] and Little [10] state, language portfolios promote reflection and being responsible for one's learning process as the learners choose the pieces of evidence that they include in their portfolios

Some of the possible ways the EPOSTL has been implemented as a dialogic reflection tool in language teacher education are to include it in different stages of the teacher training courses either before the teaching practice, during the school placements or after the teaching practice. It can be used to plan the teacher training courses in lectures to link the theoretical principles behind the descriptors and some practical applications in the class and in seminars to analyze the contents and structure before school placements, and to promote collaboration from the school mentors [11] During the teaching practice, student teachers may select, together with their university tutors or with their mentors, which elements of the EPOSTL they concentrate on, either while observing lessons or when preparing and conducting them [12]. Finally, after the teaching experience, as to reflect on it, its outcomes and results and the overall progress of the student teacher or to decide what goes into the dossier and why [13]. The Portfolio can be applied mainly for self-assessment in formative evaluation processes and a move towards another ai: to promote learner's autonomy. It may even be used for certification assessment as Hanna Schauber [14] explains, although the EPOSTL is not primarily supposed to be used by third parties as David Newby [15] states and we have mentioned before.

Self-assessment requires deep reflection; therefore, it is closely connected with it. The self-assessment sections of the EPOSTL were used in Turkey to determine self-efficacy and the results showed that student teachers found it a very valuable tool as the descriptors reflect quite well their self-efficacy levels [16]. In another case study, students appreciated the value of the EPOSTL for self-assessment as they were able to see the correlation between the descriptors and the dossier and other European documents such as the ELP and the CEFR and as a professional portfolio to reflect on their skills once they become in-service teachers [17].

2.2 As a Tool to Promote Autonomous Learning

Learner autonomy and independent learning have been given a predominant role in teaching and learning studies for many years. Autonomy in learning can be defined as the ability to set one's own learning goals and make decisions about the learning process and to attain those goals, this has changed education from being teacher-centered to becoming learner-centered. As Hanna Komorowska [10] states, becoming autonomous is a gradual 5-degree process where the teacher and the learner are interdependent at the beginning and progress to reach a point where the learner becomes independent. The teacher's role is to create independent language learners, and ultimately independent language users and consequently, the EPOSTL includes a series of descriptors about the teacher's competence in promoting learner's autonomy, on catering for the students' abilities, needs, and interest, on the one hand, and since it is a self-assessment tool, the student teacher himself becomes more autonomous, on the other hand.

Portfolios are part of the autonomous learning process due to their nature. Portfolios can be used for assessment, to showcase one's work, as development tools, for reflection or hybrid in their nature and have different formats: digital a blog, use e-portfolio sites or i-Webfolios [18]. The learner decides what to include in the portfolio and why in an autonomous process guided and/or monitored by the tutors and mentors, but the learner takes the responsibility for his learning and assesses his competence level. The dossier in the EPOSTL is meant to connect the theoretical knowledge which lies behind the descriptors and the practical applications reflected in the material, lesson and activities included in it. In an international context, digital portfolios may be instrumental in the compilation of a teacher's performance and work, both for academic and professional purposes as they can be easily produced and shared.

2.3 As a Communication Tool

When thinking about teacher education, pre-service teaching practice is given paramount importance: we become competent professionals only through professional practice and the best method to be introduced to the profession is with an initial phase of supervised practice in communities of practice [19]. Almost every teacher education program should include, according to several EURICYDE reports, a period of mentored classroom observation and practice of different lengths, either while an instructional period or consecutively. The variety of syllabuses and training programs offers a picture that is too rich in differences [20].

Mentors play a pivotal role in this apprenticeship period, but this is quite commonly the weakest part of the system. First, in some cases, mentors are not selected among the most competent professionals but appointed by educational institutions or take up this responsibility voluntarily. In this context, it is very often the case that novel teachers who lack experience or some other teachers who are not trained in teacher assessment act as mentors. Secondly, they are not always compensated for their efforts and end up being demotivated and not cooperating as such as it would be desirable. Finally, coordination between the theoretical part of the student teacher's education (universities and colleges) and the educational centres where the student teachers do their placements is missing.

Michael Fenner and David Newby [11] reflect on how the EPOSTL offers the three people involved (the student teacher, the academic tutor, and the school mentor) a framework to overcome these difficulties. The list of descriptors presents some elements to focus on when designing the theoretical content of the course and on what to observe, put into practice and provide a plan for feedback sessions. As Schauber points out, the EPOSTL facilitates interaction between the student and the mentor when used regularly for the supervision, observation of the student teacher and to provide structured feedback [12].

Furthermore, language teaching benefits of multilateral agreements enabling student teachers to do part of their teaching education in a different country. This means two different training and placement contexts and good coordination and some common ground to work on is essential. Bilingual teaching training programmes may benefit from the use of the EPOSTL as well as it can be implemented to have common content and a shared tool to work with in both countries [21].

3 Challenges in the Implementation of the EPOSTL

The implementation of the EPOSTL in different teacher education programmes has revealed some possible shortcomings and difficulties for its application. David Newby [11–13] himself mentions some of them in his works and offers some answers.

The most common feeling reported is being overwhelmed by the number of descriptors. The original portfolio consisted of about 400 descriptors, which were later reduced to 193. Newby justifies the need for so many items in the scope of the descriptors: it tries to cover some general competencies every teacher should have but also specific competencies needed in language teaching. On the other hand, some competencies seem to be missing according to others, for example, cultural competencies [22] or linguistic competencies [23].

A further difficulty was understanding the wording in the descriptors as some student teachers found them too abstract and difficult to understand [24] or even seem contradictory, such as descriptors 2 and 5 from the Resources section [23].

Marking progress by means of bars, without references to what is expected to be achieved or standard grading has also been cited as a problematic issue or found it too subjective [13]. Some implementation cases have included some form of grading with a 5-point scale [17]. As Newby [15] states, gaining teaching competencies is a lifelong process, therefore, there is not a good or maximum level to achieve, and the bar system allows the user to record his progress on different dates throughout his professional life. Apart from this fact, the EPOSTL is intended to be used for self-assessment purposes, not for summative evaluation or external assessment.

Regarding the use of EPOSTL as a communication tool for students, tutors and mentors. Most users have found it very useful as a framework but the implication and collaboration from school mentors must be improved yet. Some students got little or no support from their mentors [11], in other cases, the communication between the university and the placement centres failed even when using the EPOSTL or was not used systematically [12, 13].

Some of the difficulties found when working with the can be overcome when there is some previous team work to select of the descriptors to focus on, reflection what practical aspects they refer to and connect theoretical and aspects to the teaching practice. However, there are no concretions about how to do it within the Practicum in a teacher education course.

4 Conclusions and Future Possibilities

The EPOSTL has proved to be a valuable instrument in language education both for pre-service and for in-service teachers, however, some things have to be improved. New teachers often feel that their training courses did not equip them with the essential knowledge nor skills to be professionally competent [25].

First of all, mentoring should be structured and organized in order to improve their qualifications and involvement, as well as recognizing the importance and value of their work.in this line, Mike Grenfell proposes "defining the competencies of the "European Language Mentor" …a model of mentoring …to establish a European Mentor team" [26:169–79].

A second reason to include the EPOSTL in language teacher education is that given the internationalization of the education and professional world, what is also needed is a tool which can be shared in transnational programmes, transparent regarding the contents covered and the competencies achieved and which is internationally recognizable and recognized.

We consider that the EPOSTL can become a very valuable instrument for dialogic reflection and communication for school mentors and student teachers, as well as for academic tutors, and in transnational programmes, bearing in mind the fact that is must be through a carefully designed and systematic implementation in language teacher education are essential to enhance its value as a framework for reflection and cooperative work, especially in the practice stage of the courses.

To overcome the difficulties above mentioned, future implementations in language teaching training courses should be carefully planned and ideally include the following elements:

- an introduction to the EPOSTL, its rationale and structure to make sure the student teachers and the mentors know what it is and how to use it,
- a selection of the sections of the EPOSTL to focus on along the teacher education process: what the academic contents should cover; discussions to relate theory and practice, and dialogic reflection on what to include in classroom observation, what to put into practice during school placements; involving the student teacher, the academic tutor, and the school mentor,
- structured feedback and assessment (self-assessment, peer-to-peer, from the tutor and from the mentor) during and after the teaching practice,

- a teacher portfolio to document the teacher's performance and competence,
- finally, encourage the use of the EPOSTL for in-service teachers to reflect and document their performance and development throughout their career.

Although there have been some attempts to transform the EPOSTL, or parts of it, into a digital tool, none of them has fulfilled the requirements above, having been restricted to the creation of a digital portfolio based on the Dossier or student-written blog. A well-structured instrument to be used at different stages of the teacher education process and in which all the participants (student teacher, university tutor and school mentor) can share their reflections and feedback is the way forward.

International student mobility programmes, such as Erasmus+, or teacher education courses which include transnational components would highly benefit from the use of such a tool as they would have a structured framework to collaborate and co-work.

References

1. Richards, J.C., Jones, T.S.: Approaches and Methods in Language Teaching. Cambridge University Press, Cambridge (2014)
2. Ekbatani, G.E., Pierson, H.E.: Learner-Directed Assessment in ESL. Lawrence Erlbaum Associates Publishers, Mahwah (2000)
3. Van Ek, J.A., Trim, J.L.M.: Threshold Level 1990. Council of Europe, Strasbourg (1991)
4. Council of Europe: Common European Framework of Reference for Languages: Learning, Teaching, Assessment. Cambridge University Press, Cambridge (2001)
5. Little, D.: The European Language Portfolio: Where Pedagogy and Assessment Meet. Council of Europe, Strasbourg (2009)
6. Kelly, M., et al.: European Profile for Language Teacher Education: A Frame of Reference. Office for Official Publications of the European Communities, Luxembourg (2005)
7. Newby, D., Allan, R., Fenner, A.: European Portfolio for Student Teachers of Languages: A Reflection Tool for Language Teacher Education-European Centre for Modern Languages (2007)
8. Komorowska, H.: Learner autonomy and implications for the EPOSTL. In: Newby, D. (ed.) Insights into the European Portfolio for Student Teachers of Languages (EPOSTL). Cambridge Scholars Publishing, Newcastle upon Tyne (2012)
9. Schön, D.A.: The Reflective Practitioner: How Professionals Think in Action. Routledge, Abingdon (2017)
10. Little, D.: The Common European Framework and the European Language Portfolio: involving learners and their judgements in the assessment process. Lang. Test. **22**(3), 321–336 (2005)
11. Fenner, A.-B., Jones, B., Newby, D.: Using the European Portfolio for Student Teachers of Languages (EPOSTL). Council of Europe Publishing, Strasbourg (2011)
12. Ingvarsdóttir, H.: The EPOSTL in Iceland: getting the mentors on board. In: Fenner, A.-B., Jones, B., Newby, D. (eds.) Using the European Portfolio for Student Teachers of Languages (EPOSTL), Council of Europe Publishing, Strasbourg (2011)
13. Nihlén, C.: What goes into the EPOSTL Dossier and Why? In: Fenner, A.-B., Jones, B., Newby, D. (eds.) Using the European Portfolio for Student Teachers of Languages (EPOSTL), Council of Europe Publishing, Strasbourg (2011)
14. Schauber, H.: Using the EPOSTL for dialogic reflection in EFL teacher education. Gist Educ. Learn. Res. J. **11**, 118–137 (2015). ISSN 1692-5777

15. Newby, D. (ed.): Insights into the European Portfolio for Student Teachers of Languages (EPOSTL). Cambridge Scholars Publishing, Newcastle upon Tyne (2012)
16. Bergil, A.S., Sariçoban, A.: The use of EPOSTL to determine the self-efficacy of prospective EFL teachers: raising awareness in English language teacher education. Dil ve Dilbilimi Çalışmaları Dergisi **13**(1), 399–411 (2017)
17. Mirici, I.H., Hergüner, S.: A digital european self-assessment tool for student teachers of foreign languages: the EPOSTL. Turk. Online J. Educ. Technol.-TOJET **14**(1), 1–10 (2015)
18. Cimermanová, I.: Digital portfolio in building teaching efficacy of pre-service teachers. J. Lang. Cult. Educ. **3**(1), 57–68 (2015)
19. Gassner, O., (ed.): Strategies of Change in Teacher Education-European Views: Conference Proceedings ENTEP/BLK Conference, 18–19 January 2002, Feldkirch, Austria. Pädagogische Akademie (2002)
20. Díez Gutiérrez, E.J., Romero, L.D., Salas, A.F.F.: Hacia un nuevo modelo de formación y acceso a la función docente: propuestas a debate. Revista Docencia e Investigación (2018)
21. Jones, B.: The use of the EPOSTL in a bilateral teacher-education programme. In: Fenner, A.-B., Jones, B., Newby, D. (eds.) Using the European Portfolio for Student Teachers of Languages (EPOSTL), Council of Europe Publishing, Strasbourg (2011)
22. Byram, M.: Reflecting on teaching 'culture?' In language teacher education In: Newby, D. (ed.) Insights into the European Portfolio for Student Teachers of Languages (EPOSTL). Cambridge Scholars Publishing, Newcastle upon Tyne (2012)
23. Burkert, A., Schwienhorst, K.: Focus on the student teacher: the European portfolio for student teachers of languages as a tool to develop teacher autonomy. Innov. Lang. Learn. Teach. **2**(3), 238–252 (2008)
24. Orlova, N.: Challenges if integrating the EPOSTL into pre-service teacher training. In: Fenner, A.-B., Jones, B., Newby, D. (eds.) Using the European Portfolio for Student Teachers of Languages (EPOSTL), Council of Europe Publishing, Strasbourg (2011)
25. Mehlmauer-Larcher, B.: The EPOSTL: promoting language teacher learning in the context of field experiences. In: Newby, D. (ed.) Insights into the European Portfolio for Student Teachers of Languages (EPOSTL). Cambridge Scholars Publishing, Newcastle upon Tyne (2012)
26. Grenfell, M.: The EPOLSTL and the european profile for teacher education. In: Newby, D. (ed.) Insights into the European Portfolio for Student Teachers of Languages (EPOSTL). Cambridge Scholars Publishing, Newcastle upon Tyne (2012)

Using English in Teaching Romanian Language for Foreign Students

Ramona Mihăilă(✉)

Dimitrie Cantemir Christian University, Bucharest, Romania
ramona.mihaila@ucdc.ro

Abstract. Romanian is a Romance language and it is not a widely spoken language. Twenty seven universities out of over eighty public and private universities in Romania offer the preparatory year for learning Romanian for foreign citizens. The present article focuses on methods of using English as a tool in teaching Romanian language for foreign students. It also provides recent theoretical literature and practice in the area of teaching methodology and approaches the issues arising while teaching Romanian language, using traditional methods: the direct method, grammar-translation, audio-lingual, the structural approach, total physical response, communicative language teaching, the silent way, community language learning, immersion, task-based language learning, the natural approach, the lexical syllabus, just to mention few of them suggested by Dave Willis.

Keywords: English · Teaching methods · Romanian language · Foreign students

1 Introductory Matters

1.1 Romanian Language in International Contexts

Romanian is a Romance language, less widely spoken than the other languages of the same family, Spanish, French, Italian, Portuguese. However, the Institute of Romanian Language established in 1999 as a specialized body subordinated to the Ministry of National Education has as its main mission added on the official website "to promote the Romanian language, culture and civilizations abroad, through the specific means of the education system. The Romanian language has a vast dimension, apart from its status of the official language of Romania, respectively of the official language of the European Union.

The Institute of the Romanian Language ensures the preservation, development and affirmation of the linguistic and cultural identity, by promoting the Romanian language in academic and school settings around the world": Armenia (Yerevan Valeri Brusov State University), Austria (University of Vienna), Azerbaijan (University of Foreign Languages, Baku), Belgium (Vrije University of Brussels), Bulgaria (University of SF. Kliment Ohridski, Sofia, University of Veliko-Turnovo), Canada (York University, Toronto), China (University of Foreign Languages, Beijing), Croatia (University of Zagreb), Finland (University of Tampere, University of Turku), France (Aix-Marseille

Á. Herrero et al. (Eds.): ICEUTE 2020, AISC 1266, pp. 173–180, 2021.
https://doi.org/10.1007/978-3-030-57799-5_18

University, Sorbonne Nouvelle University, Paris, University of Strasbourg), Germany (Humboldt University, Berlin, University of Leipzig, University Johannes Gutenberg, Mainz, University of Regensburg), Greece (University Democritus of Tracia, Komotini), Hungary (E.L.T.E. University, Budapest, University of Szeged), Israel (University of Tel Aviv), Italy (University L'Orientale, Naples, Study University, Padua, University of Pisa, Sapienza University, Rome, University of Torino, University of Udine), Kazakhstan (Karagandy State University), Lithuania (University of Vilnius), Republic of North Macedonia (St. Cyril and Methodius University, Skopje), Republic of Moldavia (Comrat State University, ATU of Gagauzia), Morocco (Mohammed V University, Rabat), Poland (Jagiellonian University, Krakow, Adam Mickiewicz University, Poznan), Portugal (University of Lisbon), Serbia (University of Novi Sad), Slovakia (Comenius University, Bratislava), Slovenia (University of Ljubljana), Spain (Autonomous University of Barcelona, University of Granada, University of Salamanca, University of Seville), Sweden (University of Lund), Turkey (University of Ankara), Switzerland (University of Geneva), United Kingdom (Oxford University), United States of America (Columbia University, New York, Arizona State University), Vatican (Pontifical Oriental Institute).

1.2 Romanian as a Foreign Language in Romanian Spaces

Twenty-seven universities out of over eighty public and private universities in Romania offer a preparatory year for learning Romanian to the foreign citizens. The program of each university is evaluated every five years in order to be accredited by the Romanian Agency for Quality Assurance in Higher Education and the curriculum of the program is almost the same for all the universities; it has compulsory and elective courses for two semesters. The specific standards going to be fulfilled by the universities refer to the qualifications of the teaching staff, number of courses and seminars, number of students in a group, number of credits for each subject, ways of admitting and evaluating the students, logistics.

The program Romanian language for foreign students was introduced to Dimitrie Cantemir Christian University (name hereafter DCCU) 3 years ago and it was accredited by Romanian Agency for Quality Assurance in Higher Education two years ago. The students enrolled for the Romanian language program at DCCU came from European countries (Italy, Spain, Albania, Greece) and non-European countries (South Korea, Iran, Iraq, Turkey, China, Cameroon, Nigeria, India, Syria, Ukraine, Congo, Morocco). The Romanian program's curriculum follows the standard one imposed by the national agency, thus the students must study: Introduction to Romanian language, Romanian culture and civilization, Romanian phonetics, vocabulary, grammar, oral and written communication, writing and composition, Romanian language for specific purposes: social sciences, engineering sciences, mathematics and nature sciences, sports and physical education, humanities and the arts, biology and bio-medicine. The imperative condition for a teacher in this programs is to have already graduated the Romanian major and defended the Ph.D. In the Romanian academic system, a student at the Faculty of Foreign Languages studies two foreign languages (major, minor) and a student at the faculty of Letters studies Romanian (major) and a foreign language (minor). Ileana Oana Macari asserts that "an English teacher teaches Romanian better" being aware that this statement makes people rate her "as inflated and self-assertive. That is only natural, since traditionally it is a Romanian teacher's job to teach Romanian, while an English teacher 'ought to' stick to English." [1].

2 Ways of Teaching Romanian to Foreign Students

2.1 Students' Educational and Social Background

The students who enroll for the Romanian language program have different reasons in learning Romanian. Some of them come to Romania to attend a faculty with Romanian programs (it is mandatory for the students to have already got a certificate of Romanian language, level B1, in order to be admitted for a faculty where the classes are taught in Romanian), others want to attend a program taught in English, but they need to know Romanian for their everyday life, others are refugees and they want to settle down in Romania, others are religious missionaries in the world and it is advisable for them to learn the language where they accomplish their duty for few years, others want to work in Romania and so on. Being an academic program all the candidates need to have already got their baccalaureate diploma. Aurora Tatiana Dina also underlines this fact asserting that "teaching the courses of Romanian language as a foreign language offers large and qualitative pedagogical approaches. The graduates of the preparatory year obtain a certified diploma and their attending the courses is fully acknowledged at an academic level [2].

Thus the groups of students have different educational and cultural backgrounds. The age of the students is very different, too. So, first of all, the professors of Romanian language have to solve the problem of the group homogenization. Then, they have to assess the students' English level. Sometimes the students attending the preparatory year of Romanian finally improve their English and learn some Romanian always comparing it to English notions because they talk, listen, speak English more than Romanian, They start knowing each other, coming to university for at least 6 h per day, socializing in their spare time and speaking English all the time. They never speak Romanian each other outside the classroom, even if they are friends or stay together in the dormitory. When they meet Romanian native students in the campus or dormitory they keep speaking English, so they use Romanian especially while writing their homework or going shopping (they need to read labels, ingredients, prices, or talk to the sellers in the open market etc.) Thus, we may talk about formal, non-formal and informal teaching as they are described by the provisions of the European Union, respectively EPALE – Electronic Platform for Adult Learning in Europe. We may gradually get them involved in our interaction with our foreign students.

The formal education relies on the professor and the university and finally the student gets an official degree. The non-formal education, also called "parallel education", "non-school education", "extracurricular activities", takes place outside the academic structure, but in organized structures and informal ones (we go with our students out of the classroom, for officially organized sightseeing tours, trips, visits to the museums etc.). To sum up, in accordance with EPALE, "we can say that languages can be learned through three types of education: formal education, which is compulsory and takes place in schools, non-formal education, which is what you learn on your own, in order to expand knowledge, and the informal education that includes learning a language unintentionally by living with a host family in another country, learning from them and their surroundings and that is what the collaborative consumption community makes" [3].

2.2 Syllabus Design and Course Books

The professors have also to pay attention the way they design their courses syllabus, as according to Dave Willis the language learners "face three tasks". They must acquire an enormous body of knowledge, they must store it in such a way that they can act on it automatically, and they must use the language with which they are familiar as a basis for exploring the further possibilities which exist in the language. In order to help learners achieve this, the syllabus designer must first specify syllabus content as economically as possible. Almost any language course specifies what the designers consider necessary for the learners to learn and know at a certain stage of language development, even though they cannot guarantee when and if learners will acquire what is presented to them [4]. The Romanian authors of different articles mentioned in the present text and listed in the references section have written their studies after having been teachers for the Romanian language programs for foreign students and they shared their experiences in different Romanian universities. Cristina Eugenia Burtea-Cioroianu asserts that "the methodology of teaching Romanian to foreign students involves countless aspects of both classical and modern approaches. In this type of teaching approach one must take into consideration certain effective and subtle stratagems in order to make the learning difficulties of lexical and morpho-syntactic structures of the Romanian language more accessible to the foreign students" [5] while Aurora Martin analyzes the stereotyped approaches in teaching Romanian [6].

As for the manuals for Romanian as a foreign language following the European Framework of Reference for Languages, there have been published a lot of more or less useful books, some of them have bilingual versions, English- Romanian, French-Romanian, Turkish-Romanian and as Roxana-Magdalena Bârlea considers "the main challenge concerning course books is that most of them focus mainly on grammar and that the types of class activities proposed are structural exercises. Certainly, there are some exceptions where authors try to introduce action-based activities, oral comprehension, etc." [7] while Florica Faur considers "conversation as a didactic dialogue, achieved through a succession of questions and answers, still one of the most effective methods for learning a language" [8]. Teaching Romanian for foreign students refers also to the socio-cultural aspects of reaching objectives and specific competences as Anca Păunescu enumerates "the system of attitudes, values and behaviors, getting a set of knowledge, listening comprehension and speaking skills, intellectual capacities, rational practice of language [9]. In his exquisite study, "Romanian-Language Teaching: A Century of Pedagogical Materials Published outside Romania," Charles M. Carlton tries to find out common elements of the typical Romanian textbook like teaching materials and concludes that "certain features have no unique connection with Romanian: thy are part of the complex elements which comprise manuals whatever the language they purport to present" [10]. Thus, he finally catalogues them in accordance with the genre, format, history and culture, language and its external aspects, magnitude, the academic calendar, sequencing, esthetics, maps, audio, popular language, workbooks and texts.

2.3 Traditional and Modern Methods in Teaching Romanian Through English

As I have already mentioned the educational and cultural backgrounds of the students enrolled for the Romanian as a foreign language are very different and only few of them have graduated humanities fields so it is vital for the professor to start teaching basic grammar notions. The Romanian grammar is not at all like the English one. The fact that the students try all the time to understand Romanian following English grammar rules is an impediment. They also try to make connections even to their mother tongue and many times in their search of finding similarities they cannot concentrate on logical comprehension.

This situation repeats itself in terms of reading, too. The students try to pronounce Romanian words as if they are English words. The phonetical rules of these two languages are totally different. The traditional translation method is also difficult to use, even when we deal with short sentences. The students tend to use the word order of the English language. We may say that the professors of Romanian teach or remind the student notions of the English grammar and only then they go further teaching Romanian. The students always want to add information about the grammar of their mother tongue and explain their attempts in finding similarities with Romanian or with English. In terms of vocabulary, Onorina Botezat asserts that "discovery strategies in acquiring new vocabulary comprise determination strategies, based on retrieving the meaning from contextual clues, with the help of a dictionary and word lists and social strategies that rely on instructions, mutual projects, and communication with native speakers. On the other hand, consolidation strategies include memory, cognitive and metacognitive strategies" [11]. Laura Ioana Coroamă and Eliana-Alina Popeţi comment on the situation when English language functions as a communication language in teaching Romanian, and practically "Romanian will remain at the stage of a material that is either read, written or listened to, things will become complicated, and Romanian will not reach the status of a language that students can communicate in, or this is the level they need. While working with foreign students we have noticed that some of those who have advanced knowledge of English tend to filter the entire educational process of learning Romanian through this language. At a given moment, they will not be able to translate simple vocabulary items rendering certain features of Romanian, like inflected forms, pronominal clitics, strong and weak forms of pronouns, etc. and this will bring about many issues as they will be compared to English grammar" [12].

Reviewing other research literature concerning methods of using English in teaching Romanian we may add suggestions recommended by Linda Hodges who tries to make teaching more efficient by flipping the classroom and offering students "recorded lectures outside of class and homework-like activities in the class" [13].

The online methods are characterized as being very efficient by few researches who described them both at the theoretical and practical level. Thus, Silvia Monica Feraru, Horia-Nicolai Teodorescu, Marius Dan Zbancioc who propose a web-based resources for languages and language technology e-learning SRoL web site in order "to provide a free Romanian database for students and researchers, for linguists, for teachers, in view of teaching, learning and analysis the Romanian language sounds. The database includes the pronunciation corpus and related documentation. The database contains among others, sections with: recordings of syllables and words pronounced in various contexts, like accentuated word, interrogative sentences, exclamations, various emotions conveyed by the speaker, etc.(…) files of sounds, syllables and words pronounced by persons with various pathologies; visual tools that exemplify and fully explain spelling for a large variety of voices and contextual and emotional states. (…) Not only is it the first for the Romanian language, but its multidimensionality also makes it somewhat unique and novel in concept for language learning and teaching in general" [14]. Talking about the effective implementation of eLearning tools for the Romanian language and literature classes for university students at the Faculty of Philosophy, University of Novi Sad, Ivana Janjic and Virginia Popović notice that for students "the eLearning teaching process is more interesting, they have more learning materials, and most important, everything is available in the form most suitable for them. Based on the survey whether the implementation of eLearning has effective influence the process of studying most of the students (82%) found it very helpful. All this have influence on improving the relationship between students and professors" [15]. But Cristina Mihaela Nistor remarks that "since technology encourages on-line communication to the detriment of face-to-face interaction, teachers may have a hard time proving the benefits of more traditional methods in studying a foreign language [16].

3 Conclusions

Teaching Romanian for foreign students is a challenge both for novice and experienced professors because it offers them a lot of unexpected situations. It mostly depends on the enrolled students for the respective year of study and on the professor's skill to firstly homogenize the group and then make the students interact with each other paying attention from the very beginning to terms of intercultural communication language, like euphemisms, racism, sexism, and political correctness [17].

With the multi-language, multi-culture and mixed and shifted populations, on one hand and the continuous progress and development of software and media, on the other hand, teachers must be duly equipped to meet [18] and face the challenges. While Ana Cristina Lemnaru reflects on the teacher's role as the mediator for every level of learning "by experience, reflection, generalization and application. It is therefore necessary to design new methods and modern tools for learning the Romanian language, articulated with the cultural dimension and at the same time, focused not on the history of literature, nor grammar, but on the practical dimension" [19], Cristina Dafinoiu mentions how "Romanian language teachers have found themselves in the position to teach their mother tongue from a completely different perspective, something not easily achieved. And, from this viewpoint, many have rightly asked themselves about the magic formula by which the foreigner, regardless of country of origin, age, sex, occupation, social status or religion, could learn Romanian quickly and easily [20].

Acknowledgement. The present paper is part of the Erasmus+ Project VIR_TEACH: A VIRtual Solution for a comprehensive and coordinated training for foreign language TEACHers in Europe (Ref. 2018-1-ES01-KA203-050045).

The European Commission's support for the production of this publication does not constitute an endorsement of the contents, which reflect the views only of the authors, and the Commission cannot be held responsible for any use which may be made of the information contained therein.

References

1. Macari, I.O.: An English teacher teaches better Romanian. In: Developing a Pan-European Network of Language Resource Centres for Less Widely Used Less Taught Languages, The Proceedings of the eEuro inclusion International Conference "Equal Chances for Any European Language", CDRMO, Iaşi, 30 September–2 October, pp. 66–69 (2005)
2. Dina, A.T.: Successful approach for teaching Romanian as a foreign language. Proc. - Soc. Behav. Sci. **70**, 1032–1037 (2013)
3. *** EPALE – Electronic Platform for Adult Learning in Europe. https://epale.ec.europa.eu/en/node/20687
4. Willis, D.: The Lexical Syllabus. Collins ELT (1990)
5. Burtea-Cioroianu, C.E.: Methodological aspects of teaching Romanian as a foreign language. In: Proceedings of the 5th International Multidisciplinary Scientific Conference on Social Sciences and Arts, vol. 18, pp. 373–380. SGEM, Sofia (2018)
6. Martin, A.: Reprezentare identitară feminină în cultura română. Editura Pro Universitaria, Bucharest (2018)
7. Bârlea, R.M.: Teaching Romanian as a foreign language. Formal and non-formal education perspectives. In: Grosu-Rădulescu, L.M. (ed.) Multilingual Education, vol. 27, pp. 111–123. Springer, Cham (2018)
8. Faur, F.: Elemente de structură în predarea limbii române ca limbă străină (Structural Elements in Teaching Romanian as a Foreign Language). Language and Literature Studies, Foreign Languages Learning, pp. 107–112. Editura Universităţii Vasile Goldiş, Arad (2019)
9. Păunescu, A.: Interactive group methods in teaching Romanian as a foreign language. In: The Proceedings of the International Conference Globalization, Intercultural Dialogue and national identity. Section: Language and Discourse, vol. 1, pp. 128–133. Arhipelag XXI Press Tîrgu-Mureş (2014)
10. Carlton, C.M.: Romanian-language teaching: a century of pedagogical materials published outside Romania. Mod. Lang. J. **68**(4), 354–369 (1984)
11. Botezat, O.: Various approaches in teaching vocabulary. Ann. "Dimitrie Cantemir" Christ. Univ. Linguist. Lit. Methodol. Teach. **18**(2), 104–121 (2019)
12. Coroamă, L.I., Popeţi, E.A.: Using English in Teaching romanian for foreigners: what balance between students' linguistic Portfolio and communicative efficiency. Quaestus Multidisc. Res. J. **9**(2), 271–276 (2016)
13. Hodges, L.: Making our teaching efficient: flipping the classroom. In: Tomorrow's Teaching and Learning. National Teaching and Learning Forum, 25/5. https://onlinelibrary.wiley.com/doi/epdf/10.1002/ntlf.30033
14. Feraru, S.M., Teodorescu, H.N., Zbancioc, M.D.: SRoL - web-based resources for languages and language technology e-learning. Int. J. Comput. Commun. Control **5**(3), 301–313 (2010)

15. Janjic, I., Popović, V.:. Effective implementation of eLearning - Romanian language and literature classes for university students in Serbia. In: Proceedings of the 10th International Conference on Virtual Learning Virtual Learning – Virtual Reality. Models & Methodologies, Technologies, Software Solutions Phase II - Period 2010–2020: e-Skills for the 21st Century, pp. 447–456. Editura Universităţii din Bucureşti, Bucharest (2015)
16. Nistor, C.M.: E-learning and social networking: friends or foes in studying Romanian as a foreign language? In: Conference Proceedings of "eLearning and Software for Education" (eLSE), pp. 183–187. "Carol I" National Defence University Publishing House, Bucharest (2016)
17. Mihăilă, R.: Intercultural communication language: euphemisms, racism, sexism, and political correctness. Ann. "Dimitrie Cantemir" Christ. Univ. Linguist. Lit. Methodol. Teach. **18**(2), 82–92 (2019)
18. Botezat, O., Mihăilă, R., Barros-Del Río, M.A.: Education management: good practice and modern technologies in teaching foreign languages. In: Proceedings of International Conference "Creativity and Innovation from the Sustainable Development Education Perspective." UNESCO Chair Culture of Peace and Sustainable Development Education, pp. 128–138. Editura Universitară, Bucharest (2019)
19. Lemnaru, A.C.: New methods in teaching Romanian language for foreign students. In: Proceedings of the 5th International Conference EDU-WORLD - Education Facing Contemporary World Issues. Procedia - Social and Behavioral Sciences, 15 April 2013, vol. 76, pp. 451–455 (2012)
20. Dafinoiu, C.: Adapting methods of teaching Romanian as a foreign language to the target group. J. Humanist. Soc. Stud. **7**, 93–105 (2016)

The Relevance of an Intercultural Approach in Teaching English for Academic Purposes

Paola Clara Leotta(✉)

University of Catania, Piazza Università 2, 95100 Catania, Italy
pcleotta@unict.it

Abstract. Transnational education is one of the many facets of inclusive education, with its relevance of English as the natural Lingua Franca in a vast range of contexts. The University of Catania, Italy, is an interesting case-study of a highly international University, which welcomes students from all over the world, and where transnational education has become an ordinary practice. In the light of my experience of English language teaching, I would like to report both on the cognitive value of an intercultural, learner-centered approach and on the various factors that concur to make mixed classes a motivating practice. Based on the hypothesis that people who speak different languages have different worldviews, my teaching proposal aims at fulfilling two objectives: understanding the perspectives of other cultures, and comparing the language and culture of others with students' own.

Within a transnational education project, the teacher's role is that of a cultural mediator who takes students' needs into account and practices scaffolding as a guidance on the development of cognitive and social skills. Finally, some excellent sources to help teachers connect culture with the English language teaching process will be presented, as this cultural mediation inevitably leads to the practice of translation.

Keywords: Intercultural approach · Teaching English · Translation studies

1 Translation Studies in Intercultural Language Teaching

For the last 50 years, translation has been to all extents and purposes considered as an autonomous discipline which has shaped its specificity and scientific domain by drawing on elements from neighbouring disciplines, linguistics, anthropology, and communication studies being the most common.

This contact has contributed to the growth of Translation Studies and to its "coming of age" in the academic arena, but it has also given rise to much innovative thinking in the practice of translating.

The current view of Translation studies is that it comprises the study of the process and product of translation as a form of linguistic and intercultural mediation in a specific socio-cultural and ideological setting. Many of the developments within Translation Studies owe a debt to research within the descriptive tradition and to the insights of the scholar and translator James S. Holmes [1], who divided the discipline into Descriptive, Theoretical and Applied branches.

Á. Herrero et al. (Eds.): ICEUTE 2020, AISC 1266, pp. 181–186, 2021.
https://doi.org/10.1007/978-3-030-57799-5_19

A factor that has contributed to the development of "Applied TS" is the move away from prescriptive and evaluative positions to descriptive ones, which, in recent years, has affected almost every linguistic approach to translation, and has shifted the attention from some sort of ideal notion of translation, basically conceived as the equivalent reproduction of a source text towards which all translators should aspire, to a more negotiable and contingent one, based on the observation of how translations are actually produced in specific situations and with specific purposes. This shift in perspective can be mainly attributed to the advent of computer corpora. A pioneer in this field, Mona Baker [2] has demonstrated the potential of these new tools with evidence from monolingual comparable corpora comprising texts originally written in English and texts translated into English from various languages. Her findings have shown that "translation universals" can be identified in a translated text, whatever the original language. These consist in abstract categories of regularly-occurring patterns of translational behaviour, generated by the process of mediation, such as explicitation, simplification, normalization and the levelling out of differences between spoken and written modes.

The use of multilingual parallel corpora is a motivating and fulfilling activity. Even if it is impossible to teach students how to transfer the "creative spirit" from language to language, learners can observe a translator's performance and compare their own competence in a stress-free environment; they can even construct their own hypotheses and general guidelines regarding possible and viable "dynamic equivalents" [3] on the basis of natural patterns of language at discourse levels beyond the word.

Also, students can compare not only a source text in English and a target text in Italian, but also two target texts, i.e. either two translations in two different languages of the same text, or two translations of the same text carried out by different translators, or again two translations done at different points in time by the same or different translators, as well as studying cases of intralingual translation (especially in the case of scientific texts and their popularized versions).

Moving from a pragmatic, empirical position towards a more scientific, collaborative discourse [4], the practice of teaching translation inevitably leads to a linguistic awareness that improves both the quality of teaching and that of students' learning in the English language classroom at an undergraduate level [5, 6].

As we can see, the introduction of a diachronic rather than a purely synchronic dimension within Translation Studies opened up entirely new prospects for the discipline. For instance, it emerged that Translation plays a significant role in human communication, and cultures tend to translate differently at different historical moments [7].

It is a cross-cultural issue, the channel through which various nations and cultures exchange.

And because of this tremendous interest in translation in the global world, urgent needs arise to include translation courses in the EFL teaching programme among university curricula. By adopting a transdisciplinary perspective, we will bring together the various fields of scholarly enquiry and practice that make a valuable contribution to diversifying the uses of translation in education.

Just to have a historical perspective: when modern languages came to be broadly established as part of the University curriculum in Europe and North America over the course of the 19th century, both the rationale for their inclusion and the pedagogy employed were modelled on the study of the ancient classical languages.

The purpose of language study was to gain privileged access to prestigious foreign cultures. Accordingly, the grammar-translation method was the preferred pedagogy which studied the grammatical rules and structures of foreign languages. It required not only a word-for-word substitution, but also a cultural understanding of the way people in different societies thought.

Although intellectual movements challenged the grammar-translation emerged in Europe in the late 19th and early 20th centuries, [8, 9] the approach continued to dominate the study of foreign languages in higher education in Europe and North America until the 1940s [10] and in many other parts of the world it continues to be employed widely.

Another model which appeared later was the text-based translation model which focused on texts rather than words or sentences in the translation process.

2 Approaches and Methodologies

What I would like to focus our attention on is what I define as a "challenge" of teaching translation within a syllabus of ELT and, in the light of my experience of ELT at the University of Catania, I would like to suggest approaches and methodologies for classroom activities.

High school teacher Carol McCarthy has developed the "Poetry in Translation" unit [11], through which she invites her students (in a multilingual class) to translate published poems from their home languages into English and to compare these with professional translations for differences in how language is used and the meaning it conveys. Students also research poets' lives and the cultural, social, and political context in which the poems were written. Finally, students write and translate their own poems, which is also a stimulus for the goal of improving students' writing skills.

Park [12] found that English learners developed metalinguistic awareness and engaged in collaboration to construct meaning. Additionally, translation required them to attend to not just semantics and syntax, but also the ways that language is both cultural and political.

In addition, subtitling short media materials is a potent tool in English teaching that can be used both individually and collaboratively. With a handful of open-source subtitling applications available, subtitling can be part of classroom and out-of-classroom activities. The activity as briefly presented here consists in students' producing subtitles to a media material selected by the teacher or the students. Effective performance necessarily involves dealing with technical, inter-semiotic, intercultural, textual, contrastive, and language learning challenges. In my own educational practice, I tend to select materials which have a social role to play, e.g. a short publicity media material for a students' association at our university, or even a short movie for a social campaign.

This way, translation can be used as a means to an end, a way to learn another language, and therefore focuses on the target language. It helps my students develop their knowledge about English, thus, translating into English becomes beneficial as a language-learning device.

In the meantime, learning English as a foreign language in this respect is closely tied to learning translation methods whereby students convey the meaning of the source language into the target language. When translating, students will not only learn the grammar, syntax and the lexis of both languages they are working at, but many other skills will be enhanced, such as listening, reading, comprehension, interpreting and editing. In the end, the translated text is a "sort of rewriting", as Lefevere states in 1992 [13]:

"translating is a sort of rewriting, whereby students read, analyse, transfer, and restructure the original. In this sense, the final text stands on its own; it has its own identity, but governed by the rules of the TL".

Another Modern Approach I favour in class is the Communicative approach[1] which focuses on the importance of the students' oral skills and where language is presented as a set of forms (grammatical, phonological, lexical) which have to be practised. An often-stated goal of communicative language teaching is to get learners 'beyond translating', which amounts to an acknowledgement that despite the monolingual veneer of classroom instruction, learners are inevitably constantly translating, in the sense of mentally shuttling back and forth between (or among) the language being learned and the one(s) they already know, at least at the beginning stages. In terms of impact on foreign language education, as the rationale for modern language study has changed, the goal now is having a translingual and transcultural competence.

Following on Henry Widdowson [14, p. 235]:

"But what if we were to … take a genuinely learner-centred approach by allowing learners to react to tasks in a natural way without … imposing the constraining conditions of conformity to L2 norms? What if we allowed them to draw pragmatically on all of the linguistic resources at their disposal to achieve their communicative outcomes? What if, in other words, we allowed them to be translators"?

In short, he proposes a radical new vision for language learning: instead of trying to teach a language as a set of distinct encoding rules and usage conventions, one would teach the properties of language in general as a means of conceptualization and communication which are variously realized through different languages. The objective so defined would represent an L2 not as something dissociated from the learners' own linguistic experience, but closely related to it and an additional resource in their linguistic repertoire.

Such a drastic reorientation in language study would of course involve profound changes in pedagogical practice as well as in assessment. For instance, it is essential for teachers to be sensitive to students' needs and for students to be asked to translate "real world texts". This will increase their motivation, expectation, and curiosity.

In response to these realities, Claire Kramsch and Michael Huffmaster [15, p. 118] argue that foreign language study in the 21st century must foster, beyond knowledge of

[1] Communicative approach refers to classroom activities in which students use language as a vehicle of communication, and the main purpose is to complete some kind of task. Students are required to use any and/or all the language that they know to develop their language skills.

the vocabulary and rules of syntax of a standard national language, "an awareness of the semiotic potential of switching linguistic codes, discourse structures and discourse modes, the symbolic potential of various media and modalities, and the way they position speakers and hearers, readers and writers". To that end, and with an expansive notion of translation that encompasses re-keyings, transpositions, and transmodalities, they describe various multilingual practices in foreign language study which they find foster such multiliteracies.

One example is translating translations, in which learners translate translations published in the language they are learning (from any language) into their own language. For this practice, the source text should ideally be a covert translation that domesticizes the original, and the target text an overt one that attempts to restore what the learner-translator sees as important elements of the original source text culture.

Another is transposing across modalities, in which learners produce two different translations of the lyrics to a song with two different musical renditions, aiming to convey verbally the different musical styles. To be effective, such practices must include a self-reflexive element, such as a self-assessment describing specific challenges faced and the rationale behind the chosen resolutions.

3 Conclusions and Future Directions

Translation in English language teaching has been used in many different ways since the days of the grammar-translation method. After its near complete rejection in the name of native-speaker authenticity and priority given to the spoken mode of communication, as well as in response to the globalization phenomena, it has slowly made a return and in the name of increasing multilingualism, multiliteracy and multimodality, the role of translation in language study is being reassessed.

As a classroom activity [16], translation makes students feel they are acting as real cultural mediators, actually bridging the gap between members of different socio-cultural backgrounds that do not share the same knowledge. At the same time, it provides students with ample scope for exploring the similarities and differences between languages and cultures, and sensitises them to the compensatory strategies that need to be activated to achieve pragmatic equivalence and cross-cultural understanding that make up the goal of translation.

In fact, the ultimate goal of Intercultural Language teaching is to help learners transcend their singular world view through the learning of a foreign *linguaculture*, leading them progressively towards intercultural competence.

Within this context, as English language teachers, we have to reassess the role of translation and incorporate it increasingly into instruction as an ideal pedagogical method to promote the new goals of language study in the globalized, multilingual world of the 21st century.

References

1. Holmes, J.S.: Translated! Papers on Literary Translation and Translation Studies. Rodopi, Amsterdam (1988)
2. Baker, M.: In Other Words: A Coursebook on Translation. Routledge, Abingdon (1992)
3. Nida, E.: Toward a Science of Translating. Brill, Leiden (1964)
4. Cook, V.: Multi-competence: black hole or wormhole for second language acquisition research. Underst. Second Lang. Process **25**, 16–26 (2007)
5. Buffagni, C., Garzelli, B. (eds.): Film Translation from East to West. Dubbing, Subtitling and Didactic Practice. Peter Lang, Bern (2012)
6. Gambier, Y., Caimi, A., Mariotti, C. (eds.): Subtitles and Language Learning: Principles, Strategies and Practical Experiences. Peter Lang, Bern (2014)
7. Even-Zohar, I.: The Position of Translated Literature within the Literary Polysystem. Papers in Historical Poetics, pp. 21–27. The Porter Institute for Poetics and Semiotics, Tel Aviv (1978)
8. Cook, G.: Translation in Language Teaching: An Argument for Reassessment. O.U.P, Oxford (2010)
9. Laviosa, S.: Translation and Language Education: Pedagogic Approaches Explored. Routledge, London (2014). Le Page, R. and Tabouret-Keller
10. Richards, J.C., Rodgers, T.S.: Approaches and Methods in Language Teaching, 3rd edn. Cambridge University Press, Cambridge (2014)
11. https://poets.org/lesson-plan/poetry-translation. Accessed 07 May 2020
12. Park, J.: Learning in/through collaborative poetry translation: documenting the impact of poetry inside out with high school-aged English language learners. J. Lang. Lit. Educ. **11**(2), 134–149 (2015)
13. Lefevere, A.: Translation, Rewriting and the Manipulation of Literary Fame. Routledge, Abingdon (1992)
14. Widdowson, H.G.: The role of translation in language learning and teaching. In: House, J. (ed.) Translation: A Multidisciplinary Approach, pp. 222–240. Palgrave Macmillan, Basingstoke and New York (2014)
15. Kramsch, C., Huffmaster, M.: Multilingual practices in foreign language study. In: Cenoz, J., Gorter, D. (eds.) Multilingual Education: Between Language Learning and Translanguaging, pp. 114–136. Cambridge University Press, Cambridge (2015)
16. Popovic, R.: Forging Peace through ELT – Utopia or Reality? In: Beaven, B. (ed.) IATEFL 2008 Exeter Conference Selections, pp. 196–205. IATEFL, Canterbury (2009)

Adopting the Montessori Methodology in Teaching Languages to Adult Students- Transnational Approach

Alina Doroch(✉)

SSW Collegium Balticum, ul Mieszka I 61C, Szczecin, Poland
a.doroch@vp.pl

Abstract. Teaching adults unlike educating children, is a difficult area, owing to the learners' educational and social background, compounded by their expectations and assumed goals. Bearing in mind that learning and teaching strategies may not be equally effective with all learners, it is advisable to propose such a mode of teaching that would accommodate especially those adults who face difficulties with gaining knowledge, due to factors outside the strictly educational sphere. This is where, Maria Montessori's pedagogy steps in with a proposal addressed at those persons who have so far failed to succeed in learning a foreign language in a traditional way. Her pedagogy has inspired teachers and educators all over the world and consequently, had a profound effect on the structure and quality of teaching until today. Although it has mainly dealt with educating children and young people until the age of 18, there have been attempts to apply her methodology to adult teaching. The author of this article will look at the ways her philosophy can be adopted in teaching adults, however to get a better perception of the nature of her ideas, some attention has to be drawn to what drove Maria Montessori to devoting her professional life to education.

Keywords: Montessori pedagogy · Teaching · Principles · Manual

1 Maria Montessori and Her Pedagogical Philosophy

In order to understand the educational philosophy of Maria Montessori, one has to realize that it is not really educational methodology but rather an approach or attitude. It is regarded as a philosophy towards life, characterized by a specific way of considering what life is about and what it means to be human and how life should be led to promote optimal development. Since "this philosophy is not lived as part of mainstream society," adults who wish to introduce this way of life, should find ways to sustain and reinforce their own development. Maria Montessori explained this attitude as the "spiritual preparation of the adult" to make the children develop the aspects and qualities, such as empathy, sympathy, loyalty, fairness, consideration, helpfulness, all of which the teacher, as the guide, should possess [8].

Á. Herrero et al. (Eds.): ICEUTE 2020, AISC 1266, pp. 187–195, 2021.
https://doi.org/10.1007/978-3-030-57799-5_20

1.1 Maria Montessori - The Beginnings

Maria Montessori (1870–1952) was the first woman physician in Italy who set her heart on improving the situation of the poor and subnormal children, disadvantaged both by their physical condition and the social circumstances. Having won the doctor's degree, she started working at the hospital, at the same time engaging in her private practice. As part of her research work at the psychiatric clinic of the University of Rome, in 1897 she joined the staff as a voluntary assistant visiting the Rome asylums for the insane, with the view to selecting the suitable patients for treatment at the psychiatric clinic of the University. It was at the asylum that she met the retarded children, who were locked up and neglected, with no one taking care of their educational and emotional needs. While observing them, she noticed that the children, described as "idiots", responded to the stimulation, which proved her observation that their minds were not useless but just unused [7].

While working with those children she became acquainted with the works of two renowned men, Jean-Marc-Gaspard Itard and his disciple, Edouard Seguin, whose findings profoundly influenced her own research, which led her to say "that mental deficiency presented chiefly a pedagogical, rather than mainly a medical, problem" [14]. Hence those children should be helped with the use of special educational methods at school, rather than kept in hospitals. That finding made her turn her attention to the study of education. In the years 1897–98, she committed herself to studying pedagogy, reading all available works on educational theory of the past two hundred years, which resulted in Maria forming her own theory of education. She was greatly influenced by Friedrich Froebel, the German schoolmaster who came up with the idea of the kindergarten, the philosophy of which was later popularized in other countries. Similarly to Montessori, Froebel had developed the idea that early childhood education should be realized in a specially prepared environment, the kindergarten, or the "child's garden" [5]. The concept of a specially prepared surrounding became the icon of her philosophy in guiding children through the whole period of education.

In her next iconic project, called the Children's House (Casa dei Bambini) Montessori continued this approach perceiving this as the way to educate the children by giving them a lot of freedom in exploring the environment [10]. The whole teaching process was realized in the properly arranged environment and with the use of the carefully selected and organized materials [5].

1.2 Maria Montessori International Approach

Soon Maria Montessori and her method attracted international attention. In America, where in 1913 there were already more than one hundred schools, the proponents of her educational philosophy founded the national organization, the Montessori Educational Association, with the view to promoting the method. Although her method was received enthusiastically by many, Montessori faced a critical attack from a group of some leading progressive educators associated with John Dewey's instrumental philosophy.

In his book, *The Montessori System Examined* (1914), William Kilpatrick dismissed Montessori techniques as outdated, stating that she "still holds to the discarded doctrine of formal or general discipline". Kilpatrick saw the deficiencies in Montessori methodology in the fact that "she does not provide situations for more adequate social cooperation," and as a contrast to such approach he would "put the children into such a socially conditioned environment that they will of themselves spontaneously unite into larger or smaller groups to work out their life impulses as these exist on the childish plane" [9].

That opinion had a significant influence on dissolving the enthusiasm for the Montessori methodology and the author herself, which meant that by 1918 there were only occasional references to her in the journals. Yet, five years after her death (1952) her educational methodology experienced renaissance in America again [7]. Nowadays, it is highly recognized worldwide, resulting in thousands of schools and educational institutions adopting the principles that have become the hallmark of Montessori's educational philosophy.

2 The Transnational Character of Montessori Pedagogy in Adult Education

2.1 The Montessori Principles

Maria Montessori's philosophy was immensely instrumental in the today's conception of pedagogy, not only catering for children but increasingly entering the field of adult education. There are numerous examples of how her educational method can be applied in dealing with senior students or the mentally retarded ones. Much has been said and written how the Montessori pedagogy has influenced young children as well as young people and adults all over the world with regard to their emotional and intellectual development. People who reach out to gain success in business and other disciplines requiring autonomy and independence in making well-thought choices in most cases have come from the Montessori school environment. That sort of experience had prepared them to face the world and its challenges with an open mind, relying on their decisions in all their endeavours. All of this arises from the fact that having been taught with the Montessori methodology behind, they, like all the children, had gone "through the same process of development everywhere, regardless of the place or clime" [5]. Hence, the Montessori method assumes the transnational and transcultural character, as its use can be conditioned by the cultural context, yet it is not dependent on it, nor is it determined by it. Still, different cultural settings may have a slight influence on the way tuition is conducted, which means that the method itself can be applied in any country and culture owing to the universal character of the human and child nature.

This very aspect of the universal character of Maria Montessori methodology means that educators in many countries across the world still consider her pedagogical approach as the one that shaped teaching not only children but their parents as well who appreciate the beneficial aspects of the type of schooling that children receive in the Montessori institutions.

According to Dr. Kloskowski, one of the aspects, instrumental in facilitating children's and adults' education is again the properly prepared environment. This need has been recognized in many communities, where training centres have been founded with the view to providing educational programs for the adult population. Those programs, based on Montessori methodology would be instrumental in making the students employable or getting a possibility of obtaining a better job through the rehabilitation service [6].

Other examples included the studies conducted on a group of patients with dementia, in which the Montessori-based activities were applied. On completion of the course, it was demonstrated that the patients were more actively engaged in individual and small group activities. This has led the trainers to the conclusion that the Montessori techniques have positive psychosocial effects and the patients with dementia can successfully continue learning when they are proposed with stimulating activities, adapted to their cognitive capacity and complemented by external compensatory support, such as task break down [9].

What makes the Montessori approach so successful is the fact that it rests on the premises stated below:

- Self-confidence and self-esteem;
- A sense of achievement and self-worth;
- A sense of responsibility for themselves and their actions;
- Independence and adaptability;
- Cooperation with others and a sense of community;
- Respect for the rights and a sense of community;
- Initiative and self-motivation;
- Concentration and persistence in completing the task.

All of these, though primarily associated with children's education, can be observed in any situation when learning is involved. The Montessori methodology is designed in such a way that it explores the students' desire to learn and their ability to develop their own competence. The curriculum allows them to experience and enjoy the process of learning, and consequently developing the sense of achievement and self-esteem [2].

Maria Montessori said, "[M]y vision of the future is of the individual passing from one stage of independence to a higher, by means of their own activity, through their own effort of will, which constitutes the inner evolution of the individual" [12]. Although this statement was made with reference to children and their developmental path, yet right now its meaning is applied to young adults and adults as well since the meaning is universal and even the latter ones draw on in approaching various tasks in their lives.

Naturally, many of the aspects characteristic of the Montessori approach can be applied to adult education. Although, as it has already been indicated, there exist institutions devoted to teaching senior learners, yet these cases refer to a specific model of a learner, namely the one affected with dementia. However, up to date, no organized education, entirely Montessori-related has been proposed with reference to language learning. Does it mean then that this methodology cannot be successfully implemented into the adult learning curriculum?

The answer is not that simple. Although there is general enthusiasm for the Montessori methodology, it is mainly associated with children's learning. Parents and teachers tend to view this approach through the cognitive and emotional development of their children. What is so characteristic of the Montessori approach is the genuine interest and enthusiasm of the children themselves, while they are learning. However, an adult learner is a much more difficult and demanding recipient of the learning process, which makes it quite challenging for the teacher to propose such a model of teaching that would accommodate the student, leaving him satisfied and content.

The Montessori philosophy is addressed to educate students in the most successful way in order to allow them to live in a fulfilling and satisfying way. This can be achieved by stimulating their development through subtle guidance, which leaves ample space for individual evolvement at one's own time and pace. One of the most fundamental aspects of this philosophy is the teacher's assistance, yet avoiding imposing himself. Allowing the student to do something on his/her own produces an enormous sense of achievement, while the self-correcting materials limit the teacher's interference to a minimum, which enhances the learning process since the material learned in this way will be remembered better [13].

Naturally, all these aspects can be transferred to adult teaching/learning. Even repetition in the form of producing model sentences or short dialogues, when performed in the playful atmosphere can be successfully incorporated into adult education. Moreover, when the element of competition is excluded from the lessons and replaced with cooperation and team work, the effects will be clearly visible, making learning more attractive and effective [2].

3 International Education and the Montessori Approach

3.1 Montessori and the Issue of Global Citizenship

The universal character of Montessori pedagogy means that it can be adapted to various educational spheres, of which teaching foreign languages assumes the prime position. Still, it must be emphasized that Maria Montessori did not design a specific methodology for teaching and learning foreign languages but the approach to educating learners promotes easier acquisition of the material based on its understanding owing to the relaxed and student oriented way of teaching where learners are treated with respect and given a large space to learn at their own pace in the friendly atmosphere and environment promoting and enhancing progress [11].

Since human success cannot be divorced from the knowledge of foreign languages offering the learners access to a better future through a personal and educational development, being able to communicate with people across the world, using the target language as a means of communication can be attributed to Maria Montessori's idea of development.

Learning foreign languages in a global context, making it transnational allows students across the world get to know it better and by getting in touch with other people, owing to the ability to communicate in other languages offers them a chance of enriching their lives, develop a sense of understanding and appreciation of other cultures in the multiethnic, multilingual and multicultural world. In this way the target language becomes the vehicle to explore what the world offers. This attitude seems most relevant nowadays when the European job market is open to all people, considering the fact that it is the knowledge of foreign languages that makes the learners eligible for taking advantage of the professional opportunities.

The Montessori methodology offers a chance to approach education in a more open and relaxed way, thus allowing them to use their intellectual potential as they are not bound by the constraints of the traditional schooling. Unfortunately, in many cases the latter way of approaching learning foreign languages is still employed, which does not guarantee obtaining accuracy and expertise, and consequently a success in the future. Hence the Montessori teaching offers unquestionable benefits as confronted with the traditional education since it takes into consideration the needs of an adult student regardless of the country and culture where the education is realized.

This is when the Montessori approach to education can be compared with the international programmes of which the International Baccalaureate can be most closely associated with promoting values identified with global citizenship with the view to preparing students for new challenges when faced with the increasingly global world. As it has already been mentioned with reference to the qualities represented by the graduates of Montessori schools, these persons are prepared for a number of issues that globalization brings, hence educational programmes worldwide should focus on promoting attitudes that reflect tolerance, respect, knowledge and understanding of other nations as well as the ability to work with the people coming from various, most diverse cultures and communicate with them successfully. The idea of international education is concerned with educating the global citizens of tomorrow [1] and although there is no international curriculum specifically designed in the course of the developing methodology based on Maria Montessori philosophy, there is no doubt that it is this philosophy allowing humans to develop without the constraints of the traditional education by offering the learner a lot of autonomy in learning and the choice of ways to do it [2].

The already mentioned IBO (International Baccalaureate Organization) was founded in Geneva, Switzerland, with an initial mission to prepare the internationally mobile student for university education who as a knowledgeable and inquiring citizen would strive to create a better and more peaceful world through intercultural understanding and respect. This is why the international education writes very well into the idea and philosophy of Maria Montessori who already had talked about education for peace when in 1932 she made a powerful speech at the Geneva Convention [1] clearly seeing the necessity to prepare young people for the future education and working towards developing cooperation on an international level, in this way responding to the need of global citizenship, yet to come.

3.2 Adult Learners and the Montessori Approach

The issue of the specially prepared environment in the Montessori pedagogy has often been pointed out to as the direct reason behind the success that the people taught with this methodology have enjoyed. Indeed, without going into a detailed list of the differences between the Montessori approach and the sometimes, less effective, traditional one, we learn that when students are given more freedom and initiative in learning, the results are much more promising.

This is where the Montessori methodology comes forward. What is interesting is the fact that a teacher does not have to be deemed as the Montessori educator to introduce some principles and methods pertaining to this methodology. What is more, many teachers in a number of countries engaged in traditional forms of education have already been adopting some of the principles, without even realizing that. Still, certain elements of this philosophy, owing to their specific nature, cannot be adopted in organized classes, such as the school or university, controlled and bound by the constraints and standards imposed by the educational authorities.

Consequently, the element of individual planning of the classes and organizing learning in the way most suitable for the students can be experienced only in small groups outside the teaching institutions or in one-to-one classes [2]. This type of organized education will probably be favoured by adult learners who, as opposed to young students, will be hampered by many aspects, often resulting from their age or other conditions, adversely affecting their success in learning languages. Some of these problems may be directly connected with all sorts of fears that adult students may experience. Since adult learners are very often extremely sensitive, regarding their performance within the group, those fears may severely impede their learning.

These particular anxieties seen as a universal and transnational issue hindering students' progress were addressed in a manual which was created within the framework of the international project, the MMLT - Montessori Methodology in Language Training project, funded by the Lifelong Learning Programme of the European Commission, 530963-LLP-1-2012-1-GR-KA2-KA2MP - KA2 Multilateral Projects. It offers the teachers teaching languages to adults some hints how to do it in a more effective way and thus help adult learners overcome these obstacles [3, 4].

Following surveys and discussions with both the teachers and adult students 10 main causes of stress sustained by these learners were identified. These included:

- Fear of teacher's expectations;
- Need of immediate results;
- Difficulties in memorization (caused by personal and/or psychological deficiency);
- Need of knowledge of grammar structure before starting speaking a foreign language;
- Fear of making mistakes;
- Peer pressure (from class mates);
- Time pressure due to scheduled programs;
- Sense of loyalty towards own native language and culture;
- Difficulty in using the target language avoiding literal translation;
- Lack of self-confidence.

Recognition of the above problems, both by the teachers and the students themselves across the world is part of the success since the appropriate attitude of both parties will result in better constructed and conducted classes, where students' expectations and capacities are better acknowledged and satisfied.

Low foreign language command or its lack is nowadays the factor which in many situations can exclude adult learners from the mainstream community, not allowing them to fully use the work opportunities. Hence, considering all the unquestionable merits of the Montessori methodology, this philosophy may be considered as the best way to accommodate many problems that adult learners experience.

Since the Montessori principles are known for their universality adults from various countries and walks of life can benefit from them and many aspects which made this approach well-known worldwide can be easily adapted to the adult way of setting about learning and what is more, the adults' vision of what they expect of linguistic education.

4 Conclusion

For a number of years, the Montessori philosophy has been successfully used in teaching children and adolescents. However, high school and academic teachers, who use some elements of this method unwittingly, demonstrate a rather limited knowledge regarding this approach.

Being appreciated and used in many countries and yet, at the same time demonstrating a specific character, the Montessori methodology cannot be introduced into all types of schools for adult learners, but it can be realized in many different institutions where teaching would not be subject to external rules and standardization. What Maria Montessori insisted on in her philosophy, was the element of individualization and independent learning, which permeates the whole learning process. Hence, setting oneself a goal and consequently working towards accomplishing it is a key to a success.

Many graduates from the Montessori schools where teaching languages was conducted in line with the principles promoting and allowing a wider and more comprehensive approach to human and intellectual development, emphasize that the Montessori environment and the atmosphere of tuition has had a crucial impact on their life and career from the transnational perspective and in the international context.

References

1. Brunold-Conesa, C.: International education: the international Baccalaureate, Montessori and global citizenship. J. Res. Int. Educ. 9(3), 259–272 (2010)
2. Doroch, A. Research on Montessori educational and learning methods for adults in Poland. The Polish National Report, p. 3 Poland (2013)
3. Gastaldi, L., Marsili, R.: Montessori Methodology in Language Training for adults. Lingua Più Associazione, p. 3, Cità di Castello, Italy (2013)
4. Gastaldi, L., Marsili, R.: Montessori principles for adult education and training: Is the Montessori methodology suitable for adults? The Italian experience at Lingua Piu Citta di Castello. In: Conference Proceedings. The Future of Education., 4th Edition, Florence, Italy 12–13 June 2014, p. 436, Pixel, Padova (2014)
5. Gutek, G.L. (ed.) The Montessori Method. The origins of an educational innovation: Including an abridged and annotated edition of Maria Montessori's The Montessori Method. Rowman & Littlefield Publishers, Inc., Lantham MD (2004)
6. Kloskowsky, V.J.: Didacticism: Montessori and the Disadvantaged Student. Author House, Bloomington (2008)

7. Kramer, R.: Maria Montessori: A biography. Perseus, Reading (1988)
8. Lloyd, K.M.: An analysis of Maria Montessori's theory of normalization in light of emerging research in self-regulation. ProQuest, LLC, AnnArbor (2008)
9. Polk Lillard, P.: Montessori, A Modern Approach. Schocken Books Inc., New York (1972)
10. Roland, A. Lubienski Wentworth. Montessori for the New Millenium: Practical Guidance on the Teaching and Education of All Ages, Based on the Rediscovery of the True Principles and Vision of Maria Montessori. Lawrence Erlbaum Associates, Mahwah (1999)
11. Samsonova, O, Hormazabal, H.: Montessori Approach as a Way to Overcome ESL Students' Educational Drawbacks. Conference Paper, June 2019. https://www.researchgate.net/publication/333601848. Accessed 19 Apr 2020
12. Smith, M.C., DeFrates-Densch, N.: Handbook of Research on Adult Learning and Development. Routledge, New York (2009)
13. Stark, M., Schmidt, D.: Understanding Montessori: A Guide for Parents. Dog Ear Publishing, USA (2009)
14. Taylor, G.B.: Italy, and the Italians. American Baptist Publication Society, Philadelphia (1898)

Cross-Cultural Experiences in Transnational Education: Preservice Teachers and Global Perspectives

Concetta Maria Sigona(✉) and Alba Fernández-Alonso

Facultad de Humanidades y Comunicación, Universidad de Burgos, Paseo de los Comendadores, s/n, 09001 Burgos, Spain
{cmsigona, albafa}@ubu.es

Abstract. In the European Higher Education Area, future foreign language teachers should receive training under a global perspective that integrates a global education and citizenship within a specific framework, to internationalize teacher education programs. In this sense, the development of academic coursework regulated by a policy of good practices is important for supporting global curriculum content and the inclusion of co-curricular and cross-cultural experiences. In-depth analysis at a national level demonstrates the lack of global design for the implementation of global and cross-cultural competences. Thus, the European project VIR_TEACH suggests the design of a White Paper for a comprehensive Language Teachers Training at a European scale where coordinated policy actions and a strategy for engaging policymakers and public administrators are encouraged.

Keywords: Foreign language teacher · Internationalization · Globalization

1 Introduction

For the three last decades, the mobility of future foreign language teachers, has increased thanks to a constant growth in offers from universities that provides the most varied programs not only for studying abroad but also for preservice teacher training. All this sparks a need internationalise teacher training. Furthermore, it has been shown that mobility in education is constantly increasing, which justifies the need for awareness of the internationalization and globalization of future foreign language teachers. [1]. Furthermore, as Phelps [2] states when referring to international doctoral students, as "cosmopolitan sojourners", teachers are part of a transnational context, since their work takes place in a "space where their field of activity consists of simultaneous locations—at minimum their home country and the country in which they are studying, and possibly other locales of research, activism and professional work as well" [3].

First of all, it is necessary to recognize the importance of education in understanding and solving social, political, cultural and economic issues at a global scale. From this perspective, the role of an education aimed at building values and attitudes facilitating international cooperation and promoting social transformation, becomes essential. Today's education counts with an omnipresent internationalization since it has become the final objective of the institutions dedicated to education.

Á. Herrero et al. (Eds.): ICEUTE 2020, AISC 1266, pp. 196–201, 2021.
https://doi.org/10.1007/978-3-030-57799-5_21

Prospective teachers are faced with a training process which implicitly pushes them towards an increasingly imperative internationalization. As Leutwyler, B., Popov, N., and Wolhuter, C. [4] assert, the request of higher education institutions presents programs that clash with the training of future teachers since it is aimed at national and in some cases local training.

This dichotomy represents the starting point of this study. Based on what has been already said in the European Higher Education Area, future foreign language teachers should receive training under an international perspective that integrates global education and citizenship within a specific framework that pursue the internationalization of teacher education programs.

2 Training Under Global Perspectives

The continuous internationalization that has taken place over these last decades represents the answer to the growing increase of globalization [5]. International teacher training presumes languages, cultures and societies exchange different from those of origin and becomes, in this way, a new model for a personal growth and not only since, the future teacher compares different methodologies and educational skills to compare.

It is a matter of stimulating awareness of the global and interpersonal growth in future teachers as regards the development of interpersonal skills from both a communicative and a pedagogical point of view [6]. In a study under the Erasmus+ VIRTEACH project, A VIRtual Solution for a comprehensive and coordinated training for foreign language TEACHers in Europe[1], a questionnaire "to detect current weak aspects regarding language teacher training, and to single out positive measures for the advancement of foreign language teacher education in Europe has been filled by teacher students and graduated as a teachers of foreign languages, foreign language teachers, researchers at secondary or high schools, language schools and universities, and education policy makers and public administrators" [7]. The results of this questionnaire show that, both future foreign language teachers and experienced teachers–as well as policy makers and public administrators agree with a need for an internationalization of training programs.

In 2010 Quezada, R.L. [8] supported the internationalization of training programs for future teachers as well as for counselor and administrator candidates. In this same vein, Phelps [2] reminds us that there are different theories on globalization both from a social and educational point of view.

In our current society, where intercultural and international contacts are commonplace, students must be aware of two fundamental issues. In our societies where intercultural and international contacts are commonplace, students must be aware, on the one hand, of what a misunderstanding can lead to and on the other, the implicit enrichment resulting from intercultural exchanges. The impact of international

[1] For more information on the VIRTEACH project, please visit www.virteachproject.eu. The European Commission's support for the production of this publication does not constitute an endorsement of the contents, which reflect the views only of the authors, and the Commission cannot be held responsible for any use which may be made of the information contained therein.

experience can be categorized in a global perspective and perceptual understanding [9] and international experiences for personal growth and new interpersonal relationships. Intercultural communication is deemed essential to increase their intercultural values because of the international relations and educational experience.

3 Academic Coursework Regulated by a Policy of Good Practices: The Global Curriculum Content, the Co-curricular Experiences and the Cross-Cultural Experiences

In the context of internationalization, it is of paramount importance to regulate strategies and good practices with the aim of developing a model for a "comprehensive Internationalization" as defined by the Center for Internationalization and Global Engagement (CIGE) in American Council on Education (ACE). It is "a strategic, coordinated process that seeks to align and integrate international policies, programs, and initiatives, and positions colleges and universities as more globally oriented and internationally connected" [10]. The ACE defines it as "the obligation of colleges and universities to prepare people for a globalized world, including developing the ability to compete economically, to operate effectively in other cultures and settings, to use knowledge to improve their own lives and their communities, and to better comprehend the realities of the contemporary world so that they can better meet their responsibilities as citizens." (2011) [11]. From this perspective, curricular learning experiences become essential for the development of a global awareness. Even more, "An internationalized curriculum and co-curriculum ensure that all students, including those who do not have the opportunity to study abroad, are exposed to international perspectives and can build global competence." [11].

3.1 The Global Curriculum Content and its Internationalization

How has the perception of students, future teachers and experienced teachers changed for thirty years now towards the curriculum and its orientation? Just as we talk about internationalization of students and future teachers, it is necessary to introduce the concept of internationalization of the curriculum. And, when we talk about curriculum internationalization we refer to the involvement of students, teachers, public institutions and programs centered on the introduction of global perspectives within programs aimed at both international and national audiences. Knight (2004) defines this concept as something that is interpreted in different and sometimes confusing ways since it often turns out to be ambiguous in terms of its definition and implementation.

As for the types of global curriculum and the various and diversified processes to which it has been subjected, it is also important to mention the study of Crosling, G., Edwards, R., and Schroder, B. [12], an example of implementation of internationalizing the curriculum. The internationalization of the curriculum is not a new concept, it is part of a broader process mentioned in the previous pages, that of global citizen education. Another example of internationalization of the curriculum is offered to us by Renfors [13]. In this respect, it states that this is a somewhat difficult term to conceptualize. It does not only involve students, but also–and most importantly–teachers aiming at "defining its formal aspects' [13]. Renfors also adds that "the lecturers may not always understand the meaning of internationalization of the curriculum." [13].

3.2 Co-curricular Experiences

The co-curricular activities, unlike the extra-curricular that form us on different fields than the central interest of our training, are those that complement our main activities and that provide us with specific but complete training in the specific training sector. The co-curricular experiences do nothing but motivate the students and help them in their training path aimed not only at a diversified specialization that can further support them in the search for a well-defined profession [14]. In the case of future teachers, co-curricular training is essential since it provides integrated experiences both from a practical and theoretical point of view.

3.3 Cross-Cultural Experiences

Cross-cultural experiences play the same important function as co-curricular experiences since they contribute to the development of global skills in preservice teachers. Poole C.M. and Russell III W.B. [15] in their article conclude "Since the global content courses and co-curricular cross-cultural experiences investigated in this study were found to have a significant positive relationship with preservice teacher global perspectives, colleges of education should work toward increasing preservice teachers' frequency of participation in these experiences." (p. 50). Cross-cultural experiences enrich and benefit Institutions and communities [16].

4 VirTEACH European Project and the White Paper

One of the aims of Virteach a consists of the design of a White Paper for a comprehensive Language Teachers Training at a European scale where coordinated policy actions and a strategy for engaging policy makers and public administrators are suggested. The White Paper for the coordination of policy actions, aimed at decision makers and teacher trainers to promote the internationalisation and convergence of national formats, to enhance a greater coherence between the training and practice phases and to facilitate dialogue and transparency among the actors involved. The research carried out by partners into the current literature in the fields of Language Teaching and Teachers training programmes at European and International level together with the knowledge acquired in the project implementation will lead to a White Paper (in the format of an e-book) including policy recommendations and general guidelines to public administrators, offering suggestions for improving the current educational approaches taken along the legislations and normative in the EU Member States. These contributions will also be useful for similar projects in the field of teachers training as well as long-life education and capacity building for teachers regardless the specialisation of the target group.

The tentative content of this e-book is:

- Best Practices summaries
- Fact-Sheets with Case Study
- Relevant Impact Assessment Conclusions
- Guidelines for stakeholder identifications and key messages for engagement
- Policy Recommendations.

It will be available for all society and specially addressed to policy makers to ensure the adequate and desired reception, adoption and widespread of the contents of the project, facilitating the early adoption of innovative methodologies. The White Paper will be professionally designed and edited using Open Source software (Scribus or any other alternative). It will be translated by the partners to their national language, then shared with representatives from different levels of the education systems from the countries participating in the consortium, and in participative workshops at a national level). Finally, project partners will analyze all their feedback and, if accepted, incorporate ir into the final version of the Online White Book.

5 Conclusions

In conclusion, fostering global awareness and international collaboration is highly beneficial to future language educators. Global awareness and international collaboration during training courses encourage students to see things from different perspectives and helps them to make informed decisions and acquire transferable skills that will be useful to them and will remain with them for life. Apart from communicating effectively in a foreign language, building the identity of a global citizen is of fundamental importance. In fact, from what has been said, the new form of learning would derive not only lead to the development of a critical mind, but also–and above all–to act and participate in a world that is already global. Thus helping future teachers in the development of the new skills required, both technology and criticism and communication [17].

References

1. Barros del Río, M.A., Sigona, C.M., Simarro Vázquez, M.: The international and intercultural perspective in future teachers of english (efl): exchanging experiences. Forthcoming
2. Phelps, J.M.: International doctoral students' navigations of identity and belonging in a globalizing university. Int. J. Dr. Stud. **11**(1), 1–14 (2016)
3. Gargano, T.: (Re) conceptualizing international student mobility: the potential of transnational social fields. J. Stud. Int. Educ. **13**(3), 331–346 (2009). Cited in Phelps 2016, (p.2)
4. Leutwyler, B., Popov, N., Wolhuter, C.: The Internationalization of Teacher Education: Different Contexts, Similar Challenges. Bulgarian Comparative Education Society, Sofia (2017)
5. Pais, A., Costa, M.: An ideology critique of global citizenship education. Crit. Stud. Educ. **61**(1), 1–16 (2020)

6. Unesco: Global citizenship education: preparing learners for the challenge of the 21st century. United Nations Educational, Scientific and Cultural Organization (UNESCO) (2013)
7. Barros-del Río, M.A., Mediavilla-Martínez, B.: The European Perception of Foreign Language Teacher Training. Findings on a Survey in Countries, pp. 68–69. "Dimitrie Cantemir" Christian University, Bucharest (2019)
8. Quezada, R.L.: Internationalization of teacher education: creating global competent teachers and teacher educators for the twenty-first century (2010)
9. Short, K.G.: Critically reading the word and the world: building intercultural understanding through literature. Bookbird: J. Int. Child. Lit. **47**(2), 1–10 (2009)
10. https://www.acenet.edu/Research-Insights/Pages/Internationalization/Comprehensive-Internationalization.aspx
11. https://www.acenet.edu/Research-Insights/Pages/Internationalization/Intlz-in-Action-2015-October.aspx
12. Crosling, G., Edwards, R., Schroder, B.: Internationalizing the curriculum: the implementation experience in a faculty of business and economics. J. High. Educ. Policy Manag. **30** (2), 107–121 (2008)
13. Renfors, S.M.: Internationalization of the Curriculum in finnish higher education: understanding lecturers' experiences. J. Stud. Int. Educ. **1** (2019)
14. Freeman, K., Ellis, M., Parham-Mocello, J., Walker, H.M.: Panel: supporting student co-curricular experiences. In: Proceedings of the 51st ACM Technical Symposium on Computer Science Education, pp. 641–642 (2020)
15. Poole, C.M., Russell III, W.B.: Educating for global perspectives: a study of teacher preparation programs. J. Educ. **195**(3), 41–52 (2015)
16. McKeown, S., Vedan, A., Traplin, J., Sanford, L., Bourne, C.: Cross-cultural and global competencies and their role in admissions policies and practices. Higher Education Admission Practices: An International Perspective, vol. 96 (2020)
17. Fromm, N., Raev, A.: The emergence of transnationalisation of higher education of German universities. In: Nickl, B., Popenici, S., Blackler, D. (eds.) Transnational German Education and Comparative Education Systems. Global Germany in Transnational Dialogues. Springer, Cham (2020)

Special Session: Personalization and ICT: a Path to Educational Inclusion

NetExtractor. A Semi-automatic Educational Tool for Network Extraction Conceived to Differentiate by Student Interest

Luis Miguel Cabrejas-Arce, Jorge Navarro, Virginia Ahedo(✉), and José Manuel Galán

Universidad de Burgos, Avenida Cantabria s/n, 09006 Burgos, Spain
lca0037@alu.ubu.es, vahedo@ubu.es

Abstract. Network science is an interdisciplinary field that provides a wide range of analytical and computational tools to conceptualize, develop, analyze, and understand interconnected systems. The recent technological and social media developments based on this field explain the increasing interest to include network-based concepts across all educational stages. In this work, we present a web-based application to obtain networks from novels and/or movie scripts semi-automatically. These graphs can be used as teaching examples and in assignments, thus implementing differentiation by interest instruction and facilitating the adaptation of the contents to multicultural classrooms.

Keywords: Network science · Transnational education · Differentiation by interest · Complex networks

1 Introduction

We live in an increasingly interconnected world where networks are ubiquitous. Examples include the "network of networks", i.e. the Internet, social networks such as Facebook, Instagram and Twitter, the electric grid, the cell phone network, the airline networks, etc. [1].

Remarkably, in spite of their pervasiveness, most ordinary people ignore the mechanics and formal knowledge underlying such networks. The above-mentioned gap of knowledge may stem from the lack of effective network science outreach programs and from educational systems worldwide that still need to be adapted to provide students with the necessary skills to meet the challenges of the 21st century: the Big Data revolution and the increasing complexity of problems that science has to deal with in all fields [2].

In his 'millennium' interview on January 23, 2000 (San José Mercury News), Stephen Hawking stated in reference to the 21st century: "I think the next century will be the century of complexity". We are surrounded by complex systems composed by many interconnected components whose collective behavior cannot be inferred from the individual knowledge of their parts. Many of those systems have existed since the dawn of time and many others are being created and broadened by us. The differential aspect is that at present, we are developing the tools to study and disentangle their intricacies: network science and its analytical methods.

Á. Herrero et al. (Eds.): ICEUTE 2020, AISC 1266, pp. 205–214, 2021.
https://doi.org/10.1007/978-3-030-57799-5_22

Network science—the science of connectivity—is an interdisciplinary field of research on connected systems that offers a powerful approach for conceptualizing, developing, analyzing and understanding solutions to complex social, health, technological and environmental problems [3]. In fact, behind each complex system, one can identify an intricate network that encodes the interactions between the system's components [4]. Outstandingly, network science is also characterized by its empirical, data-driven approach, which relies on a highly sophisticated quantitative (mathematical/computational) core.

Besides the interest to make network-science knowledge generally accessible, so that everybody can have an intuition of what is happening behind the scenes in the networks they are part of, we would also like to elaborate on its pedagogical potential. Network science offers an exciting supplement to standard mathematics curricula in schools. Examining problems from the network perspective is helpful to show that mathematics is about abstract thinking and formalization rather than just calculations, and it allows to do so through everyday examples closely connected to students' daily experiences [5]. The inclusion of the network paradigm in the curricula of the different educational stages may help students to approach problems differently and acquire an analytical mindset.

Lastly, we would like to give some thought to the present and future multicultural societies, and to how network science could act as an enabling tool for transcultural education. In particular, we consider that the application of network analysis tools to literature pieces and/or to screenplays impregnated with the ideology, values and main cultural traits of a given region/country, could help students with different backgrounds better grasp the concepts and applications of the various network science curricula.

The extraction of social networks from literary works has gained popularity in recent years [6]. Nonetheless, thus far, we have not found any studies that explore the connection between culturally aware AI systems for literary network extraction and pedagogical approaches. Hence, we decided to develop an application that brings together both dimensions and that may also facilitate cultural integration.

The main objective of our application (NetExtractor) is the obtention of the social network that codifies the interactions between all characters in a literary work and/or screenplay. The additional functionalities implemented include visualization tools and personalized topological analyses. For a more detailed explanation, please refer to Sect. 4: NetExtractor application and its features.

The remainder of this paper is organized as follows: Sect. 2 succinctly presents the basics of network science and covers in a little more detail their pedagogical dimension. Section 3 reviews some previous works on networks and education. Section 4 presents the NetExtractor application and its functionalities, and, finally, Sect. 5 presents the concluding remarks and prospective educational uses of NetExtractor.

2 Theoretical Background

2.1 Network Science

A network is a catalog of a system's components, often called *nodes* or *vertices* and the interactions between them, called *links* or *edges*. This network representation provides a common language to study systems that may differ significantly in nature, appearance, or scope [4].

The scientific study of networks is an interdisciplinary field that combines knowledge from mathematics, physics, biology, computer science, the social sciences, and many other areas [1]. It is precisely because of its interdisciplinary nature that knowledge about networks is somewhat dispersed, existing no consensual curricula for network studies: different instructors have adopted different curricular designs focusing on different areas depending on their background, knowledge and objectives. According to the systematic analyses of network science course contents from Sayama [7] and Gera [8], the key aspects covered by most networks courses are:

1. Network basic definitions and properties: degree, clustering coefficient, diameter, density, shortest paths, node similarity, and homophily.
2. Centralities: degree, closeness, betweenness, eigenvector, Katz, PageRank, hubs and authorities.
3. Network types.
4. Synthetic network models: Erdős–Rényi random networks, Watts-Strogatz small-world networks, the Barabási–Albert preferential attachment growing model and its variants, the Malloy-Reed configuration model, and the random geometric model.
5. Community structure.

Notwithstanding, in overall terms, the desirable knowledge for a successful user-level management and understanding of networks would cover the concepts and metrics necessary for network characterization at different levels: individual (node) level, intermediate (mesoscale) level and whole-network level.

2.2 Educational and Pedagogical Dimension of Networks

Nowadays, children have regular contact with networks such as Facebook, Instagram, Google, YouTube, etc. from a very early age, thus being familiar with their intuition and dynamics. As a consequence, they are able to grasp the concepts from network science with minimal effort, which renders the impartation at schools of short network outreach sessions meaningful. In addition, the pervasiveness of networks offers an ideal means to motivate students to pursue mathematics and science in greater depth [5].

From all the above, there is growing interest in the inclusion of network-related curricular content throughout the different stages of the educational process. Actually, various pilot programs have already succeeded in this endeavor, see for instance the examples of (i) NetSci High, a NSF-funded educational outreach program that has been bringing network science research to regional US high schools since 2010 [2], the outreach activities conducted both at local schools and at Somerville College in Oxford for students aged 13–16 [5], and the replication of the Oxford experience in Madrid - Spain- [9].

Network science teaching, just like all teaching-learning processes, can make use of a wide range of channels and methodological approaches such as direct instruction, concept and inquiry-based teaching, case-based teaching, cooperative learning, problem-based learning, and many others. In this regard, it is important to note that new teaching strategies keep emerging every day to meet the challenges of twenty-first-century education: growing societal press for standards and accountability, increased student diversity and fundamental changes in technology and globalization [10]. From that pool of new pedagogical strategies, worthy of special mention are (i) curriculum design for student learning and (ii) instructional differentiation, both very much interlinked.

With respect to curriculum design, it suggests adopting the learner's point of view to accomplish three main objectives: (1) tailoring the curriculum to the particular abilities, readiness and prior knowledge of students, (2) connecting specific contents and learning activities to students' interests and (3) providing them with appropriate challenges suited to their age and socio-economic environment [10].

As for differentiation, it consists of adjusting the curriculum, teaching strategies, assessment strategies and the classroom environment to meet the needs of all students [11]. More specifically, differentiation implies a paradigm change; it requires to move from the traditional teacher-centered standardized approach to student-centered whole class teaching [10]. Given that our 21st-century classrooms are full of differences: in gender, in culture, in cognitive levels, in abilities, in intelligence, in learning styles, in languages and in interests, differentiation is needed more than ever to accommodate such range of diversity by providing personalized opportunities to learn [10].

Because of the above, a variety of frameworks have emerged to analyze and manage the differentiated classroom. One of the most famous among them may be the one provided by Carol Ann Tomlinson [11]. Tomlinson's is a six-part framework covering students' academic readiness, their interests, and their learner profiles as well as how to adapt content, processes and products to meet different student needs. Particularly worthy of further attention is the notion of *differentiation by interest*, which according to Tomlinson means connecting school tasks to students' interests in order to facilitate the learning process. For such an approach to be successful, it is necessary to identify the initial interests of the students and to help them create new ones by presenting them with different ideas, topics and out-of-the-box thinking [10].

Coming back to network science teaching, we find that its contents are very well suited for the implementation of *differentiation by interest*, as thanks to networks' ubiquity, it is relatively easy to find networks aligned with the personal interests of each and every one of the students. Furthermore, allowing students to work on a network of their choice may be extremely useful to reach across cultural differences, since putting into context the concepts studied in class into a network that is familiar to you or that is permeated by your native culture, facilitates understanding and boosts social integration. In this respect, it is interesting to note that even though differentiation by interest may be extremely relevant in multicultural contexts, in general, it is not a frequent differentiation technique [12].

3 Previous Research and Application Examples

After reviewing some relevant scientific advances dealing with networks and education, we present here a brief summary of several of the most inspiring application examples as well as the lessons learned.

In [8], Gera presents his approach for bringing together active learning and network science in a way that engages students and propels their learning: allowing the students to pick individual networks of their interest that they will analyze throughout the course. Eventually, the variety of networks analyzed by all the students is discussed jointly. Concerning analysis tools, Gera suggests using the three resources most commonly used in network science [13–15]: Gephi [16] for visualization purposes, and either Python [17] or R [18] for detailed network analysis.

In [19], the authors review the most relevant initiatives over the last 25 years to use network science as a tool for instruction. Especially worthy of mention is the booklet entitled "Network Literacy: Essential Concepts and Core Ideas", the result of a year-long process which involved several brainstorming sessions on the question: "What should every person living in the twenty-first century know about networks by the time he/she finishes secondary education?" reaching professional network science researchers, educators, and high school students. Such a booklet has already been translated into 20 languages [20].

In [21], Tanizawa elaborates on the need for browser-oriented network teaching enabling tools, since in many of the stages of the educational process where networks can be taught, teachers may not be programming experts. Only in undergraduate courses requiring high programming skills, the teaching style should be command-line oriented.

Further examples of network outreach initiatives are the *Network of Thrones* [22] and *Who Is the Most Important Character in Frozen? What Networks Can Tell Us About the World* [23], the latter aimed at young minds. In the *Network of Thrones*, the authors parsed the third book in the series: *A Storm of Swords,* to turn the *Game of Thrones* world into a social network. The result was a network of 107 vertices that represent the characters joined by 353 integer-weighted edges, in which higher weights correspond to stronger relationships between those characters. The weight of the edge between two characters was established by increasing its value when their names (or nicknames) appeared within 15 words of one another. Notably, some manual validation and cleaning were necessary for the process of building the network. Once obtained, the network was analyzed to determine the importance of the main characters in accordance with the most common centrality measures, and to identify the different communities in the network by means of Louvain algorithm [24].

For its part, in the work entitled *Who Is the Most Important Character in Frozen? What Networks Can Tell Us About the World*, the authors illustrate through the story of Frozen the most renowned centrality measures (degree centrality and eigenvector centrality) and elaborate on two critical notions in network science: (1) the versatility of networks, i.e., that in spite that they can be used to model varied phenomena in which the real components of the network can be very different, all these networks can be analyzed using the same mathematical tools; and (2) the importance of the question you want to answer to, as depending on the question, the same system/story/phenomena can be modeled differently.

In Fig. 1 we present a self-sourced example of the network of interactions in the film "Joker", to visually illustrate what we mean by a network of interactions and intending to facilitate the understanding of the above-mentioned examples.

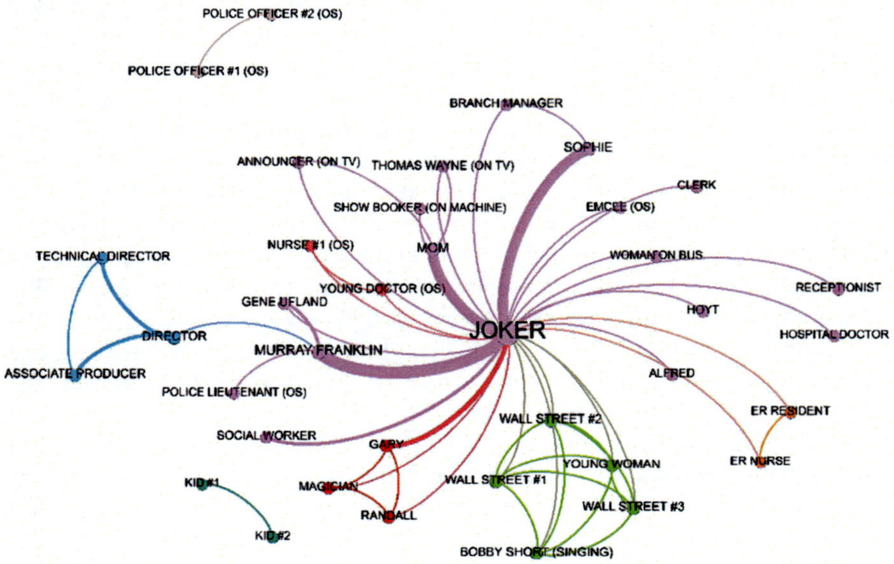

Fig. 1. An illustrative example of the network of interactions in the film "Joker" as retrieved from the script available at https://www.imsdb.com/scripts/Joker.html. The network has been obtained with NetExtractor and the visualization has been made with Gephi [16].

4 NetExtractor Application and Its Functionalities

In view of all the above, we concluded that there is an urgent need for the democratization of network knowledge among all members of society. To achieve such a goal, along with outreach programs targeting the different stages of the educational process, it would be necessary to develop browser-oriented enabling tools, since primary and secondary school teachers do not need to be programming experts.

It was precisely for that reason that we decided to create NetExtractor, a web application for the general public, accessible at http://netextractor.herokuapp.com/, and conceived to extract the network of interactions between characters from literary pieces (in ePub format) and/or from film scripts available at https://www.imsdb.com/. NetExtractor has been developed in Python and the source code is fully available at https://github.com/jng0020/NetExtractor. Our application meets several objectives: (i) it fills the existing gap regarding network science enabling tools, providing teachers in particular and society in general, with a simple semi-automatic tool to understand the basics of network science; (ii) it supports the differentiation by interest pedagogical approach, as it allows the student to choose the book or the film of her interest from which to extract the network of interactions; (iii) it is also intended to facilitate teaching and integration in multicultural environments, as the possibility to select the book/film from which to extract the network, is also a way to adapt the contents to the student's background.

NetExtractor is a semi-automatic tool that incorporates multiple functionalities ranging from network extraction and visualization to network analysis, community analysis, role detection and prediction of node gender and ethnicity. The rationale behind making it semi-automatic is to be found in the difficulties that arise during the identification of the different characters in novels/scripts. In Dekker et al. [6], the authors evaluate several natural language processing tools for the automatic extraction of social networks from novels (Named Entity Recognition tools -NER-) and conclude that among the issues that complicate NER in novels, the greatest challenge is coreference resolution, for which the best solution thus far consists in manually compiling a list of aliases for each character with the aid of external resources or annotators. In light of the foregoing, NetExtractor was designed to solve ambiguities and coreferences manually.

NetExtractor is structured around different windows through which we can explore the multiple functionalities that it incorporates. The flowchart of the application can be seen in Fig. 2.

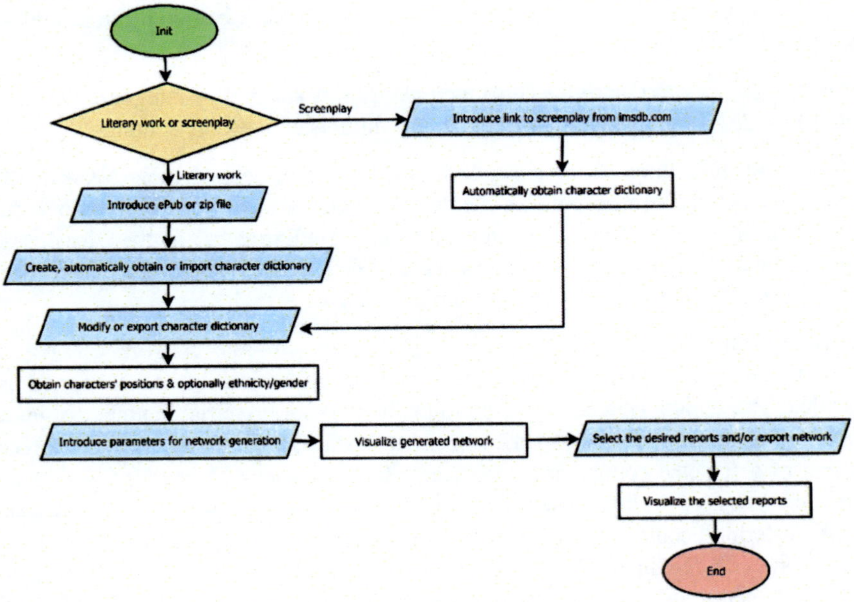

Fig. 2. Flowchart of the NetExtractor application. Actions requiring a decision from the user appear in blue and those done automatically appear in white.

A general overview of the main functionalities that NetExtractor incorporates is provided below:

Network Creation. The user can choose between introducing an ePub or the URL of the script (note that only scripts available at https://www.imsdb.com/ are supported). Depending on the input, the tool creates undirected weighted networks following different approaches. In the case of novels, two characters are considered to have an interaction if their names are mentioned within a given number of words (radius) of one another. The weight of the link is the total sum of interactions between the two characters under consideration. As for screenplays, two characters interact if they appear in a scene together, being the weight of the link again the sum of interactions.

Regarding the obtention of the dictionary of characters, for scripts it is done automatically. However, for ePubs the user is given different possibilities: (i) manually creating a dictionary of characters, (ii) uploading a pre-existing dictionary (from Wikipedia, certain websites, etc.) and/or (iii) automatic creation of the dictionary.

In a further step, the user is asked/allowed to solve coreferences and/or possible ambiguities: the same character having different names/nicknames, different characters with the same name, etc., which are fixed manually by selecting the names to join from a checklist. Characters can also be deleted.

Once the dictionary of characters has been created and cleaned up, the appearances (positions) of the different characters are obtained. In addition, prior to the obtention of the network, the user is asked to introduce a threshold: the minimum number of appearances; those characters with a number of appearances below the threshold will not be included in the network.

Ethnicity and Sex Prediction. This functionality was attained by integrating Ethnea + Genni [25, 26], both available at http://abel.lis.illinois.edu/resources.html.

Network Visualization. Once created, the network is visualized using Netwulf [27], one of the most performant tools for interactive network visualization in Python to date, and which has successfully been integrated into NetExtractor. In the visualization window the user can also export the network as .GML, .GEXF and/or .NET, in order to analyze/visualize it externally (with Gephi, R, etc.).

Network Analysis and Role Detection. The analysis of the network can be customized in the report generation window, which has intentionally been designed as a checklist. The options available are the following: (i) basic metrics (number of nodes, number of edges); (ii) distance metrics (diameter, radius, eccentricity, average distance, number of connected components); (iii) centrality measures (degree, eigenvector centrality, closeness centrality, betweenness centrality, PageRank; (iv) clustering measures (local clustering coefficient, average local clustering coefficient, transitivity); (v) community analysis (*k*-clique, Girvan-Newman, Louvain, Clauset-Newman-Moore); and (vi) role detection in accordance with Guimerà & Amaral [28] on the results from the different community detection algorithms implemented.

5 Concluding Remarks

NetExtractor has been conceived as a semi-automatic enabling tool for network science outreach, being mainly aimed at people with little programming knowledge. As it operates on books/scripts of the user's choice, it incorporates both differentiation by interest and integrative teaching approaches, being thus particularly well-suited for multicultural environments.

From a technical point of view, future work lines that would improve NetExtractor could focus on: (i) the inclusion of centralization measures in the network analysis section; (ii) the broadening of the community detection algorithms implemented; and (iii) the development of the required functionality to create and handle dynamic networks, i.e., networks that evolve throughout time, which would allow to study the temporal evolution of the interactions over the course of the story.

Acknowledgements. The authors acknowledge financial support from the Spanish Ministry of Science, Innovation and Universities (excellence networks HAR2017-90883-REDC and RED2018-102518-T), and from the Junta de Castilla y León - Consejería de Educación through BDNS 425389 and the predoctoral grant awarded to Virginia Ahedo (supported by the European Social Fund). In addition, the authors would like to especially thank Dr. Luis Izquierdo for his insightful suggestions to improve the manuscript and Dr. Álvar Arnaiz-González, Yi Peng Ji and Alicia Olivares-Gil for their advice and help with some programming work.

References

1. Newman, M.E.J.: Networks, 2nd edn. Oxford University Press, Oxford (2018)
2. Cramer, C., Sheetz, L., Sayama, H., Trunfio, P., Stanley, H.E., Uzzo, S.: NetSci high: bringing network science research to high schools. In: Mangioni, G., Simini, F., Uzzo, S.M., Wang, Dashun (eds.) Complex Networks VI. SCI, vol. 597, pp. 209–218. Springer, Cham (2015). https://doi.org/10.1007/978-3-319-16112-9_21
3. Cramer, C.B., Porter, M.A., Sayama, H., Sheetz, L.: Network Science In Education. Springer, Cham (2018)
4. Barabási, A.-L.: Network Science. Cambridge University Press, Cambridge (2016)
5. Harrington, H.A., Beguerisse-Díaz, M., Rombach, M.P., et al.: Teach network science to teenagers. Netw. Sci. **1**, 226–247 (2013). https://doi.org/10.1017/nws.2013.11
6. Dekker, N., Kuhn, T., van Erp, M.: Evaluating named entity recognition tools for extracting social networks from novels. PeerJ. Comput. Sci. **5**, e189 (2019). https://doi.org/10.7717/peerj-cs.189
7. Sayama, H.: Mapping the curricular structure and contents of network science courses. In: Cramer, C.B., Porter, M.A., Sayama, H., Sheetz, L., Uzzo, S.M. (eds.) Network Science In Education. SCI, pp. 101–116. Springer, Cham (2018). https://doi.org/10.1007/978-3-319-77237-0_7
8. Gera, R.: Leading edge learning in network science. In: Cramer, C.B., Porter, M., Sayama, H., Sheetz, L., Uzzo, S.M. (eds.) Network Science In Education. SCI, pp. 23–44. Springer, Cham (2018). https://doi.org/10.1007/978-3-319-77237-0_2
9. Sánchez, A., Brändle, C.: More Network Science for Teenagers. **18**, 1–4 (2014). http://arxiv.org/abs/1403.3618
10. Arends, D., Kilcher, A.: Teaching for Student Learning. Routledge, New York (2010)

11. Tomlinson, C.A.: How To Differentiate instruction in Mixed-ability Classrooms, 2nd edn. Association for Supervision and Curriculum Development, Alexandria (2001)
12. Adami, A.F.: Enhancing students' learning through differentiated approaches to teaching and learning: a Maltese perspective. J. Res. Spec. Educ. Needs **4**, 91–97 (2004). https://doi.org/10.1111/j.1471-3802.2004.00023.x
13. Cherven, K.: Mastering Gephi Network Visualization. Packt Publishing Ltd., Birmingham (2015)
14. Csárdi, G., Nepusz, T.: The igraph software package for complex network reasearch. InterJ. Complex Syst. **1695**, 1–9 (2006)
15. Kolaczyk, E.D., Csárdi, G.: Statistical Analysis of Network Data with R. Springer, New York (2014)
16. Bastian, M., Heymann, S., Jacomy, M.: Gephi: an open source software for exploring and manipulating networks. In: Proceedings of the Third International Conference on Weblogs and Social Media, pp 361–362. AAAI Press, Menlo Park (2009)
17. Oliphant, T.E.: Python for scientific computing. Comput. Sci. Eng. **9**, 10–20 (2007). https://doi.org/10.1109/MCSE.2007.58
18. R Core Team: R: A Language and Environment for Statistical Computing. R Foundation for Statistical Computing, Vienna, Austria (2013)
19. Cramer, C.B., Sheetz, L.: Secondary student mentorship and research in complex networks: process and effects. In: Cramer, C.B., Porter, M.A., Sayama, H., Sheetz, L., Uzzo, S.M. (eds.) Network Science In Education. SCI, pp. 141–157. Springer, Cham (2018). https://doi.org/10.1007/978-3-319-77237-0_9
20. Sayama, H., Cramer, C., Porter, M.A., et al.: What are essential concepts about networks? J. Complex Networks **4**, 457–474 (2016). https://doi.org/10.1093/comnet/cnv028
21. Tanizawa, T.: Network science in your pocket. In: Cramer, C.B., Porter, M.A., Sayama, H., Sheetz, L., Uzzo, S.M. (eds.) Network Science In Education. SCI, pp. 189–199. Springer, Cham (2018). https://doi.org/10.1007/978-3-319-77237-0_12
22. Beveridge, A., Shan, J.: Network of thrones. Math Horizons **23**, 18–22 (2016). https://doi.org/10.4169/mathhorizons.23.4.18
23. Holme, P., Porter, M.A., Sayama, H.: Who is the most important character in frozen? What networks can tell us about the world. Front Young Minds **7** (2019). https://doi.org/10.3389/frym.2019.00099
24. Blondel, V.D., Guillaume, J.-L., Lambiotte, R., Lefebvre, E.: Fast unfolding of communities in large networks. J. Stat. Mech: Theory Exp. **2008**, P10008 (2008). https://doi.org/10.1088/1742-5468/2008/10/P10008
25. Torvik, V.I., Agarwal, S.: Ethnea – An instance-based ethnicity classifier based on geo-coded author names in a large-scale bibliographic database. In: International Symposium on Science of Science. Library of Congress, Washington, D.C. (2016)
26. Smith, B.N., Singh, M., Torvik, V.I.: A search engine approach to estimating temporal changes in gender orientation of first names. In: Proceedings of the 13th ACM/IEEE-CS Joint Conference on Digital Libraries - JCDL 2013, p. 199. ACM Press, New York (2013)
27. Aslak, U., Maier, B.: Netwulf: Interactive visualization of networks in Python. J. Open Source Softw. **4**, 1425 (2019). https://doi.org/10.21105/joss.01425
28. Guimerà, R., Amaral, L.A.N.: Cartography of complex networks: modules and universal roles. J. Stat. Mech: Theory Exp. **2005**, P02001 (2005). https://doi.org/10.1088/1742-5468/2005/02/P02001

Comparative of Clustering Techniques for Academic Advice and Performance Measurement

María Teresa García-Ordás[1], José Antonio López-Vázquez[2], Héctor Alaiz-Moretón[1], José Luis Casteleiro-Roca[2(✉)], David Yeregui Marcos del Blanco[3], Roberto Casado-Vara[4], and José Luis Calvo-Rolle[2]

[1] Department of Electrical, Systems and Automatic Engineering, University of León, Campus de Vegazana s/n, 24071 León, Spain
{mgaro,hector.moreton}@unileon.es

[2] Department of Industrial Engineering, University of A Coruña, CTC, CITIC, Avda. 19 de febrero s/n, 15405 Ferrol, A Coruña, Spain
{jose.lopez,jose.luis.casteleiro,jlcalvo}@udc.es

[3] IE University, Maria de Molina, 28006 Madrid, Spain
dmarcos@faculty.ie.edu

[4] BISITE Research Group, IoT Digital Innovation Hub, Edificio Multiusos I+D+i, University of Salamanca, 37007 Salamanca, Spain
rober@usal.es

Abstract. This article presents an innovative proposal for improving personalized student performance counselling. The methodology implemented applies clustering techniques in order to obtain group profiles of students with similar features. The research has been performed utilizing anonymized real academic grades from student data sets of the Polytechnic School of the University of A Coruña. The ultimate purpose for the proposed tool is to be dynamic and adaptive to different data sets. Therefore, only the most representative, universal variables are considered. Overall, three techniques have been evaluated for clustering, with two additional ones for dimensional reduction with very promising results.

Keywords: Clustering · Dimensional reduction · Academic performance · Academic advice

1 Introduction

It is widely known that the European Higher Education Area (EHEA) compels Universities of EU Member countries to implement quality assurance systems for performance tracking purposes [34]. In consequence, University Schools have the responsibility to implement and improve their internal quality systems [16]. Such praiseworthy initiative poses nonetheless new challenges to the Higher Education

Á. Herrero et al. (Eds.): ICEUTE 2020, AISC 1266, pp. 215–226, 2021.
https://doi.org/10.1007/978-3-030-57799-5_23

Institutions, to which solution the present research paper tries to contribute in the following pages [22,23,27].

Once the challenge has been presented, the authors considered that one approach with great potential would be to develop a tool for faculty and staff to measure and analyze student performance in order to provide an enhanced, more personalized follow-up.

To that end, the first step could be student identification by means of a previously defined pattern. In such case, ideally it would be possible to predict even if roughly, a student performance, based on her belonging to the defined pattern.

In relatively recent works, it was possible to obtain a model with the aim to accomplish prediction tasks, based on a data set of previous students. The applied techniques ranged from the traditional ones to the most updated, sophisticated ones [2,7–9,12,13,15,18,24,26,32,33]. Regarding the more traditional ones, the results were not particularly satisfactory due to a lack in the number of previous cases; hence not implementing satisfactory generalizations [1,3–6,10,11,19–21,25,28,31].

The present research tests a different approach, based on clustering techniques. Specifically, with the available data set, the aim is to create groups of students with similar features. Taking into consideration that the developed tool should be ideally adaptive, with the ability to create ad-hoc models, an additional task is to identify the more representative and significant courses so that it is not necessary to consider the whole .

As introduced in the abstract section, for the present research the authors have used a real, anonymized data set of academic grades from the Polytechnic School of the University of A Coruña, Spain. The data covers 7 academic years, specifically from 2001/2002 to 2008/2009.

The rest of the document is structured as follows: Sect. 2 briefly describes the study case. The applied techniques to achieve the fault detection single-class categorization are addressed in Sect. 3. While Sect. 4 elaborates on the experiments and achieved results. Finally, the conclusions and future works are exposed in Sect. 5.

2 Case Study

The data set for this research are the real, anonymized academic records of the Polytechnic School of the University of A Coruña, specifically from the Electrical Engineering degree.

It is structured in 25 courses divided into three academic years for a total of 236 academic credits. The first and second year involves 9 courses, with the third and final one adding the last 7.

The final dataset is achieved through a two different step methodology:

Initially, all the students enrolled in the degree from the 1996/1997 to the 2008/2009 academic years were considered. A statistical analysis gathered the following main findings:

- Only 20% of female students.
- 0.5% of the students come from outside Galicia region.
- 91.65% of the enrolled students come from secondary education.
- The average enrollment age was 20.86.
- The average graduation age was 26.40.
- An average of 5.96 academic years were needed to graduate.

The second step was to fine-tune the data set by removing two groups of students: those enrolled before the 2001/2002 academic year due to the lack of information in the academic records, as well as students who eventually dropped off without graduating.

2.1 Dataset

The final data set used in this research includes the following variables:

- Number of examination sittings to pass each course.
- The final grade for each course.
- Academic background: Secondary School or vocational education and training.
- Academic Grade for University Admission.

3 Used Techniques

3.1 Error Metrics

All the clustering methods included in the present document have been reviewed by means of three unsupervised metrics: Silhouette coefficient, Calinsky-Harabasz and Davies-Bouldin.

Silhouette. The Silhouette coefficient measurement is defined by the following Eq. (1) for each point.

$$\frac{x - y}{max(x, y)} \tag{1}$$

where x represents the mean intra-cluster distance and y corresponds to the mean nearest-cluster distance, i.e., the separation between a given sample and the nearest cluster to which the said sample is not pertaining.

It is important to consider that there is a prerequisite to apply the Silhouette coefficient: the total number of labels must be between two and the number of samples minus one.

If the condition is satisfied, the Silhouette is calculated as the mean of all the mean silhouettes through the different clusters.

Calinski-Harabasz. Given N, defined as the total number of observations and K as the amount of clusters, the Calinsky-Harabasz index is defined as: (2):

$$C = \frac{\frac{BGSS}{K-1}}{\frac{WGSS}{N-K}} = \frac{N-K}{K-1}\frac{BGSS}{WGSS} \tag{2}$$

where

$$BGSS = \sum_{k=1}^{K} n_k ||G^k - G||^2 \tag{3}$$

(where G^k is the dispersion of the barycenters for every cluster, G the barycenter of the complete dataset and n_k the number of samples for cluster C_k)

$$WGSS = \sum_{k=0}^{K} WGSS^k \tag{4}$$

$$WGSS^k = \sum_{i \in I_k} ||M_i^k - G^k||^2 \tag{5}$$

(for M_i^k representing the coefficients corresponding to the i-th row in the data matrix for cluster C_k, and I_k being the set of the indexes of the observations for cluster C_k).

Davies-Bouldin. The Davies-Bouldin metric definition is the mean value for all the clusters, of the samples M_k (See 6).

$$DB = \frac{1}{K}\sum_{k=1}^{K} M_k \tag{6}$$

With the aforementioned formula being correspondent to the following:

$$DB = \frac{1}{K}\sum_{k=1}^{K} max_{k' \neq k} \left(\frac{\delta_k + \delta_{k'}}{\Delta_{kk'}} \right) \tag{7}$$

$$\Delta_{kk'} = d(G^k, G^{k'}) = ||G^k - G^{k'}|| \tag{8}$$

3.2 Cluster

In the current section, the authors review the clustering algorithms that have been used to classify the data:

K-Means. This algorithm cluster purpose is to separate the data into different group sets while minimizing the inertia [17]. To that end, it is necessary to specify the number of clusters, usually called centroid μ_j, representing the mean of all its elements. See following Eq. 9.

$$\sum_{i=0}^{n} \min_{\mu_j \in C}(||x_i - \mu_j||^2) \tag{9}$$

Spectral Clustering. The Spectral Clustering algorithm [30] main characteristic is the use of similarity graphs to divide a dataset. Once the graph is obtained, the adjacency matrix and the degree matrix are generated, indicating the relation between the samples and the number of connections respectively.

Subsequently, the associated Laplacian matrix is calculated, extracting the adjacency matrix from the degree matrix. The final step involves running K-means over the eigenvectors of the Laplacian matrix so that the samples are arranged into clusters. The introduction of K-Means, implies that it is required to specify the number of centroids beforehand.

Agglomerative Clustering. Part of the so called Hierarchical Clustering algorithms family, the Agglomerative Clustering [14], builds clusters by a succession of mergers or splits. It starts with just one cluster per sample. With each iteration, the two most similar clusters merge, with this process continuing until all samples become members of one unique cluster. The most common way for graphically representing this family of algorithms are dendrograms.

3.3 Dimension Reduction

Principal Component Analysis: Usually referred to as PCA; it is a linear dimensional reduction algorithm based on the extraction of information from a high-dimensional space [29]. Specifically, PCA projects this high-dimensional space into a lower-dimensional sub-space. In order to preserve the basic components carrying more information, the algorithm fixes the features with a higher variation capacity within the data set and subsequently removes the redundant parts with a lower variation capacity.

Linear Discriminant Analysis: Abbreviated as LDA; it is a technique for dimensional reduction which can be summarized in five steps:

1. Compute the d-dimensional mean vectors for the different classes from the dataset.
2. Compute the in-between-class and within-class scatter matrix.
3. Compute the eigenvectors ($e_1, e_2, ..., e_d$) and corresponding eigenvalues (λ_1, $\lambda_2, ..., \lambda_d$) for the scatter matrices.

4. Sort the eigenvectors by decreasing eigenvalues and select the k eigenvectors with the largest eigenvalues to form a $d \times k$ dimensional matrix W (where every column represents an eigenvector).
5. Use the generated $d \times k$ eigenvector matrix to transform the samples into the new subspace. This can be summarized by the matrix multiplication: $Y = X \times W$ (where X is a $n \times d$-dimensional matrix representing the n samples, and Y are the transformed $n \times k$-dimensional samples in the new subspace).

4 Experiments and Results

The PCA is able to reduce an elevated number of features preserving the variance retrieved in the transformation with the selected number of features. For the research explained in the present article, a PCA reduction has been carried out to determine the minimum possible number of features while still preserving enough percentage of information and thus establishing a starting point for the clustering analysis. The following Fig. 1 shows a plot with the percentage of retrieved information depending on the number of features selected. It is worth noting that at least 20 features from the original dataset are needed to preserve a 70% of the dataset information.

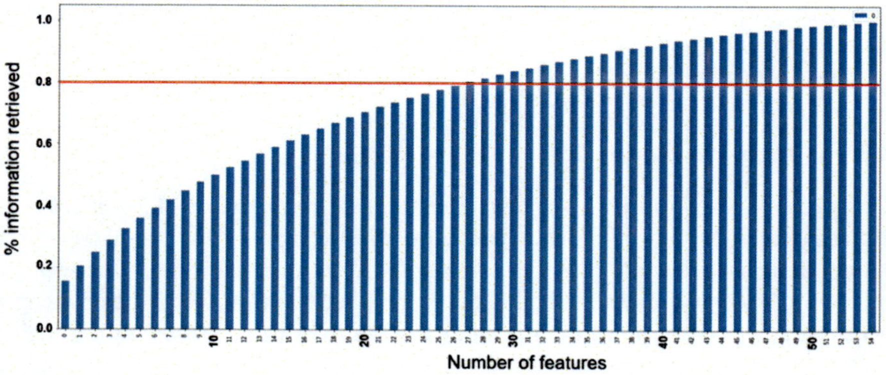

Fig. 1. Percentage of information retrieved when reducing the dataset by implementing PCA.

The LDA, in contrast to PCA, requires a supervised dataset with all elements labelled.

In order to label the entire dataset, the authors have evaluated different clustering algorithms by means of a hyperparameters study to determine the best configuration. The different experiments and the best result for each one are shown in the following Table 1.

Table 1. Hyperparameter selection for all different clustering methods using all the features.

Clustering	Hyperparameter	Best value
K-Means	Number of clusters	4
Agglomerative	Number of clusters	4
Spectral clustering	Number of clusters	2

The hyperparameters have been obtained after evaluating each cluster with the three metrics introduced in Sect. 3.1 (Silhouette, Calinski-Harabasz and Davies-Bouldin) and subsequently choosing the best score. Once the clustering process has finalized, each element in the cluster is considered as located inside one class. Afterwards, a LDA dimensionality reduction algorithm is launched, assigning relevance coefficients to each class, therefore determining which of the features is the most relevant.

As previously introduced and according to the results obtained in the PCA dimensional reduction study, the 20 most important features have been selected for each grouping method after applying LDA to reduce our dataset to a 2D dimensionality while preserving 70% of the original information.

The results showed that Agglomerative clustering just needed 20 features while with K-Means it was necessary to select 30 whereas and in Spectral clustering 25 features were needed to preserve the 70% threshold of the dataset variance.

The following Table 2 introduces the most relevant features for each clustering technique, showcasing that some features like TeoriadeCircuitosCalif, MaquinasElectricasIN_C, FisicaCalif, etc. were never taken into account whereas FísicaN_C, ExpresionGraficaN_C, QuimicaCalif, etc. were included in all the clustering methods evaluated.

Subsequently, a new evaluation of the best hyperparameters have been done in order to ensure the best performance of the clustering techniques.

Table 3 shows the best configuration when the most representative features are selected for each clustering and the associated scores obtained.

The metrics show a better performance for both agglomerative and Spectral Clustering, although it is very important to point out that K-Means was able to retain the same number of clusters after feature reduction. Since it is widely known that the more clusters detected by the algorithm, the worse the metrics turn out, the authors conclude than K-Means is the best option, considering the dataset. The main reason supporting the assessment is that K-Means was able to categorize the data into 4 classes with acceptable value metrics, which increased the overall impact of the experiment.

As a final step, a new LDA reduction was applied in order to evaluate the 2D distribution of each cluster. The results are shown in the following Fig. 2: K-Means has more than 2 clusters while the other two clustering methods (Agglomerative and Spectral Clustering) found just two different groups.

Table 2. Most relevant features extracted with LDA for each clustering method

Feature	K-Means	Agglomerative	S. Clustering
Nota acceso	YES	NO	NO
FísicaCalif	NO	NO	NO
FísicaN_C	YES	YES	YES
MatemáticasICalif	NO	NO	YES
MatemáticasIN_C	YES	NO	NO
TeoríadecircuitosCalif	NO	NO	NO
TeoríadecircuitosN_C	YES	NO	YES
FundDeInformaticaCalif	YES	NO	NO
FundDeInformaticaN_C	NO	NO	NO
ExpresionGraficaCalif	NO	NO	NO
ExpresionGraficaN_C	YES	YES	YES
EstadisticaCalif	YES	NO	YES
EstadisticaN_C	NO	NO	NO
QuimicaCalif	YES	YES	YES
QuimicaN_C	YES	YES	YES
MaterialesElectricosyMagneticosCalif	YES	YES	NO
MaterialesElectricosyMagneticosN_C	NO	YES	NO
TeoríadeMecanismosyEstructurasCalif	NO	NO	YES
TeoríadeMecanismosyEstructurasN_C	NO	YES	YES
ElectrometriaCalif	NO	NO	YES
ElectrometriaN_C	NO	NO	NO
ElectronicaIndustrialCalif	YES	YES	YES
ElectronicaIndustrialN_C	NO	NO	NO
InstalacionesElectricasCalif	NO	NO	YES
InstalacionesElectricasN_C	YES	NO	NO
RegulacionAutomaticaCalif	NO	NO	YES
RegulacionAutomaticaN_C	YES	YES	NO
MáquinasElectricasICalif	NO	YES	YES
MáquinasElectricasIN_C	NO	NO	NO
CircuitosElectricosCalif	YES	NO	YES
CircuitosElectricosN_C	NO	NO	NO
ProgramacionIndustrialCalif	YES	NO	NO
ProgramacionIndustrialN_C	YES	YES	YES
ProgramacionIndustrialN_M	YES	NO	NO
MatematicasIICalif	YES	YES	YES
MatematicasIIN_C	NO	YES	YES
MatematicasIIN_M	NO	NO	NO
MecanicadeFluidosYTermodinámicaCalif	NO	NO	NO
MecanicadeFluidosYTermodinámicaN_C	YES	YES	NO
TransportedeEnergiaElectricaCalif	YES	NO	NO
TransportedeEnergiaElectricaN_C	YES	NO	NO
CentralesElectricasCalif	NO	NO	YES
CentralesElectricasN_C	YES	NO	NO
MaquinasElectricasIICalif	YES	NO	NO
MaquinasElectricasIIN_C	NO	NO	YES
OficinaTecnicaCalif	YES	YES	NO
OficinaTecnicaN_C	YES	NO	NO
AdministraciondeEmpresasyOrganizacionIndustrialCalif	YES	YES	NO
AdministraciondeEmpresasyOrganizacionIndustrialN_C	NO	NO	YES
AplicaciondeMaquinasElectricasCalif	NO	YES	YES
AplicaciondeMaquinasElectricasN_C	NO	NO	NO
ConstruccionIndustrialCalif	YES	NO	YES
ConstruccionIndustrialN_C	YES	YES	NO
ProyectoFindeCarreraCalif	YES	YES	YES
ProyectoFindeCarreraN_C	NO	NO	NO
Acceso	NO	YES	YES

Table 3. Hyperparameter selection for all different clustering methods using the most relevant features and their scorings.

Clustering	Hyperparameter	Best value	Silhouette	Calinski-Harabasz	Davies-Bouldin
K-Means	Clusters	4	0.1059	20.2477	2.4335
Agglomerative	Clusters	2	0.2584	28.7999	1.7843
Spectral clustering	Clusters	2	0.4383	10.3803	0.5714

Once all the relevant information and results have been gathered, the authors can conclude that K-Means is the technique with a better 2D data reconstruction due to a lower number of overlapped groups.

In conclusion, K-Means showed the best performance due to its high performance achieved by dividing the data set into four different groups with a very low class overlay, as shown in Fig. 2. Even with an initial disadvantage because of a larger number of clusters, K-Means obtained very relevant and valuable scores in comparison with the easier clustering based on just two groups.

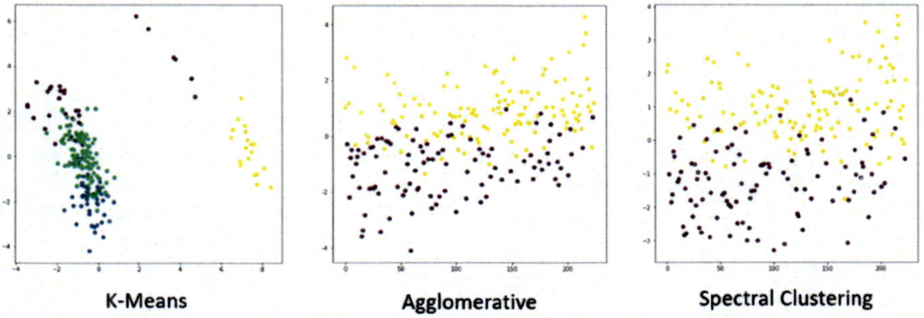

Fig. 2. 2D representation of the dataset for each clustering technique.

5 Conclusions and Future Works

The new European Higher Education Area guidelines have introduced stricter requirements for quality assurance and tracking. This situation has increased the challenges that Universities face in order to keep up with the new standards.

In the present article, the authors aim to contribute to the aforementioned situation by developing an evaluation method for students performance, which could be very useful in order to design improved University courses and degrees.

The unsupervised learning techniques present the ability to cluster the students profiles in order to understand the most important courses when the segmentation is successful.

In the present article, three clustering techniques have been implemented. The best results have been achieved with Kmeans due to the fact that it can process fours groups, even after the process of feature reduction.

Additionally, the other two clustering techniques implemented (Agglomerative and Spectral) could identify which were the most important features and which had a lesser impact when the clustering was performed.

The authors believe that a promising area for future works could be the evaluation of additional clustering techniques in order to discover which are the most suitable ones for better understanding the behavior of student performance and the main subjects related to this fact.

References

1. Alaiz-Moretón, H., Calvo-Rolle, J., García, I., Alonso-Alvarez, A.: Formalization and practical implementation of a conceptual model for PID controller tuning. Asian J. Control **13**(6), 773–784 (2011)
2. Anderman, E.M., Gimbert, B., O'Connell, A., Riegel, L.: Approaches to academic growth assessment. Br. J. Educ. Psychol. **85**(2), 138–153 (2015)
3. Calvo-Rolle, J.L., Machón-Gonzalez, I., López-Garcia, H.: Neuro-robust controller for non-linear systems. Dyna **86**(3), 308–317 (2011)
4. Calvo-Rolle, J.L., Quintian-Pardo, H., Corchado, E., del Carmen Meizoso-López, M., García, R.F.: Simplified method based on an intelligent model to obtain the extinction angle of the current for a single-phase half wave controlled rectifier with resistive and inductive load. J. Appl. Logic **13**(1), 37–47 (2015)
5. Casteleiro-Roca, J., Calvo-Rolle, J., Meizoso-López, M., Piñón-Pazos, A., Rodríguez-Gómez, B.: Bio-inspired model of ground temperature behavior on the horizontal geothermal exchanger of an installation based on a heat pump. Neurocomputting **150**, 90–98 (2015)
6. Casteleiro-Roca, J., Quintián, H., Calvo-Rolle, J., Corchado, E., Meizoso-López, M., Pinón Pazos, A.P.: An intelligent fault detection system for a heat pump installation based on a geo-thermal heat exchanger. J. Appl. Logic **17**, 36–47 (2016)
7. Casteleiro-Roca, J.L., Barragán, A.J., Segura, F., Calvo-Rolle, J.L., Andújar, J.M.: Fuel cell output current prediction with a hybrid intelligent system. Complexity **2019**, 1–10 (2019)
8. Casteleiro-Roca, J.L., Calvo-Rolle, J.L., Méndez Pérez, J.A., Roqueñí Gutiérrez, N., de Cos Juez, F.J.: Hybrid intelligent system to perform fault detection on BIS sensor during surgeries. Sensors **17**(1), 179 (2017)
9. Casteleiro-Roca, J.L., Jove, E., Gonzalez-Cava, J.M., Pérez, J.A.M., Calvo-Rolle, J.L., Alvarez, F.B.: Hybrid model for the ANI index prediction using remifentanil drug and EMG signal. Neural Comput. Appl. **32**, 1–10 (2018)
10. Casteleiro-Roca, J.L., Pérez, J.A.M., Piñón-Pazos, A.J., Calvo-Rolle, J.L., Corchado, E.: Modeling the electromyogram (EMG) of patients undergoing anesthesia during surgery. In: 10th International Conference on Soft Computing Models in Industrial and Environmental Applications, pp. 273–283. Springer (2015)
11. Cecilia, A., Costa-Castelló, R.: High gain observer with dynamic deadzone to estimate liquid water saturation in PEM fuel cell. Revista Iberoamericana de Automática e Informática industrial **17**(2), 169–180 (2020)
12. Cook, W.D., Tone, K., Zhu, J.: Data envelopment analysis: prior to choosing a model. Omega **44**, 1–4 (2014)
13. Crespo-Ramos, M.J., Machón-González, I., López-García, H., Calvo-Rolle, J.L.: Detection of locally relevant variables using SOM-NG algorithm. Eng. Appl. Artif. Intell. **26**(8), 1992–2000 (2013)
14. Defays, D.: An efficient algorithm for a complete link method. Comput. J. **20**(4), 364–366 (1977). https://academic.oup.com/comjnl/article-lookup/doi/10.1093/comjnl/20.4.364
15. Fernández-Serantes, L.A., Vázquez, R.E., Casteleiro-Roca, J.L., Calvo-Rolle, J.L., Corchado, E.: Hybrid intelligent model to predict the SOC of a LFP power cell type. In: International Conference on Hybrid Artificial Intelligence Systems, pp. 561–572. Springer (2014)
16. Ferreira, F.H.G., Gignoux, J.: The measurement of educational inequality: achievement and opportunity. World Bank Econ. Rev. **28**(2), 210–246 (2014)

17. Forgy, E.W.: Cluster analysis of multivariate data: efficiency versus interpretability of classifications. Biometrics **21**, 768–769 (1965)
18. Freeman, S., Eddy, S.L., McDonough, M., Smith, M.K., Okoroafor, N., Jordt, H., Wenderoth, M.P.: Active learning increases student performance in science, engineering, and mathematics. Proc. Natl. Acad. Sci. **111**(23), 8410–8415 (2014)
19. Garcia, R.F., Rolle, J.L.C., Castelo, J.P., Gomez, M.R.: On the monitoring task of solar thermal fluid transfer systems using NN based models and rule based techniques. Eng. Appl. Artif. Intell. **27**, 129–136 (2014)
20. Ghanghermeh, A., Roshan, G., Orosa, J., Calvo-Rolle, J., Costa, A.: New climatic indicators for improving urban sprawl: a case study of Tehran city. Entropy **15**(3), 999–1013 (2013)
21. Gomes, I.L.R., Melicio, R., Mendes, V.M.F., Pousinho, H.M.I.: Wind power with energy storage arbitrage in day-ahead market by a stochastic MILP approach. Logic J. IGPL (2019, in press)
22. Grissom, J.A., Kalogrides, D., Loeb, S.: Using student test scores to measure principal performance. Educ. Eval. Policy Anal. **37**, 3–28 (2015)
23. Jove, E., Antonio Lopez-Vazquez, J., Isabel Fernandez-Ibanez, M., Casteleiro-Roca, J.L., Luis Calvo-Rolle, J.: Hybrid intelligent system to predict the individual academic performance of engineering students. Int. J. Eng. Educ. **34**(3), 895–904 (2018)
24. Jove, E., Blanco-Rodríguez, P., Casteleiro-Roca, J.L., Quintián, H., Moreno Arboleda, F.J., LóPez-Vázquez, J.A., Rodríguez-Gómez, B.A., Meizoso-López, M.D.C., Piñón-Pazos, A., De Cos Juez, F.J., Cho, S.B., Calvo-Rolle, J.L.: Missing data imputation over academic records of electrical engineering students. Logic J. IGPL **34**, 895–904 (2019)
25. Jove, E., Gonzalez-Cava, J.M., Casteleiro-Roca, J.L., Méndez-Pérez, J.A., Antonio Reboso-Morales, J., Javier Pérez-Castelo, F., Javier de Cos Juez, F., Luis Calvo-Rolle, J.: Modelling the hypnotic patient response in general anaesthesia using intelligent models. Logic J. IGPL **27**(2), 189–201 (2018)
26. Kokkinos, C.M., Kargiotidis, A., Markos, A.: The relationship between learning and study strategies and big five personality traits among junior university student teachers. Learn. Individ. Differ. **43**, 39–47 (2015)
27. López-Vázquez, J.A., Orosa, J.A., Calvo-Rolle, J.L., Juez, F.J., Castelerio-Roca, J.L., Costa, A.M.A.: A new way to improve subject selection in engineering degree studies. In: International Joint Conference: CISIS 2015 (2015)
28. Marrero, A., Méndez, J., Reboso, J., Martín, I., Calvo, J.: Adaptive fuzzy modeling of the hypnotic process in anesthesia. J. Clin. Monit. Comput. **31**(2), 319–330 (2017)
29. Martínez, A.M., Kak, A.C.: Pca versus lda. IEEE Trans. Pattern Anal. Mach. Intell. **23**(2), 228–233 (2001)
30. Ng, A.Y., Ng, A.Y., Jordan, M.I., Weiss, Y.: On spectral clustering: analysis and an algorithm. In: Advances in Neural Information Processing Systems, vol. 14, pp. 849–856 (2001). http://citeseerx.ist.psu.edu/viewdoc/summary?doi=10.1.1.19.8100
31. Quintián, H., Calvo-Rolle, J.L., Corchado, E.: A hybrid regression system based on local models for solar energy prediction. Informatica **25**(2), 265–282 (2014)
32. Quintián, H., Casteleiro-Roca, J.L., Perez-Castelo, F.J., Calvo-Rolle, J.L., Corchado, E.: Hybrid intelligent model for fault detection of a lithium iron phosphate power cell used in electric vehicles. In: International Conference on Hybrid Artificial Intelligence Systems, pp. 751–762. Springer (2016)

33. Tomás-Rodríguez, M., Santos, M.: Modelling and control of floating offshore wind turbines. Revista Iberoamericana de Automática e Informática Industrial **16**(4), 381–390 (2019)
34. http://www.ehea.info/ (2017). Accessed 19 Mar 2017

English for Specific Purposes and Dyslexia at Higher-Education Level: Overcoming the Challenges of Vocabulary Acquisition

Alba Fernández-Alonso(✉) and Concetta Maria Sigona

Facultad de Humanidades y Comunicación, Universidad de Burgos, Paseo de Los Comendadores, s/n, 09001 Burgos, Spain
{albafa, cmsigona}@ubu.es

Abstract. The specific learning difficulties (SpLD) are evolutionary disorders whose origins go back to biological and genetic alterations. Dyslexia is one of these disorders that occur when children learn to read. Due to the growing increase of dyslexic students in the university area, it is imperative and urgent to redevelop teaching materials that can respond to their specific academic needs. On the one hand, knowing the characteristics of dyslexic students enable the mitigation of the harmful effects of this difficulty. On the other, it leads to the improvement of the quality of the university life of dyslexic students thanks to the educational responses specifically adapted for them. With this in mind, our proposal aims to implement appropriate strategies for the acquisition of Business English vocabulary, a challenge for dyslexic students due to the intrinsic features of this type of English for specific purposes (ESP). This proposal is aimed not only at providing adequate professional training, but also at improving their meaningful learning process.

Keywords: Dyslexia · Business english · Vocabulary · Higher education

1 Dyslexia and Language Acquisition: Brief Literature Review

Dyslexia is a learning difficulty that directly involves reading, in some cases writing and, in others, phonological alterations related to the problem of perception of some phonemes. Some students might find difficult to recognize certain sounds within a word or to represent sounds graphically. A study carried out to evaluate the phonological, auditory, visual, cerebral, magnocellular theories on dyslexia reveals that the phonological theory prevails as a basic problem of dyslexic students. However, this study also signals the presence of sensory problems and motor disturbances as secondary problems for many others [1]. In addition, the lack of automation in reading, due to the difficulty of discrimination of the graphic signs, accompanied by their pronunciation, is the cause of this problem which, as a consequence, pours into the final objectives achieved by students suffering from dyslexia. As far as reading is concerned, one should not overlook the fact that dyslexic students are not always able to read in a correct and precise way [2]. The degree of difficulty that learners of a foreign language

Á. Herrero et al. (Eds.): ICEUTE 2020, AISC 1266, pp. 227–235, 2021.
https://doi.org/10.1007/978-3-030-57799-5_24

may encounter depends not only on the difficulties found by the learner, but also on the characteristics of the foreign language studied [3]. Just as these students read and write with difficulty in their mother tongue, they find the same difficulties in learning a foreign language albeit to different degrees. Dyslexic students generally have difficulties in acquiring a foreign language due to a lack of phonetic awareness in their mother tongue. Other problems, related to a slower processing of information, can be found in memorizing vocabulary and concepts, and deciphering speech in a foreign language [4].

Moreover, it should be highlighted that students with dyslexia should be granted same academic opportunities as students who do not have this difficulty, therefore, when exposed to learning a foreign language, they should be put at ease and work in a supportive, motivating and inclusive environment with individual support. All this in order to take advantage of their rights as students and obtain a high-quality teaching [5]. As stated in Recital 13 of the Council Recommendation of 22 May 2019 on a comprehensive approach to the teaching and learning of languages in the European Union:

> "The European Pillar of Social Rights states as its first principle that everyone has the right to quality and inclusive education, training and life-long learning in order to maintain and acquire skills that allow full participation in society and successful transitions in the labor market. Multilingual competence is one of the key competences that could foster employability, personal fulfillment, active citizenship, intercultural understanding and social inclusion; it is defined as 'the ability to use different languages appropriately and effectively for communication'".

In connection therewith, Daloiso and Melero [6], advocate for a multicultural learning environment and emphasize the need for a diversity of languages in the European space: "The European Union, since its formation, has always promoted only multilingualism and multiculturalism, considering them founding principles of European citizenship, but also equal educational and social opportunities, with particular attention to people with Special Educational Needs (SEN)".

2 Dyslexia in the Higher Education Environment

Since dyslexia does not end with the end of school years and its presence at university has been increasing in the last years, reworking teaching materials that can meet the particular needs of the students has been deemed of extremely importance and urgency. Being aware of the needs of dyslexic students allows, on the one hand, to alleviate the harmful effects of this difficulty and, on the other, to improve the quality of university life of the students thanks to the increasingly appropriate educational responses to this type of student. Furthermore, Corredor et al. [7], support the importance of higher education structures to detect cases of dyslexic students and identify their specific difficulties, so as to provide support aimed at improving their learning. Our proposal is precisely designed for this purpose, as we will try to put into practice strategies suitable for dyslexic students to both provide appropriate training and improve their significant learning. In addition, major international organizations have been promoting inclusive education that enables children, young people and adults to be trained qualitatively and fairly within an inclusive institution. From this point of view, one of the objectives defined by UNESCO (2015) an referring to the quality of education, cannot be omitted: "to improve all aspects of the quality of education and ensure excellence of all so that recognized and measurable learning outcomes are achieved by all, especially in literacy, numeracy and essential life skills" [8].

As regards specific studies on dyslexia in a university context, [7], affirm: "A literature review shows that university students with dyslexia may have not been diagnosed and/or treated before starting their studies at university, though they are able to compensate with certain strategies."

As far as the scientific literature is concerned, it is important to point out what MacCullagh et al. [9] say about the studies carried out on the university experiences of dyslexic students. This study sheds lights on how the learning strategies developed by dyslexic students at university benefit then both academically and personally speaking.

3 Major Challenges of Dyslexic Students Learning English as a Second Language

Having said the previous, we could first seek an answer to the following questions: "What is the approach of dyslexic students to foreign language learning in the university environment? And, what kinds of strategies are recommended for achieving meaningful learning? For dyslexic students, precisely because of their difficulties in reading, phonologic discrimination and/or writing, learning a foreign language, in our case English in particular, turns out to be a huge effort. The curricular adaptation, in terms of attention to diversity, of the English language program is considered essential and necessary in order to guarantee progress in learning English. It is important to resort to the use of strategies and tools that highlight what they can do to enhance their self-esteem and encourage them to feel an inclusive part within the classroom. In this proposal, we stick to putting this into practice by following the guidelines of the inclusion principles.

In applying these inclusive principles, we will provide dyslexic students with a mentoring system that will respond to their requests and suggestions and provide them with a close relationship in which they will be offered a natural and close treatment. In class, teachers will use simple and direct language for explanations and provide synonyms for expressions that may be more complicated. All this will be developed within a framework of respect and inclusion that meets the specific needs of each student.

In addition to the curricular adaptation of the content, we will also adapt the design of our proposal following the guidelines of the study of Rello, L. and BaezaYates, R. [10], on the impact of typefaces on readability by people with dyslexia.

According to these authors, the most suitable fonts for people with dyslexia are Helvetica, Courier, Arial, Verdana and Computer Modern Unicode. On the other hand, the study advises against the use of words in italics. Based on the results of this research, we will use the Helvetica font and avoid the use of italics, even when quoting Anglo-Saxon terms in Spanish language explanations.

4 Adapting a Business English Academic Proposal to the Needs of Dyslexic Students

4.1 Justification

Globalization and the internalization of markets and the dominance of the United States as an economic power, have transformed the English language into the lingua franca of the financial environment [11], which implies an increased demand for subjects such as business English in university degrees in the economic sciences. In recent years, Spanish has ceased to be considered a hermetic language and its vocabulary has been influenced by the inclusion of terms from English due to the increasing globalization of the economy and the influence of the United States worldwide [12]. The permeability of languages has led to the production of new concepts with corresponding terms in the English economic language that are used in different languages. Many of these new concepts and their coined terms have been exported into Spanish, sometimes adapting to our language or remaining in English.

From our point of view and our experience as English teachers, we have observed that the inclusion of these new concepts in Spanish can be a source of confusion and, at times, lead to problems of understanding and clarity in teaching and subsequent assimilation of terminology in the economic field within the business English curriculum.

Aware of these difficulties, we have committed ourselves to the project of a series of activities with the working languages in Spanish and English that arise from the proposal of Fernández Alonso and Alcalde Peñalver [11]. The proposal presented below is a new version adapted to the needs of students with dyslexia in the university environment. The suggested activities pay particular attention to the following aspects: neologisms, false friends and common metaphors in financial jargon. Currently, there are no concluding studies about the prevalence of dyslexia in adults, but a tentative figure estimates that between 5% and 10% of the whole population is considered to have dyslexia [13]. Therefore, we believe it is highly probable that our classrooms at university will also be prone to including a considerable number of students with dyslexia, who will need to be provided with the pertinent adaptations to face their different learning patterns.

4.2 Theoretical Basis for the Adaptation of the Proposal

Although dyslexia does not have a neurobiological origin –and therefore does not depend directly on the mother tongue [14] – its manifestations may vary depending on the characteristics of the languages in which the student has a certain familiarity and communicative competence. Brunswick [15] emphasizes the difficulty that students with dyslexia can experience in their daily academic routines between two different languages such as Spanish –with a transparent spelling system– and English –with an opaque spelling– [16]: While dyslexic readers of all orthographies have difficulty converting written letters into their corresponding sounds, those who need to translate between the spoken and written forms of complex, deep, irregular languages will be the ones who experience the greatest reading difficulties. Dyslexic readers of deep spelling systems will be identifiable on the basis of their extremely slow and impaired Reading, spelling and phonological processing. Dyslexic readers of shallow languages, however, may cope adequately with their daily reading [...], but their performance is characterized by reading that is slower (but no less accurate) than expected, poor spelling ability and effortful phonological processing.

Since the impact of student performance can be influenced by the use of one language or another, Spanish students with dyslexia have more difficulty in learning English, which becomes a double challenge when it comes to tackling a professional and academic variety of English, such as Business English. Given the circumstances, the differences between students with no difficulties and students with dyslexia can be accentuated if no adjustments are made in daily classroom activities. Therefore, we believe that this proposal is not only vital to help students with dyslexia to achieve the objectives of having an optimal command of the Business English jargon and consolidate good communication skills, but also to help to eliminate the gap in their learning.

When adapting the proposal, we have paid attention to the needs that the student with dyslexia may find in the acquisition of Business vocabulary in English based on the general difficulties of learning English and the requirements of the program. Therefore, we have developed a series of strategies that will serve as a basis for adapting each of the activities:

- Vocabulary learning support with explanations in their mother tongue and writing [17]. In this way, the student's mother tongue becomes a vital tool for understanding, strengthening and using the mnemonic rules that facilitate the memorization and learning of vocabulary [18].
- Design of a small group of words for each exercise grouped by characteristics they have in common (grammatical categories, suffixes, prefixes, Latinisms, etc.).
- Implementation of mind maps to show the student in a single image the thoughts and ideas that his/her brain connects to a particular concept. This visual strategy provides the student with support for the acquisition of terminology and provides the connections needed to store new information [19]. In designing these activities, we also considered the main characteristics of the financial jargon in English highlighted by Alcaraz and Hughes [20]. Among these, we find the metaphorical component, false friends and neologisms. The metaphors characteristic of this area are based on the colloquialism that, among others, promoted by the founder of The Economist over 160 years ago. On the other hand, neologisms feed on the dynamism and globalization of the corporate sphere, a reality that forces language to adapt to the new realities that are emerging in order to fulfill its function as a communicative tool. Finally, in relation to false friends, it is necessary to underline that, although they are often easily detectable, it is necessary to clarify the translation of some of them into Spanish, such as the case of marginal or commodity.

The implementation of these three activities designed for a 90-min lecture session will help our students to develop a critical eye for using the correct terminology. This proposal has been designed for the students of the subject English for Tourism II, included the syllabus of the Degree on Tourism by the University of Burgos. After having taught this subject in the past years and witnessed the difficulties that dyslexic students face when learning a new language –sometimes even leading them to the development of a counterproductive insecurity in the academic environment– we have come up with this proposal we will start implementing firmly in the next academic winter semester 2020/2021. Our observations these past years have enabled us to design this proposal based on a hands-on approach that we believe will substantially help dyslexic students to acquire Business English vocabulary in an easy, integrating and transparent manner. As their final project, the students are asked to create a business plan that concisely describes a business idea of their own. For this purpose, they will need to use Business English jargon in connection with the recent developments in the field of their idea such as the project's economic feasibility, expected financial results, advertising and marketing options, potential customers or sustainability among others. For many of these students, this session will be their first direct contact with the Business English language. Furthermore, this set of activities will also contribute to greater transparency in the communication of this specialized language allowing the elimination of all the obstacles that students with dyslexia might encounter in their learning process. After having completed the activities, students with dyslexia will receive an anonymous questionnaire in which they will be asked to assess the adaptation and an evaluation of the aspects that could be implemented for possible improvement.

5 Adapted Proposal

5.1 Exercise on Neologisms

Students will work with current articles obtained from frequent sources of business advice and will be able to observe their use in a specific context.

- Look for examples of the use of the following English terms in Spanish print texts (Diario Expansión, El Economista, 5 Días, etc.). Explain their meaning in English, look for its possible equivalent in Spanish and whether its use is more widespread in English or Spanish.
- List of terms: quantitative easing, swaps, haircut, commodities, bonuses.
- Adaptation: we provide students with a definition in their own language, Spanish, to help them understand and facilitate the possible development of mnemonic rules. For their part, the students with dyslexia should proceed in the same way as their peers without adaptation.

5.2 Exercise on False Friends

The intention of this exercise is to make students become familiar with the meaning of terms that may prove to be false friends in the business environment and prevent their misuse.

- Explain the meaning of the following terms that appear in English and a possible equivalent in Spanish, as well as the resources used to search for this information.
- List of terms: quantitative easing, swap, haircut, commodities, bonus.

- Adaptation: Following Schneider and Crombie [21], we classify related words by grammatical categories using different coding system based on colors to identify each category. The students will also be provided with different examples in context that illustrate the meanings of these worlds.

5.3 Exercise on Metaphors

Metaphors are linguistic mechanisms that allow us to perceive reality in a different way and encourage the development of a more naturalized discourse in the target language.

- Explain and translate the titles of the related articles that include the following metaphors: knock yourself down, pay freeze, to keep your head above everything, ready to use.
- Adaptation: we change metaphors for others who share a common term and allow the student to create a mental map that facilitates their memorization by putting the main idea in a central image. The metaphors chosen for this adaptation share the term "money": to shovel the money out of the door, to pump money, a pot of money, to see the color of money, money laundering. To complete this exercise, the students will use the free online tool CmapTools, developed by the Institute for Human and Machine Cognition (IHMC) of the University of West Florida in the United States, and will classify the different metaphors into branches representing nominal verbal structures or sentences.

6 Conclusions

We believe that the design of these activities will guide our dyslexic students to better understand the complexities and problems associated with Business English terminology within a quality and highly meaningful learning environment. Its design is not only an example of the characteristics of this professional and academic language, but also a tool for inclusion for students with dyslexia in the English class.

The design of the adapted activities is located in an active learning framework in which the student ceases to be a mere spectator in the classroom to become the protagonist of their own learning, aware of their own difficulties and the tools they have to overcome them. In addition, this active learning strategy will allow the students to bring and share in the classroom their learning style, experiences and feelings. It will provide all the information and resources within its reach –some old and some new— to create a class that is truly their class [22].

Finally, the results of the surveys of students with dyslexia will shed light on possible improvements to this proposal, which we consider to be open and flexible and intends to be fully adapted to their needs within the English classroom environment.

References

1. Ramus, F., Rosen, S., Dakin, C.S., Day, L.B., Castellote, J.M., White, S., Frith, U.: Theories of developmental dyslexia: insights from a multiple case study of dyslexic adults. Brain **126**, 841–865 (2003)
2. Cuetos Vega, F., Soriano-Ferrer, M., Rello, L.D., Despiste, N.: ni Pereza: Todas las Claves para Entender el Trastorno. La Esfera de los Libros (2019)
3. Bianchi, V., Favilla, M.: Dislessia e Apprendimento di Lingue Seconde: Problemi da Considerare per la Costruzione di un Cillabo per L'insegnamento dell'italiano L2 ad Apprendenti Dislessici, pp. 149–165 (2018)
4. Babuder, M.K., Jazbec, S.: Exploring pre-service and in-service teachers' perceptions about early foreign language learning and dyslexia. Porta Linguarum: Revista Internacional de Didáctica de las Lenguas Extranjeras **32**, 87–102 (2019)
5. Nijakowska, J., Tsagari, D., Spanoudis, G.: English as a Foreign Language. Teacher Training Needs and Perceived Preparedness to Include Dyslexic Learners: The case of Greece, Cyprus, and Poland. Dyslexia, **24**(4), 357–379 (2018)
6. Daloiso, M., Melero Rodríguez, C.A.: Foreign languages and special educational needs. In: Rodríguez, M., Alberto, C. (ed.) Languages in Italy, Languages in Europe: Where we are, where we go. Edizioni Ca'Foscari, Venice, pp. 119–316 (2016)
7. Corredor, C.M., Ferreira, S.C.O., Monsalve, A.M.S., Currea, A.M.: Piloting a self-report questionnaire to detect reading-writing difficulties in students from two colombian universities. In: 2019 IEEE 19th International Conference on Advanced Learning Technologies (ICALT), pp. 311–313 (2019)
8. UNESCO: Education for All 2000–2015: Achievements and Challenges. United Nations Educational, Scientific and Cultural Organization, Paris (2015)
9. MacCullagh, L., Bosanquet, A., Badcock, N.A.: university students with dyslexia: a qualitative exploratory study of learning practices. Challenges Strategies Dyslexia **23**(1), 3–23 (2017)
10. Rello, L., Baeza-Yates, R.: Good fonts for dyslexia. In: Proceedings of the 15th International ACM SIGACCESS Conference on Computers and Accessibility, pp. 1–8 (2013)
11. Fernández, A.A., Alcalde, P.E.: Terminología y Traducción para el Mundo Empresarial, pp. 125–132. Formulación de los Nuevos Espacios Docentes. Tecnos, Madrid (2017)
12. Pizarro, I.: Análisis y Traducción del Texto Económico. Netbiblio, La Coruña (2010)
13. Siegel, L.S.: Perspectives on dyslexia. Paediatrics Child Health **11**(9), 581–587 (2006)
14. Daloiso, M.: Supporting Learners with Dyslexia in the ELT Classroom. Oxford University Press, Oxford (2017)
15. Brunswick, N.: Unimpaired Reading Development And Dyslexia Across Languages. In: Brunswick, N., McDougall, S., de Mornay Davis, P. (eds.) Reading and Dyslexia in Different Orthographies. Psychology Press, New York (2010)
16. Cuetos, V.F.: Reading Systems in Transparent Orthographies: Evolution of Profound Dyslexia in Spanish- Cognitive **14**(2), 133–150 (2002)
17. Rondot-Hay, M.: Teaching english to the dyslexic student. New Standpoints, The Modern Language Journal **28**(3), 2–4 (2006)
18. Nijakowska, J., Kormos, J., Hanusova, S., Jaroszewicz, B., Kálmos, B., Imrene Sarkadi, A., Smith, A.M., Szymańska-Czaplak, E., Vojtkova, N.: Dyslexia for Teachers of English as a Foreign Language. Erasmus+ Programme, KA2 Strategic Partnership, Project number: 2014-1-PL01-KA200-003578. University of Lodz, Faculty of Philology, Department of Pragmatics, Poland (2013)

19. Herrera, S., Murry, K.: Mastering ESL and Bilingual Methods: Differentiated Instruction for Culturally and Linguistically Diverse (CLD) students. Pearson, Boston (2005)
20. Alcaraz Varó, E., Hughes, B.: Diccionario de Términos Económicos. Financieros y Comerciales, Ariel, Barcelona (2008)
21. Schneider, E., Crombie, M.: Dyslexia and Foreign Language Learning. David Fulton Publishers, London (2003)
22. Schwartz, S., Pollishuke, M.: Aprendizaje Activo: Una Organización de la Clase Centrada en el Alumnado. Narcea, Madrid (1998)

Educational Inclusion: Teachers' Technical Needs and Supports

Beatriz F. Núñez[1](✉), Antonella Valenti[2], and Sonia Sapia[2]

[1] Universidad de Burgos, C/Villadiego 1, 09001 Burgos, Spain
bnunez@ubu.es
[2] Universitá della Calabria, Via Pietro Bucci 7/11/B, Aravaca di Rende, Italy
sonia.sapia@unical.it

Abstract. The Erasmus + Project "European diversity design for inclusive education" (EURODDIP-e), co-financed by the European Union, aims to provide education professionals with the knowledge, tools and devices necessary to assess the individual needs of the Students with difficulties in the upper limbs and personalize the Information and communication technologies (ICT) resources (interface, software, buttons, etc.). A Questionnaire Teacher (QT) with $\alpha = 0.75$, has been prepared to find out the needs of professionals in the teaching-learning process, taking into account the characteristics of the students and the existing technological possibilities, both in paper format and by using Google Forms is anonymous, and collects sociodemographic data, as well as 27 other items (16 yes/no answers, 10 Likert-type answers and the last open-ended answer. 1237 teachers from public and private schools in Belgium, Italy, Portugal and Spain, have responded. Items related to the training and performance of teachers in their teaching with inclusion students and use of ICT are analyzed. The need for training and greater resources is unanimous. English, French and Spanish versions in https://riubu.ubu.es/handle/10259/5262. It is the beginning for the elaboration of a methodology for the detection of capacities for all partner countries.

Keywords: Teachers' needs · Educational inclusion · ICT

1 Introduction

At the beginning of the current decade Vega-Godoy [25] argued the need to carry out profound changes in the curriculum, methodology and organization of schools, breaking with the traditional educational scheme that considers that all students are equal and consequently everyone has to do the same thing at the same time. It also maintained the existence of doubts and negative attitudes, related to ignorance of the subject, lack of socialization and sensitivity to special educational needs. He alludes to Marchesi and Martín [17] when commenting on the obstacles of the inclusion process and the fear of the unknown and not knowing how to face and manage differences on the part of teachers, in such a way that they experience a low expectation towards students with special educational needs what influences their learning rhythms.

Á. Herrero et al. (Eds.): ICEUTE 2020, AISC 1266, pp. 236–244, 2021.
https://doi.org/10.1007/978-3-030-57799-5_25

Martinez-Serrano [18] comments on the existence of studies such as those of Nketiah-Amponsah et al. [21] that demonstrate that there is no positive effect on student learning when using the applications and technological resources that exist in the market. Therefore, it is important to discriminate and recommend some devices over others to carry out academic tasks.

Hatlevik [13], Gudmundsdottir and Hatlevik [12], when talking about digital teaching competence, comment that it is intrinsically related to variables such as: the training of their teachers, perceived self-efficacy, pedagogical competence and the motivation of their teaching activity, these last two are alluded to by Deng, et al. [6], Arancibia, Comino and Casanova [1] and Karaseva, Ruulmann-Vengerfeld and Siibak [14].

On the other hand, Ekberg and Gao [8] affirm that teachers find difficulties in new teaching situations, even if they have previous training. In this sense, OECD [21] in a study on 14 countries, confirmed that Spain, Australia and Uruguay have made an effort to provide more devices (computers and tablets) and to increase teacher training, whereas Spain has a lower level than the OECD, average regarding the use of these tools with the appropriate methodologies.

For Garrido-Miranda [11] it is a challenge to train teachers who help future generations to develop capacities to cope in a very complex society, as well as, globalized, competitive, complex, inequitable and mediated by the use of languages, environments and digital interactions, as alluded to by Den Exter, Rowe, Boyd and Lloyd [5], Lim, Chai and Churchill [16], Newell, Pembroke and Boyd [20]. From all this arises the importance of initial and ongoing teacher training to incorporate digital technologies in their teaching.

In the II Plan of Attention to Diversity in Education in Castilla y León 2017–2022 [2], six strategic lines of action are proposed in order to ensure that the current school is gradually becoming more inclusive, that is, capable of offering real learning opportunities to all students. We emphasize that “teachers in inclusive education systems, in addition to identifying the educational needs of students, also need to know how effective their teaching is, as well as what to do to ensure that each student learns as much as possible” (p. 13).

The designed lines are based on the principles of equity, inclusion, standardization, proximity, universal accessibility and design for all, participation, efficiency and effectiveness, awareness, coordination, and prevention. In the objectives established for each of them, the teaching staff appears as a direct or indirect protagonist to achieve their conclusion. The objectives that we can find distributed in the different strategic lines of this plan, in which the teachers are directly involved, are:

Promoting inclusive culture in educational centers is intended to promote the training of all education professionals in inclusive methodologies (first strategic line achieves the improvement of the prevention, detection and early intervention processes of students’ educational needs (the line two) that is concretized in reinforcing the knowledge of professionals to address prevention and intervention actions in the face of possible specific support needs and in improving early detection of specific educational support needs, as objectives.

Implement technological instruments that facilitate a better understanding of the reality of students with specific need for educational support, within the proposed fifth line).

According to this II Diversity Attention Plan [2], teachers need more than knowledge in a subject, they need to know how students learn, how to understand individual differences, how to adapt their teaching to those differences, they need experience, practical knowledge, they need to know how to make adaptations of materials, evaluation, projects and activities, how to structure the tasks so that students with different levels can participate in them and how to manage the work of the classroom so that it becomes a place where there are multiple activities. Therefore, among the priorities proposed by this II Plan are the early detection of specific educational support needs in order to make decisions regarding the implementation of new curricular and/ or methodological measures, teacher training, participation of the educational community and improvement of inter-institutional coordination when necessary; and promote a line of training for all members of the educational community. Likewise, it emphasizes the lack of initial teacher training in basic principles of attention to diversity as a current weakness and considers training in inclusive education as the main focus. Other weaknesses included refer to material and personal resources in schools, lack of incentives for teachers who investigate, saturation of administrative tasks for teachers and failures in exchange of information between professionals, among others. Although, it is worth highlighting the motivation that teachers have, which demands permanent training and participates in innovative initiatives.

The ideas and principles of educational inclusion cannot be put into practice without the participation of teachers, who have to assume, together with families, the responsibility to promote, manage and implement them (Vélez, Tárraga, Fernández, and Sanz) [26] and that is why, one of the main concerns is the initial and continuous training of teachers, it is necessary to achieve an inclusive educational system, since it is essential that all teachers have knowledge that allows them to address the specific educational needs present in the classroom, since they are responsible for ensuring that every student, with any type of need, acquires the same learning results as the rest of the students (Moreno) [19]. It is essential to develop good initial teacher training in attention to diversity, regardless of the educational level at which they carry out their work, so it is important to know what training teachers receive and what their evolution has been.

According to Escudero [10] (p. 4) "what matters is not so much the quantity as the quality of the training, its valuable contributions to the teaching staff, the construction of school organizations effectively focused and committed to the learning of students and the same teaching staff".

The project "European diversity design for inclusive education" (EURODDIP-e) 2018-1-ES01-KA201-050300, co-financed by the European Union, aims to provide education professionals with the knowledge, tools and tools devices necessary to assess the individual needs of students with difficulties in the upper limbs and customize ICT resources (interface, software, buttons, etc).

In the project, the QT questionnaire has been designed and elaborated to obtain information about the needs that teachers have when it comes to attending and including the students of attention to diversity in their classrooms, and trying to personalize the educational resources to the maximum. The questionnaire has been applied in the partner countries: Belgium, Italy, Portugal and Spain, where differences in educational legislation have been observed in relation to the implementation of inclusion for students with disabilities, whose actions are in line of Sustainable Development Goals number 4 (SDG4) "ensure inclusive, equitable and quality education and promote lifelong learning opportunities for all" UNESCO (p.2) [24].

2 Method

The QT questionnaire has been applied during the months of May and September 2019, is divided into four parts: in the first part, referring to socio-demographic data (sex, age, center, course in which he teaches, years of professional experience, other studies, etc.), the second related to the profile of initial training during the years of teaching, being its educational practice and professional experience, which make up the third and fourth part Aspects of personal considerations are collected.

The formats used for the preparation and completion of the QT questionnaire were paper and online, using the Google Forms tool, offering teachers greater freedom and convenience when completing it. It is anonymous, with identifying data and 27 items (16 of answer yes/no, 10 of answer type Likert scale, from 1 to 5 being: Nothing agree, Little agree, Indifferent, Agree and Totally agree respectively). In all the blocks there are the options Do not know, Do not answer, and the last open answer item.

The sample consists of 1237 responses from the four partner countries: Belgium, Italy, Portugal and Spain.

To carry out the statistical treatment of the data, the statistical program SPSSv24 has been used.

3 Results

The reliability of the scale, measured by Cronbach's Alpha [α], identifies internal consistency, since it analyzes to what extent partial measurements obtained with the different items are consistent with each other and, therefore, representative of the possible universe of items that could measure the construct. In our case $\alpha = 0.75$, which indicates good reliability.

We are going to expose, in the first place, some sociodemographic data.

As can be seen in Table 1, the country where the most women have responded is Portugal, and the one where the most men have answered is Spain.

In relation to the age of the teachers who have answered the questionnaire, in Spain the highest percentage is between 25 and 35 years (37.04%), in Portugal between 45 and 55 years (50.68%), in Italy 41.8% have between 45 and 55 years, while in Belgium, between 35 and 45 years (27.27%) and between 45 and 55 years (28.34%). It is observed that in Spain it is the country with the youngest teaching staff.

Table 1. Results of the QT questionnaire in relation to the gender variable and in the countries where it has been applied.

	ITALY	SPAIN	PORTUGAL	BELGIUM
Female	71.1%	55.1%	81.2%	70.59%
Male	27.2%	44.9%	18.4%	29.41%
No reply	1.3%	0%	0.2%	0%
Not specified	0.4%	0%	0.2%	0%
Total	100%	100%	100%	100%

When asked about the completion of academic training, we find that in Spain it is 43.52%, in Portugal 60.06%, in Belgium 59.36% and in Italy it is around 80%, whit more than 20 years ago they finished their studies.

It is important to highlight the teaching experience, specifically, the professionals with more than 20 years and the percentages of the countries are: 67.17% Portugal, 30.56% Spain, 51.87% Belgium and more than 70% in Italy.

With regard to the workplace being public, statistics reveal that in Spain it is 59.57%, 94.86% in Portugal, 97.8% in Italy and 96.26% in Belgium.

We highlight some items, from the QT questionnaire, related to ICT and disability, which allows us to place ourselves in a wide range of actions, according to Portugal, Italy, Belgium and Spain.

95.7% of Spaniards consider specific training necessary for each type of attention to diversity (high abilities, immigration, disability, etc.), while in Portugal it is 75.5%, in Italy it is 72.4% and the 51.35% in Belgium.

Regarding what they think about whether it is necessary to know resources related to each type, the positive responses (agree and totally agree) in Spain is 87% of the teaching staff, equivalent to Belgium with 87.16%., while for Portuguese teachers it is 69.3% and for Italian teachers it is 84.5%.

In relation to the need to coordinate with teachers from other centers that have students with similar characteristics, the results of the survey show that it is Italy, with 84.5% the one that contributes the highest percentage followed by Belicaa with 81.08% and after Spain 75.9%, the lowest being Portugal with 66.2%.

One of the important questions is whether teachers consider it necessary to expand their training in attention to diversity. In Portugal 84.2% estimates, while in Spain it is 93.7, in Belgium 89.86% and in Italy 71.5%, the percentage being lower and denying the need 13%, while the doubt is expressed by 15.5%.

Training and information on ICT is an important aspect to consider and the results of the survey in the participating countries regarding the response is not: 7.3% in Portugal, 3.5% in Italy, in Spain 1.2% do not know or does not answer and in Belgium it is 26.36% they are in no, do not know or do not want to answer.

The last item contains final comments such as: "all the weight falls on the tutor", "the teacher lacks support and reinforcements", "he tries to cover everything without succeeding", and they demand that "the Administrations do what is necessary to equip the centers and to the teachers with the necessary and sufficient resources to be able to give attention to quality diversity, as well as to all the students who need it".

4 Conclusions

Vega-Godoy [25] stated that changes in the initial predisposition and attitudes of teachers towards diversity are fundamental to carry out inclusion, and improvements in classroom practices can be made, if provided to teachers conceptual and methodological tools that allow them to modify their previous schemes, their symbolic representations, replacing the homogenizing views of teaching with new approaches, focused on diversity.

We join Martinez-Serrano [18], commenting that the use of ICT in educational practice is conditioned by very heterogeneous factors, as well as that its appearance, in relation to new methodologies, has emerged in a short time. All this has led to new training needs in ICT by the teachers and the students.

Various reports and investigations by Enochsson and Rizza [9], Liaw [15], Tondeur, Roblin Couple, van Braak, Voogt and Prestridge [23], warn us that the initial teacher training offers few experiences to learn to use ICT in the teaching-learning process, as well as the influence of low levels of self-efficacy in its pedagogical use by teachers.

Caballero and Díaz [4] consider educational inclusion, and even attention to diversity, to be very important, considering it “a fundamental pillar”, but in many cases they do not know how to take their teaching to inclusion, due to training needs, lack of resources, coordination between members of the educational community, etc. These aspects coincide with a large group of teachers participating in the study, understanding school inclusion from a merely ethical perspective, as an objective that must be met.

Initial training received by teachers and previous experience with students with disabilities are the two main tools available to the teaching staff to deal with the education and inclusion of these students (Caballero and Díaz) [4]. In the data analysis, around 18% of the surveyed teachers do not know or have not received training in attention to diversity, with the consideration of working to promote inclusion in their classrooms and centers.

The vast majority of teachers demand more training on methodologies, design and strategies for inclusive practices, evaluation procedures, technical aids, internal and external resources, resources for collaboration between teachers and knowledge about the relationship with teachers. students and families.

The percentage of participants who consider that initial training, both theoretical and practical, should be expanded in terms of attention to diversity. Knowledge on how to educate and treat students with disabilities is essential so that inclusive values are present in the classroom and shared by all students (Caballero and Díaz) [4], likewise, it is exposed in the II Plan Attention to Diversity [2] need more than knowledge in a subject, they need experience and practical knowledge to know how to make adaptations of materials, evaluation, projects and activities.

Respondents stated that they agree on the need and importance of receiving more specific training on each type of diversity that exists, with the aim of being able to detect needs and attend all of them in the best possible way, an aspect that has yielded significant percentages. The “early detection of specific needs for educational support in order to make decisions regarding the implementation of new curricular and/ or methodological measures" is one of the priorities of the II Plan for Attention to Diversity [2] (p. 25), along with the training of teachers and all members of the educational community.

On the other hand, sometimes, it is the teachers who have to find the necessary means to carry out their teaching activity, which they usually find in training, acquired thanks to their initiative and personal disposition (Caballero and Díaz) [4]. A high percentage of participating teachers have continued to train with respect to students with specific needs, in all the countries participating in the survey.

In some cases they have carried out the training when they meet unknown needs or those who do not know how to work well, since a high number of participants admit that they need help in those cases or courses in which they change the type of student care to diversity, being less for women than for men.

When it comes to putting into practice and carrying out the different adaptations and adjustments necessary, most respond in the affirmative, with a minimum percentage of the teachers in private schools responding that they do not need help.

On some occasions it is considered that it is better for students with special needs to receive support outside the classroom, in this regard Yadarola [27] comments that "it may lead them to be more dependent on support professionals, to seek to take charge of the student integrated, disintegrating it" (p. 2).

One aspect of debate is the lower ratio in inclusion cases, since having inclusion in the classroom makes it more complicated for teachers to attend individually and with quality to students, as well reflected in the LOMCE [3] in its article 20, giving importance to the tutor providing personalized attention, making early diagnoses and establishing reinforcement mechanisms to achieve school success.

In order to serve these students, it is important to know resources related to each type of attention to diversity, an aspect in which differences between public and private centers are not highlighted.

Most of the teachers who participated consider that the involvement of professionals with the attention to diversity students should be encouraged, since as they have commented, to achieve a good teaching for this student "much more efforts are necessary".

We join the exposition by UNESCO [24] that alludes to factors such as capacities and attitudes of teachers, infrastructures, pedagogical strategies and curriculum design that facilitate or inhibit inclusion. In the same document, we are shown that the formulation of inclusive and equitable policies requires that it be recognized that the difficulties faced by students arise from aspects of the educational system itself, which includes the way in which educational systems are organized. today, the forms of instruction provided, the learning environment, and the ways in which student progress is supported and evaluated (p. 13).

Garrido-Miranda [11] concludes with the need to promote changes in pedagogical beliefs from which the form of teaching for learning is conceived, on the other hand, by exemplifying a variety of formative and contextualized uses of available digital technologies that offer greater cognitive potential.

The research by Messiou and Ainscow, in 2015, mentioned in UNESCO [24] states that teacher development should take place where they carry out their activity, in addition, it must create spaces to observe, plan, share ideas and resources, that is, cooperate and develop their experience within the school itself, make connections with existing knowledge and, not least, develop a common language of practice and how to improve it.

Durán and Giné [7] have related inclusion with issues such as assuming values of respect for differences, valuing each of the students, equity, etc.; the recognition of rights, increasing the participation of students in the curriculum, in culture and in the community, avoiding any form of exclusion in educational centers; transform the cultures, regulations and practice of schools so that they respond to the diversity of the needs of the students; the presence, participation and success of all students exposed to any risk of exclusion; development and training of schools, since what is important is the capacity of the center to welcome, value and respond to the needs of the students; and, the integral development of all the students.

References

1. Arancibia, M., Cosimo, D., Casanova, R.: Percepción de los profesores sobre integración de TIC en las prácticas de enseñanza en relación a los marcos normativos para la profesión docente en Chile. Revista Ensaio: Avaliaçao e Políticas Públicas em Educaçao, **26**(98), 163–184 (2018). https://doi.org/10.1590/s0104-40362017002501119
2. BOCYL: Acuerdo 29/2017, de 15 de junio, de la Junta de Castilla y León, por el que se aprueba el II Plan de Atención a la Diversidad en la Educación de Castilla y León 2017–2022. BOE 19 de junio de 2017 (2017). https://www.educa.jcyl.es/es/resumenbocyl/acuerdo-29-2017-15-junio-junta-castilla-leon-aprueba-ii-pla
3. BOE: Ley Orgánica 8/2013, de 9 de diciembre, para la mejora de la calidad educativa. BOE 10 de diciembre de 2013. Recuperado de https://www.educastur.es/-/lomce-ley-organica-para-la-mejora-de-la-calidad-educativa (2013)
4. Caballero, F., Díaz, V.: Investigación sobre las necesidades formativas de los docentes en la educación de estudiantes con discapacidad. Fundación ONCE. (2014)
5. Den Exter, K., Rowe, S., Boyd, W., Lloyd, D.: Using Web 2.0 technologies for collaborative learning in distance education- case study from an Australian University. Future Internet, **4** (1), 1–19 (2012)
6. Deng, F. et at.: The relationships among Chinese practicing teachers' epistemic beliefs, pedagogical beliefs and their beliefs about the use of ICT. J. Educ. Technol. Soc. **17**(2), 245–256 (2014)
7. Durán, G., Giné, C.: La formación del profesorado para la educación inclusiva: Un proceso de desarrollo profesional y de mejora de los centros para atender la diversidad. Revista Latinoamericana de Educación Inclusiva **1**, 153–170 (2017)
8. Ekberg, S., Gao, S.: Understanding challenges of using ICT in secondary schools in Sweden from teachers' perspective. Int. J. Inf. Learn. Technol. **35**(1), 43–55 (2018). https://doi.org/10.1108/ijilt-01-2017-0007
9. Enochsson, A., Rizza, C.: ICT in initial teacher training. Res. Rev. (2009). https://doi.org/10.1787/220502872611
10. Escudero, J.: La formación contínua del profesorado de la Educación Obligatoria en el contexto español. Profesorado. Revista de Curriculum y Formación de Profesorado, **21**(3), 1–20 (2017)

11. Garrido-Miranda, J.M.: Intención y prácticas TIC en formadores de docentes: congruencias, colisiones y autoeficacia. Estudios pedagógicos (Valdivia), **44**(3), 253–269 (2018). https://dx.doi.org/10.4067/S0718-07052018000300253
12. Gudmundsdottir, G.B., Hatlevik, O.E.: Newly qualified teacher' professional digital competence: implications for teacher education, doi: org/https://doi.org/10.1080/02619768.2017.1416085, European Journal of Teacher education, 41(2), 214-231 (2018)
13. Hatlevik, O.E.: Examining the Relationship between Teachers' Self-Efficacy, their Digital Competence, Strategies to Evaluate Information, and use of ICT at School. Scand. J. Educ. Research, **61**(5), 555–567 (2016)
14. Karaseva, A., Pruulmann-Vengerfeldt, P., Siibak, A.: Relationships between in-service teacher achievement motivation and use of educational technology: case study with Latvian and Estonian teachers. Technol. Pedagogy Educ. **27**(1), 33–47 (2017)
15. Liaw, E.: Application of computer-mediated communication on teacher efficacy during school-based field experience. Teach. Teacher Educ. **65**, 81–90 (2017)
16. Lim, C., Chai, C., Churchill, D.: Leading ICT in Education Practices: A Capacity-Building Toolkit for Teacher Education Institutions in the Asia-Pacific. Microsoft, Singapore (2010)
17. Marchesi, A., Martín, E.: Calidad de la enseñanza en tiempo de cambio. Alianza, Madrid (1998)
18. Martínez-Serrano, M.C.: Percepción de la Integración y uso de las Tecnologías de la Información y la Comunicación (TIC). Estudio sobre docentes y alumnos de educación primaria. Información tecnológica, **30**(1), 237–246 (2019). https://dx.doi.org/10.4067/S0718-07642019000100237
19. Moreno, M.I.: Análisis de la Gestión y Atención a la Diversidad en un Centro Educativo. La Ciudad Accesible. Revista Científica sobre Accesibilidad Universal, **8**, 151–189 (2016)
20. Newell, D., Pembroke, M., Boyd, W.: Crowd sourcing for conservation: Web 2.0 a powerful tool for biologists. Future Internet, **4**(2), 551–562 (2012)
21. Nketiah-Amponsah, E., Asamoah, M.K., Allassani, W.: Examining students experience with the use of some selected ICT devices and applications for learning and their effect on academic performance. J. Comput. Educ. **4**(4), 441–460 (2017). https://doi.org/10.1007/s40692-017-0089-2
22. OECD: Innovating education and educating for innovation: The power of digital technologies and skills, OECD Publishing, París (2016). https://doi.org/10.1787/9789264265097-en
23. Tondeur, J., Pareja Roblin, N., van Braak, J., Voogt, J., Prestridge, S.: Preparing beginning teachers for technology integration in education: ready for take-off? Technol. Pedagogy Educat. **26**(2), 157–177 (2016)
24. UNESCO: Guía para asegurar la inclusión y la equidad en la educación. Recuperado de (2017). http://www.unesco.org/open-access/terms-use-ccbysa-sp
25. Vega Godoy, A.: Integración de alumnos con necesidades educativas especiales: ¿Existe coherencia entre los discursos y las prácticas pedagógicas que ejercen los maestros de primaria? Estudios pedagógicos (Valdivia), **35**(2), 189–202 (2009). https://dx.doi.org/10.4067/S0718-07052009000200011
26. Vélez, X., Tárraga, R., Fernández, M.I., Sanz, P.: Formación inicial de maestros en Educación Inclusiva: una comparación entre Ecuador y España. Revista de Educación Inclusiva **9**(3), 75–94 (2017)
27. Yadarola, E.: Una mirada desde y hacia la educación inclusiva. Boletín Electrónico de IntegraRed. Recuperado de (2006). http://www.integrared.org.ar/links_internos/

Special Session: Innovation and Research Findings in Engineering Higher Education

Motivation, Technical Achievement and Communication Skills in Engineering Degree Projects Under the European Higher Education Area

Javier Rodríguez Martín(✉), Ignacio López Paniagua, Susana Sánchez Orgaz, Celina González Fernández, Carlos Arnaiz del Pozo, Ángel Jiménez Álvaro, and Rafael Nieto Carlier

Escuela Técnica Superior de Ingenieros Industriales de Madrid, Universidad Politécnica de Madrid, José Gutiérrez Abascal, 2, 28006 Madrid, Spain
javier.rodriguez.martin@upm.es

Abstract. To which extent are the final marks of an Engineering Degree Project determined by the technical quality of the work? To which extent by the horizontal skills of the student? The educational paradigm of the European Higher Education Area (EHEA) integrates both sides. Inevitably, the subjects included in the curriculum of an engineering degree are mainly technical, and prone to exercising and developing self-learning and in-depth, technical knowledge. The Degree Project, on the other side, is necessarily multidisciplinary, and usually develops loosely defined problems of wider scope. Apart from the work being supervised by a teacher, frequently projects are carried out as part of larger contexts in research teams. The Degree Project is judged by a panel, which will consider the written final report provided by the student, in combination with an oral session, including a presentation and a short turn of questions and answers on the topic. It is clear that the understanding of the work by the panel is filtered by how effectively the student is able to communicate, which will bring into the final mark a certain measure of soft skills and personality traits: self-confidence, articulation, body language, and attitude, among others. This work, on a basis of 18 degree projects, concludes that the effectiveness of the oral session is critical to obtain a high mark (9/10), while other factors, difficult to dissect, as well as certain degree of randomness, intervene on the higher marks.

Keywords: European Higher Education Area · Generic competences · Engineering degree · Oral communication skills · Written communication skills · Interpersonal skills

1 Introduction

The curriculum of the Engineering Degree at the Escuela Técnica Superior de Ingenieros Industriales de Madrid (ETSII) of the Universidad Politécnica de Madrid (UPM), includes, in its final year, a project called Trabajo Fin de Grado-Engineering

Á. Herrero et al. (Eds.): ICEUTE 2020, AISC 1266, pp. 247–256, 2021.
https://doi.org/10.1007/978-3-030-57799-5_26

Degree Project (DP). It is a lengthy work in which the student must solve a complex problem by applying and integrating several disciplines learned during the degree, as well as acquiring new competences and work methodologies. The formal outcome of the project is a written document. The work is judged by a Project Panel of three teachers after revising the document and hearing an oral presentation by the student. This system was designed to integrate horizontal, generic competences and soft skills into the marks, along the lines set by the European Higher Education Area (EHEA). This, however, inevitably leaves room for subjectivity in the evaluation of Degree Projects, and the final mark achieved by a student may be associated to a certain degree of uncertainty.

Logically, technical achievement alone might not lead to high marks, but, to which extent technical competence, generic skills and final mark are related? It is far from evident. This work is a first approach to this question. Is there a global pattern? Are generic and technical skills effectively integrated?

In marking a Degree Project, the procedures at ETSII-UPM define a methodology, which specifies, among other aspects, a list of traits to be evaluated by the Project Panels, but the question remains that how much, and in which cases, communication skills might overweigh technical soundness. This paper analyses the relation, not always evident, between motivation, attitude, technical capacity and communication skills on one side, and final mark, quality of the work and quality of the oral presentation. The base for the study is a number of Degree Projects carried out at the ETSII-Thermodynamics Group over the last years, covering a range of energy topics, and of different character.

This paper will first summarise the nature of the Degree Project and the methodology of evaluation at ETSII-UPM. Then, the characterization of project cases will be given, and analysed with graphs, each showing the relation between different traits, which will be discussed in detail. Finally, the major conclusions of the study will be summarized.

2 The Engineering Degree Project

The Degree Project is a lengthy, between 6 and 30 ECTS, comprehensive individual work that the student will carry out under the supervision of one or more teachers. The objective is that the student exercise the general competences of the degree, and apply the subjects of the curriculum [1]. In contrast to a Master's Thesis, oriented to imperfectly defined problems, research or technology integration, the Degree Project is oriented to well-established practice, and application of consolidated technology [2].

It is intended that the DP be the first contact of the student with a professional context, in the sense of wide scope problems, professional tools and practices. Thus, the project must be multidisciplinary and of sufficient complexity [3]. There exist two main project philosophies: the academic, whose objective is that the student integrate and apply the different disciplines learned during the Degree, and the professional, in which the student must produce a realist solution, both technically and economically, to a number of practical, interrelated problems, of sufficient complexity.

Formally, the project can be of three types:

- Engineering project: design/manufacturing of a prototype, facility engineering, etc.
- Technical, organizational or economic studies on equipment, systems, services, etc.
- Theoretical-experimental studies, either theoretical, computational or experimental.

Apart from the technical soundness of the work, the Panel will evaluate a list of competences, detailed in the official procedure [4], both generic and specific to the written document. Examples of these are the following:

1. The student identifies the restrictions associated with the system, component or process and integrates them into the work
2. The student analyses problems, negative and positive impacts related to the work, in the social, economic and environmental dimensions
3. The student establishes objectives and accomplishes them in the required time
4. The student shows autonomy and proactivity
5. The student is capable of integrating and searching for information additional to the given
6. The student organizes information in order to improve its understanding
7. The student adapts the style to the expectations and previous knowledge of the reader
8. The student uses graphic and other resources in order to communicate information effectively

A written document of the project must be submitted by the student to a Panel formed by three teachers, to which the student will present the work orally. The presentation will last 20 min and be followed by a turn for questions. The panel will produce the final mark.

3 Methodology

A total of 18 Degree Projects have been selected for the study. All the considered studies have been supervised by the same teacher in order to reduce uncertainty.

Each project and student has been characterized in several ways: formal and administrative context, student motivation and attitude, working methodology, technical capacity and communication skills. The result has been characterized by final mark, technical quality and presentation quality. Each of the traits has been graded.

It is worth mentioning that low values in Motivation or Dedication indicate that the student does not work regularly, with possible long periods of no dedication at all, does not attempt to solve problems or even study them, etc. High values, on the other hand, frequently mean that the student shows a personal interest in problems and questions, puts effort in studying and learning even more than the strictly needed, and, on occasion, might actively get involved in the activity of the research Group.

Capacity of analysis represents the student's ability to dissect problems, identify structure and patterns, and trace results to theory. It is understood that this trait be limited in undergraduate students, but it has been included to reflect a number of cases in the sample.

It has been observed that most students dedicate some time to planning their work, mostly in the sense of allocating time slots for their project among their other activities. A few students, however, show more developed habits, and extend planning to their technical work, for example by discriminating between levels of detail in analysing and designing, or by carrying out work package decomposition of large tasks to a certain extent (evidently, not systematically nor employing professional methodologies, but instinctively).

Table 1. Results of the characterization of Degree projects sample.

Case	Motivation	Analysis skills	Communication skills	Working methodology	Dedication	Technical quality	Presentation quality	Average subjects mark	Final mark
1	4	4	2	5	4	4	2	6.68	8.6
2	3	2	4	2	3	3	3	6.53	9.8
3	2	2	3	1	2	2	3	6.01	10.0
4	2	2	4	3	3	2	4	7.94	9.5
5	2	2	4	3	3	3	3	6.89	9.0
6	4	3	2	3	4	3	2	7.44	8.8
7	2	3	3	3	2	3	3	6.12	9.0
8	2	2	1	3	2	1	1	6.64	7.8
9	3	3	3	4	3	4	3	6.80	9.6
10	5	5	5	3	5	5	5	6.36	9.6
11	4	4	4	3	4	4	3	6.14	9.7
12	2	1	5	1	1	1	5	6.42	9.8
13	3	3	3	3	3	4	3	6.24	9.5
14	4	4	4	5	4	4	5	7.16	10.0
15	1	1	5	1	1	1	5	6.47	9.2
16	4	3	3	4	4	3	3	6.90	9.2
17	4	3	3	5	5	4	3	7.64	9.0
18	5	4	4	5	5	5	4	7.14	9.2

Two related aspects: the student's ability for communication and the quality of the final presentation, have also been studied. In general, students feel comfortable speaking in public, and some are particularly good in their choice of words, tone and non-verbal behaviour. These skills, in most cases, have led to producing adequate graphic material, a good structure of the discourse and a natural reaction to the questions of the Panel. The relative importance of communication and presentation in the final mark will be discussed in the following sections.

4 Results and Discussion

The Table 1 presents the characterization carried out over the sample of 18 Degree Projects. The average subject mark and final mark have been collected from the academic transcripts, which are expressed numerically from 0 to 10. The rest of the parameters have been graded using a five levels scale following the criteria of official procedure [4].

A first point in the research was to establish the relation between the student's academic trajectory, understood as the average mark of the student in all subjects preceding the DP, and the final mark of the Degree Project. This has been represented on the Fig. 1. A low correlation can be observed.

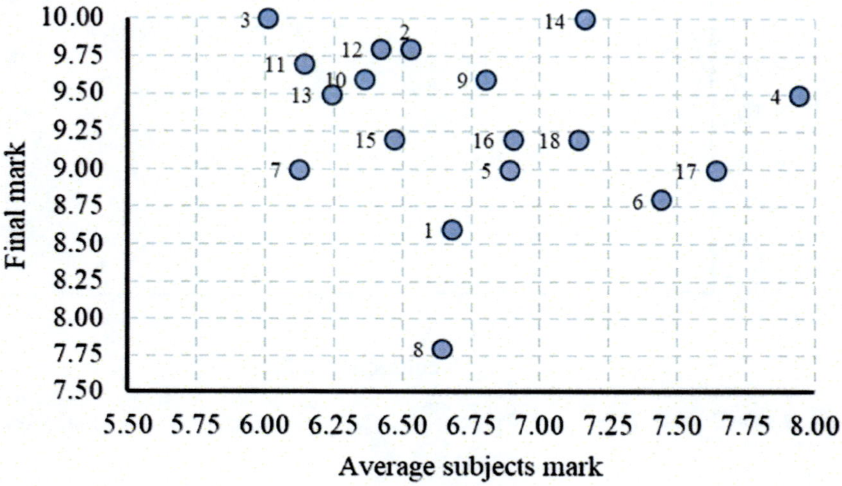

Fig. 1. Final mark of the Degree Project (DP), against the average mark of the rest of the subjects of the Degree.

It can be observed that many cases fall far from the diagonal, which would represent consistency between both. Although this could seem strange, the explanation is that the nature of the subjects of the curriculum is intrinsically very different from that of the DP. Whereas subjects imply learning new concepts, organised and focused on one topic, the Project is applied and multidisciplinary by definition. Whereas subjects are based largely on individual learning, the Project is carried out with careful supervision, and usually follows a progressive, step by step approach. Although many students may perform equally on both contexts, it is also natural that some students reverse roles. We can observe how good academic trajectories do not excel in the DP (cases 4, 6, 17), and that there are six cases with trajectories below 7, with final marks greater than 9.5.

The team had a special interest in establishing the relation between the quality of the work and the final mark, the motivation and the final mark, and the dedication of the final mark. However, our intuition was that the strongest correlation would occur between the quality of the presentation and the final mark.

As these variables are mutually dependent, it is natural to expect there be some correlation between them. For example, a student who has put a high dedication into the work, might have invested the greater part in the technical side of the project, in drafting the actual document, or in preparing the presentation, and this choice might have a more or less direct over the final mark. Indeed, the correlation can be observed in the results, which show very similar distributions for the final mark dependence with the quality of the work (Fig. 2), motivation (Fig. 3), and dedication (Fig. 4).

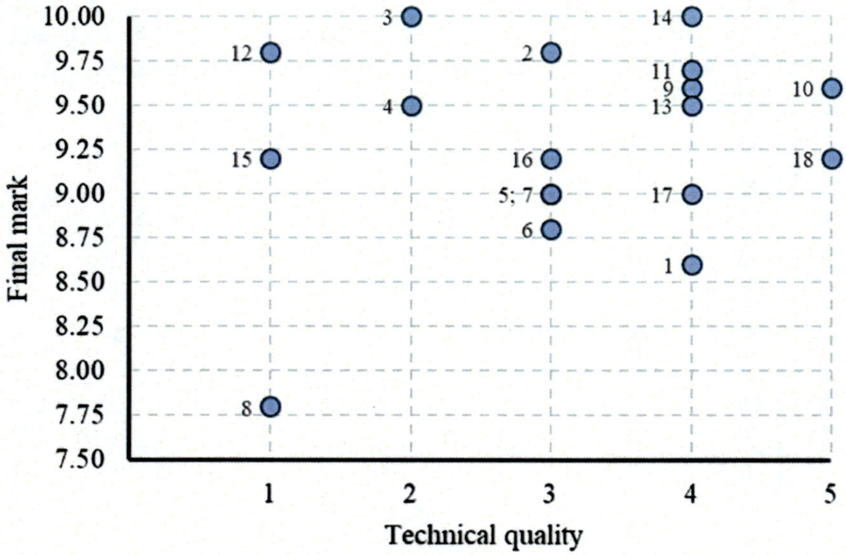

Fig. 2. Final mark of the Degree Project (DP), against the technical quality of the work.

Figure 2 represents the final mark of the Degree Project (DP), against the technical quality of the work. Although a number of points appear in the diagonal, showing certain correlation, it is remarkable the degree of dispersion in number and measure.

Figure 3 shows the relation between final mark of the Degree Project (DP) and the student's motivation. Again, cases near the diagonal show certain correlation, but motivation is not clearly defining.

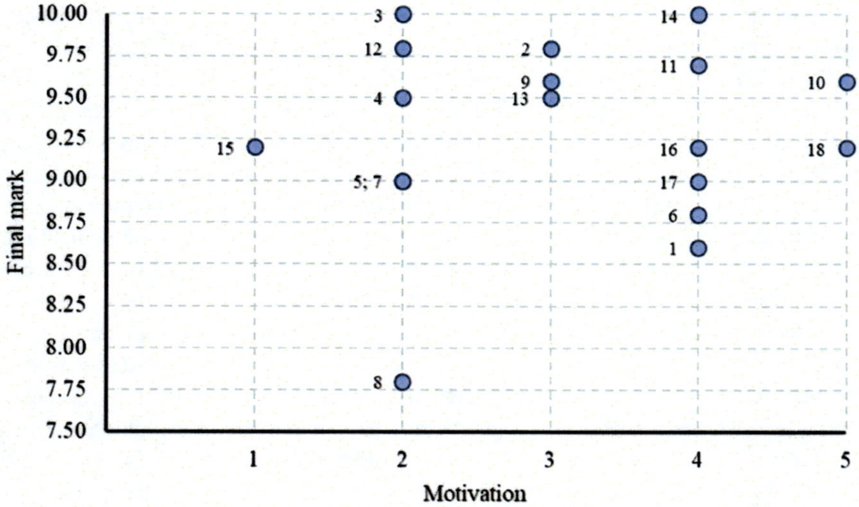

Fig. 3. Final mark of the Degree Project (DP), against the student's motivation.

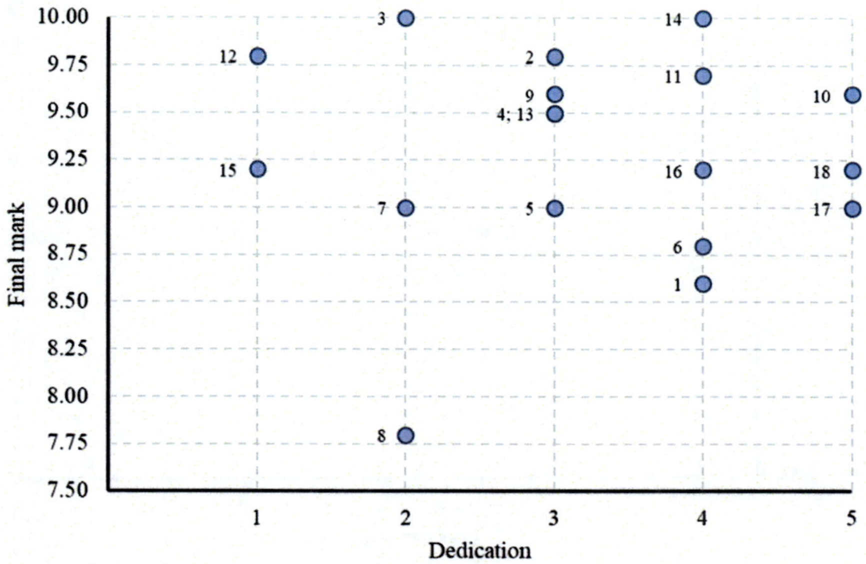

Fig. 4. Final mark of the Degree Project (DP), against student's dedication.

Final mark of the Degree Project (DP) against student's dedication have been represented on Fig. 4. Although indeed there is a certain mutual dependence, dedication does not guarantee a high final mark. Cases 1 and 17 are especially representative. It is perhaps how this dedication is reflected on the different aspects of the project that could be more telling.

However, the clearest correlation is observed between the quality of the presentation and the final mark (see Fig. 5). This deserves a further analysis and explanation. The range of final marks starts at 7.5, because projects under that level are usually prevented from being presented by their own supervisors. Good marks would start at 9, and high marks would start above 9.5, for exceptionally good projects.

On the other hand, a presentation of quality 3 is expected to be well structured, include meaningful graphs, and correct sections of result discussion and conclusions. Assessment of management aspects: dedication estimates, Gantt charts and project cost estimates are also expected. The capacity of the student to sustain a debate with the Panel on any of these aspects, with well-based arguments and scientific correction is a must. The capacity to react to broad scope questions or to develop on questions slightly off the work topic would lead to high marks. All these are included on the communication skills. The Fig. 6 shows a strong correlation between the final mark and the communication skills. The pattern is totally consistent with that of Fig. 5.

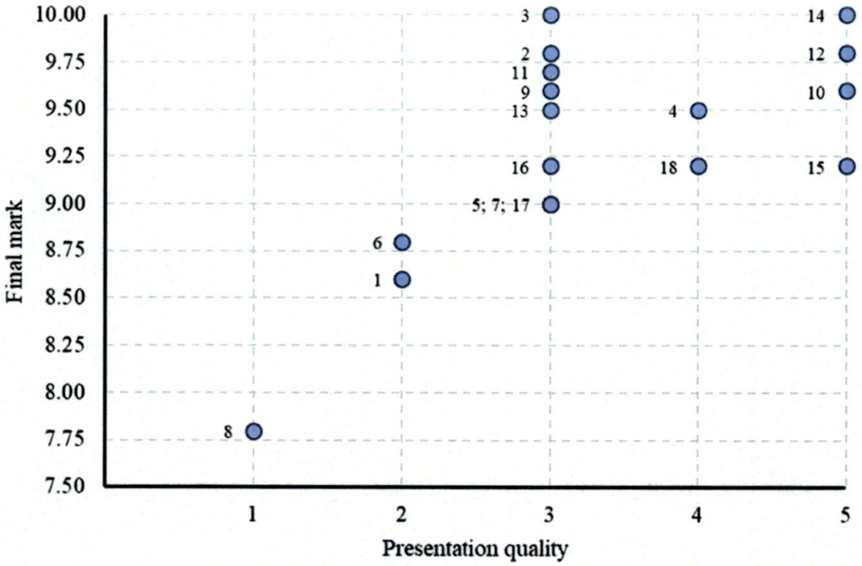

Fig. 5. Final mark of the Degree Project (DP), against the quality of the presentation to the Panel.

The Fig. 5 clearly shows that a poor presentation leads to low marks and a correct presentation (3) leads necessarily to a good mark (9), regardless any other factor. This defines a quadrant: from 3–5 in the horizontal axis and 9–10 in the vertical axis, which is guaranteed by an average or good presentation. Within this quadrant, however, there exists great variability; finding a pattern is difficult.

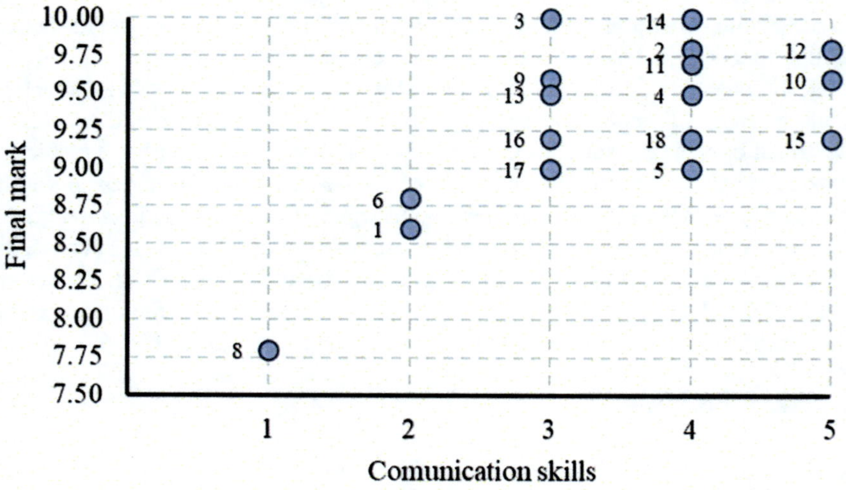

Fig. 6. Final mark of the Degree Project (DP), against the communication skills of the student.

Case 14 is especially consistent: an outstanding student, with outstanding technical and communication capacity, attitude, and sufficient self-confidence to deliver an excellent presentation, awarded with a final mark of 10. However, there are 6 more projects with 9.6 or above. Three of these, cases 9, 10 and 11, seem justified. The remaining three: 2, 3 and 12, cannot be explained by the variables observed here, and may fall within the uncertainty of the evaluation system itself. It is remarkable that the quality of the presentation of cases 2 and 3 is average and the rest of parameters is well under average, their final mark being outstanding. The pair of cases 12 and 15 in particular deserves special attention, both of them being practically identical, except for a slight difference in their motivation, even though both ranked especially low.

Some other cases are worth mentioning too. Case 17 corresponds to a specially hard-working student, with excellent attitude and organizational skills, whose work lead to a paper presented at an international congress. The quality of the presentation was average, which is the only justification for the not-exceptional final mark. In this case, a shy person with moderate or low self-confidence was penalized by the system.

Case 4, with a good presentation and a good mark, leads to further thought. This is the case of a student with a brilliant mind, and excellent technical and communication capacity. The student understood complex problems and made decisions very quickly, but in the end produced good but never excellent results. This indeed was consequence of a dedication that never exceeded average, and of an extremely high self-confidence that led the student not to follow advice from supervisors and experts.

5 Conclusions

A clear correlation between the quality of the oral presentation and the final mark of Degree Projects has been established over an 18 project sample, all supervised by the same teacher. Although out of the scope of this study, this is in agreement with the perception of the results by the projects supervised by the rest of the team.

Average presentation leads to marks equal or above to 9, regardless other factors, which indeed comes to show the importance of communication and other horizontal skills on the final mark. In fact, within this region above 9, clear patterns are very difficult to identify. In fact, the ordering of final marks between different projects is often inverted with respect to the actual traits of the students and their work. This shows a certain degree of uncertainty, logical as the sample includes projects presented on different years and judged by different panels. But it also shows the importance of other factors such as self-confidence, shyness and articulation.

This must not lead to think that the actual quality of the work is uncorrelated with the actual mark, only that there are other equally and more significant factors which may override or alter this correlation in the final mark.

A major conclusion is that, in accordance with the EHEA paradigm, horizontal competences and soft skills are indeed integrated in the Degree Project system. It is perhaps their weight in the overall evaluation, the weight of the actual work previous to the oral presentation and the controllability of the evaluation system that would require further thought.

References

1. ETSII: ANX-PR/CL/1/001-01 Guía para la elaboración y defensa del Trabajo Fin de Grado y Trabajo Fin de Máster. https://www.etsii.upm.es/
2. ENAEE: European Network for Accreditation of Engineering Education. https://www.enaee.eu/. Accessed Feb 2020
3. ETSII: Normativa de Trabajos Fin de Grado y Fin de Máster en Titulaciones Oficiales de la Escuela, July 2012. https://www.etsii.upm.es/
4. ETSII: ANX-PR/CL/1/001-03 Rúbricas para la evaluación de TFG y TFM por parte de tutores y tribunals. https://www.etsii.upm.es

Developing Engineering Skills in Secondary Students Through STEM Project Based Learning

Miguel Ángel Queiruga-Dios[1(✉)], Emilia López-Iñesta[2], María Diez-Ojeda[1], María Consuelo Sáiz-Manzanares[3], and José Benito Vázquez Dorrío[4]

[1] Facultad de Educación, University of Burgos, c/Villadiego, 1, 09001 Burgos, Spain
maqueiruga@ubu.es
[2] Department of Didactics of Mathematics, University of Valencia, Av. Tarongers, 4, 46022 València, Spain
[3] Facultad de Ciencias de la Salud, University of Burgos, Comendadores, 09001 Burgos, Spain
[4] Escuela de Ingeniería Industrial, University of Vigo, Lagoas-Marcosende, 36310 Vigo, Spain

Abstract. The development of 21st century skills allows the training of competent students prepared for work and social life. The improvement of these skills, necessary for engineering students and future engineers, can be favored with training in Secondary Education Stage. Specifically, the Project-Based Learning (PBL) methodology includes the performance of different activities that allow the acquisition of both conceptual and procedural skills. This article describes the relationship between the skills that high school students develop when they work with PBL and the competences (ABET engineering skills) they acquire. In addition, an objective evaluation of the learning outcomes of 129 Secondary students who carried out 34 projects is made and these results are related to the skills needed to achieve them.

Keywords: Skills · Secondary Education · Engineering education

1 Introduction

In the last decades of the twentieth century, the changes that were expected to occur in society and in the demands of labor, favored by the great technological and social development, brought forward, likewise, changes in Higher Education to prepare students for the new challenges of the 21st century [1–10]; training responsible, informed and committed people to improve society. Therefore, training that goes beyond the cognitive domain of disciplines and gives relevance to competences is necessary [10, 11]. These competences also facilitate access to the world of work [12], knowing that there are competences that are valuable in all careers, whatever the work area [13]. It is precisely the lack of these competences in the preparation of workforce

Á. Herrero et al. (Eds.): ICEUTE 2020, AISC 1266, pp. 257–267, 2021.
https://doi.org/10.1007/978-3-030-57799-5_27

which causes a shortage of STEM (Science, Technology, Engineering and Mathematics) professionals, as it is warned from industry [2, 8, 14–16]. It is, therefore, a political and economic objective to prevent the skills gap in people who intend to access the labor market by promoting "competences and motivations that will enable them to be flexible and adaptable in response to economic and organisational shocks" [14, p. 104], without this being detrimental to the achievement of a scientific literacy of society [17, 18]. This must imply an imperative change in the approaches of educational institutions at the primary, secondary and post-secondary levels, which, to a large extent, are products of the technological infrastructure and social circumstances of the past [19].

This article describes how STEM projects can be carried out from Secondary Education using the Project-Based Learning (PBL) methodology so as to favor the acquisition of skills that allow students to develop (obviously, adapted to their level) competences that an engineering student must have and which, as indicated, are mostly transferable to other professions. For this, the framework of skill and attribute development validated by Fisher, Bagiati and Sarma [20] is used.

In Europe, the Tuning Project addressed the mission of expressing the learning results obtained by students in competences acquired by them [21], as well as determining which competences are valuable for any student, generic or transversal (transferable skills), and which ones are more related to the specific knowledge of the disciplines. While the specific competences refer to the competence domain of the substantive content of each degree, the generic competences have an interest in preparing students for their work and social future, and are divided, in turn, into instrumental competences (cognitive, methodological, technological and linguistic abilities), interpersonal skills (social interaction and cooperation) and systemic competences (combination of understanding, sensibility and knowledge) [22].

These generic skills are strongly interlinked (to achieve systemic competence it is necessary to develop instrumental and interpersonal skills firstly, for example) and, depending on the different degrees, it is possible that some generic skills are more important than others, according to the context of professional practice [21–23], but in any case, they are developed linking to the acquisition of specific competences [21, 24]. The generic competencies defined in the Tuning Project [21], and which are subsequently reviewed in Tuning Latin America [25], serve as a reference for all university degrees. Subsequently, specific competencies have been defined for some qualifications within the framework of the Tuning Project (e.g. Civil Engineering).

There are different institutions and agencies that intend to determine what the specific competences that engineering graduates must achieve and that are necessary to achieve the accreditation of study programs, are. One of them is Accreditation Board of Engineering and Technology (ABET) [26, 27] which describes the desirable outcomes of an engineer's training, designed to encourage development of "soft" engineering skills (d, f, g, h, i y j) in addition to the "hard" skills (a, b, c, e, k); standards that have been known as EC2000 [28]: (a) an ability to apply knowledge of mathematics, science, and engineering; (b) an ability to design and conduct experiments, as well as to analyze and interpret data; (c) an ability to design a system, component, or process to meet desired needs within realistic constraints such as economic, environmental, social, political, ethical, health and safety, manufacturability, and sustainability; (d) an ability to function on multidisciplinary teams; (e) an ability to identify, formulate, and solve engineering problems; (f) an understanding of professional and ethical responsibility; (g) an ability to communicate effectively; (h) the broad education necessary to understand the impact of engineering solutions in a global, economic, environmental, and societal context; (i) a recognition of the need for, and an ability to engage in life-long learning; (j) a knowledge of contemporary issues; and (k) an ability to use the techniques, skills, and modern engineering tools necessary for engineering practice.

Another proposal is the so-called CDIO Proposal (Conceive, Desing, Implement and Operate) Syllabus, originally conceived by MIT [29, 30]. CDIO classifies learning outcomes into four categories: technical knowledge and reasoning, personal and professional attributes, inter-personal skills (teamwork and communication) and specific competencies of the engineering profession; and, from these, specify different sub-categories [30].

International Project Management Association, IPMA [31], in turn, classifies competencies into three categories: technical competencies, behavioral competencies and contextual competencies, which encompass a total of 46 competencies and whose standards are focused on improving the management and success of projects [32].

The competential organization of CDIO (Syllabus) is more rational and more detailed than ABET, but both are in tune, and the IPMA encompasses all others [33]; and, as it has been shown, in the Tuning Project only generic competencies are defined, although, as indicated, there are developments of specific competences for some branches of Engineering [25].

Among the skills of the 21st century, [20, 34] selected, from a study conducted on 135 articles, 20 of them as related to the skills that may intervene in the development of the skills of an engineer according to ABET criteria. These are: Civic Responsibility, Creativity, Critical Thinking, Cross-Cultural Skills, Disciplinary Knowledge, Ethics, Global Awareness, Humanitarianism, Interpersonal Communication, Memory, Networking, Organizational Management, Problem Solving, Public Speaking, Self-Confidence, Self-Direction, Strategy, Teamwork, Time Management and Written Communication. The development of these skills can promote the development of relevant technical and non-technical skills in the context of engineering education [20, 34].

Once the competential structure of the learning that students must acquire is defined, it is necessary to define the methodologies that lead to the acquisition of those competences [21], whereas these competences will be achieved by carrying out appropriate learning activities [22], having to be dynamic methodologies that involve the student [24, 35]. PBL is among these methodologies suitable for the achievement of competencies in engineering students [36–39]. In addition, the PBL methodology can provide students with the opportunity to apply STEM concepts [8, 40] through the activities carried out during the development of the project (design, problem solving, research, cooperative work, etc.); at the same time, students learn to work autonomously [41, 42]. On the other hand, the PBL methodology increases student performance [43–45]. Although there are different frameworks that detail the phases or stages that the PBL methodology follows, the so-called acPBL (Adaptive and Creative Problem-Projects Based Learning) [42] will be used in the context of this article, phases of which are briefly described below:

- Approach and motivation. The teacher raises the question: *What do you want to investigate?* Students have to choose a general issue for their study.
- Make teams. Students choose which classmates will form the team, taking into account those with whom they wish to work on the same subject and those with whom they can more easily meet.
- Brainstorm. Students specify what aspects of the topic they want to investigate. They make a global outline of the project by outlining what products they will generate and even what elements of dissemination they will develop. The first roles within the team are nominated, they decide who will be the coordinator and who the secretary.
- Specification. Students specify the structure of their project: *what?*, *how?*, *when?* and where? they will investigate. The tasks are also specified, and each task is assigned to a person in charge. All agreements are written on the project roadmap, which also serves as a self-assessment.
- Development. Students work according to the tasks they have defined and documented, sharing the resources they find and the content they generate. They reach a domain over the content.
- Investigation. Mastery over the content allows students to ask themselves new questions, open new lines of research and establish contact with expert researchers, institutions, industry, business, etc.; that act as external educational agents.
- Dissemination. Students design the dissemination elements they consider most appropriate: conferences to other courses, participation in scientific fairs or congresses, use of social networks and blogs, classroom exhibitions, creation of models, etc.
- Review. Although the review is carried out during the entire development of the project, once completed, the difficulties and problems arisen, alternative paths, possible improvements or changes in the project, etc. are critically analyzed.

In this methodology, the *Make Decisions* skill is especially relevant: make informed decisions using appropriate tools and resources and use multiple processes and diverse perspectives to explore alternative solutions [46], since the student has to choose: a) the research topic, related to their interests and curiosity; b) the work team, thinking that it is adequate to achieve success in the project; and c) the scope of the project, in terms of research depth and dissemination of the results, using the most appropriate means to the topic; also establish new lines of investigation.

2 Research Objectives and Methodology

2.1 Research Objectives

The objectives of this research are as follows:

- Relate skills whose development is favored with the completion of projects in Compulsory Secondary, according to the methodology described, acPBL [42], with the outcomes standards established by ABET [27].

- Show the evaluation of the projects developed by Secondary students as an indicator of achievement in the acquisition of skills.

2.2 Sample

The work sample is made up of 129 students (74 girls and 55 boys) from Compulsory Secondary School (14–16 years), forming 34 teams that developed a project following the PBL methodology described, enrolled in an educational center in the downtown area of a city in northern Spain of medium-high socio-economic level.

2.3 Development

To address the first research objective, the skills selected by [34] were considered, adding to this list the *Decision Making* skill that appears in different 21st skills frameworks [46, 47] and accreditation systems [48], and although it is related to others (e.g. Critical Thinking, Problem Solving, Self-Confidence or Strategy), strengthening it, is relevant for the future of the student as a participatory citizen [49, 50]. On the other hand, the *Humanitarianism* skill will not be considered since, usually, its development is not favored in the PBL in Secondary Education. Each team carried out a project and the following outcomes were evaluated (using a scale from 1: *large deficiencies*, to 4: *excellently achieved*): O1. *Original idea*: degree of originality of the idea on which they will investigate or in the approach they will use to address it. O2. *Cooperative work*: degree in which the student works cooperatively with the rest of the classmates, assumes responsibilities, puts effort into developing his role; and the equitable distribution of tasks. Cooperative work implies the appearance of new roles every time a new task appears. In addition, students use a roadmap where they write the agreements and the paths to follow. They also reflect on the results of their decisions. O3. *Technical report*: each team prepares a technical and scientific report explaining their research project (including what they have learned in their contact with the external educational agents, experts they have interviewed, etc.). O4. *Project presentation*: students prepare a presentation of their project and expose it in the classroom (as in a congress). In this exhibition, the communication skills, scientific rigor, the elements used and the cooperative organization among the group members are evaluated. O5. *Final product* (technical quality): students make a product that supports their presentation. It can be a set of experiments, a model, a video movie, etc.; adjusted to the type of project they carry out. O6. *Final product* (originality): Students develop a creative final product (not a copy or reproduction of usual objects or experiences). O7. *Participation in scientific fairs*: Students defend their projects in scientific fairs, for which they must develop different exhibition elements, such as posters, leaflets, stickers, etc. O8. *Social media dissemination*: students use social networks (eg, Twitter, Instagram, blogs) to share what they are learning and discovering. In addition, they look for related, current news to summarize, comment, share, etc. It is also sought, depending on the topic, to analyze the environmental impact and the social impact of the research topic.

In order to relate the results of the tasks developed during the project with the skills that are potentially being developed, 8 experts teachers from Secondary and University were interviewed. Table 1 shows the connection between these evaluated items and the skills that students have developed to generate those results, according to the experts consulted.

Table 1. Skills whose development is favored with each PBL outcome.

Skill	O1	O2	O3	O4	O5	O6	O7	O8
Civic responsibility		X		X			X	X
Creativity	X		X		X	X	X	X
Critical thinking	X	X		X	X			
Cross-cultural skills		X					X	X
Disciplinary knowledge	X		X	X	X	X	X	X
Ethics	X	X	X			X		X
Global awareness			X				X	X
Decision making	X	X	X	X	X	X		X
Interpersonal communication		X	X	X	X	X	X	X
Memory				X			X	
Networking			X					X
Organizational management		X	X	X	X	X	X	X
Problem solving	X	X			X	X	X	
Public speaking				X			X	
Self-confidence	X	X	X	X		X	X	
Self-direction	X	X	X	X			X	X
Strategy		X	X	X	X	X	X	X
Teamwork		X	X	X	X	X	X	X
Time management		X		X	X	X	X	X
Written communication	X		X				X	X

As shown in Table 1, acPBL is a framework for students to develop relevant skills. Many of the products individually favor the acquisition of all selected skills; although each of the outcomes only makes sense within the PBL framework since these are related and connected to each other.

These skills can be related to the competences of the engineer according to ABET [20, 34], as shown in Table 2.

With respect to the second objective of the research, Table 3 shows the evaluation of the products developed by the students (4 is the maximum mark). The positive evaluation, and the relationship between the evaluated outcomes and the skills, confirm an achievement in the development of these skills of the 21st century, and connecting them with ABET [20, 34], shows the achievement of these engineering competences (at the level of secondary requirement).

Table 2. Relation between ABET outcomes and core skills. Adapted from framework skills of relevance to EC2000 [20 pp. 296–297, 34].

ABET outcome	Relevant skills
(a)	Critical Thinking, Problem Solving, Disciplinary Knowledge, Self-Direction, Memory
(b)	Creativity, Critical Thinking, Disciplinary Knowledge, Decision Making, Problem Solving, Written Communication
(c)	Civic Responsibility, Creativity, Critical Thinking, Cross-Cultural Skills, Disciplinary Knowledge, Ethics, Global Awareness, Decision Making, Problem Solving, Strategy
(d)	Critical Thinking, Problem Solving, Cross-Cultural Skills, Strategy, Ethics, Decision Making, Teamwork, Interpersonal, Communication, Time Management, Organizational Management
(e)	Creativity, Problem Solving, Critical Thinking, Decision Making, Self Confidence, Disciplinary Knowledge, Self-Direction
(f)	Civic Responsibility, Ethics, Global Awareness
(g)	Interpersonal Communication, Written Communication, Public Speaking
(h)	Civic Responsibility, Critical Thinking. Cross-Cultural Skills, Ethics, Global Awareness
(i)	Self-Direction
(j)	Civic Responsibility, Cross-Cultural Skills, Ethics, Global Awareness
(k)	Critical Thinking, Self-Direction, Disciplinary Knowledge, Decision Making, Strategy, Problem Solving, Teamwork, Public Speaking, Self-Confidence, Written Communication

Table 3. Team outcomes evaluation.

Outcome	Average score	Typical deviation
Original idea	3.00	0.89
Cooperative work	3.09	0.75
Technical report	3.21	0.73
Project presentation	3.35	0.65
Final product (quality)	3.35	0.69
Final product (originality)	3.12	0.84
Social networking	3.06	0.98
Participation in scientific fairs	79.4% of the projects	

3 Discussion and Conclusions

The bibliography shows the skills and competencies that an engineer must develop and the importance of this development both professionally and socially [51–57]. As evidenced in the results of this research, linking the activities carried out during the project with the development of engineering skills according to ABET [20, 27, 34], the described PBL methodology favors that Secondary students acquire engineering skills. In addition, taking into account that most of the skills are transferable [21], the PBL methodology prepares Secondary students to develop the necessary skills in any profession [36–39], to develop professionally and for labor, while acquiring scientific literacy and preparing for a life in society.

The acPBL methodology [42] favors the development of *Decision Making* skills, which makes the student choose the best way to solve a problem in each situation [47]. However, what makes the student stay on the path to achieving objectives is self-regulation [13]. Therefore, it is necessary to introduce the techniques of self-regulated learning and metacognition in the classroom [58, 59], which will allow the students to analyze their problem solving strategies [24].

One of the skills, *Humanitarism*, is not developed, in general, in PBL, so it would be convenient to systematically incorporate in this methodology a phase in which the product has an impact on the common good or on the sustainability of the environment; It would also be very appropriate to develop a phase of service-learning, which would make each carried out project have an impact on an asset for the community [60, 61], which would allow this skill to take a more important role.

References

1. Ananiadou, K., Claro, M.: 21st century skills and competences for new millennium learners in OECD countries (2009)
2. Bybee, R.W.: The case for STEM Education: Challenges and Opportunities. NSTA Press, New York (2013)
3. Herro, D., Quigley, C., Andrews, J., Delacruz, G.: Co-measure: developing an assessment for student collaboration in STEAM activities. Int. J. STEM Educ. **4**(1), 1–12 (2017). https://doi.org/10.1186/s40594-017-0094-z
4. Kelley, T.R., Knowles, J.G.: A conceptual framework for integrated STEM education. Int. J. STEM Educ. **3**(1), 1–11 (2016). https://doi.org/10.1186/s40594-016-0046-z
5. Larson, L.C., Miller, T.N.: 21st century skills: prepare students for the future. Kappa Delta Pi Record **47**(3), 121–123 (2011)
6. National Research Council (NRC): Learning science in informal environments: people, places, and pursuits. Committee on Learning Science in Informal Environments. Bell, P., Lewenstein, B., Shouse, A.W., Feder, M.A. (eds.) Board on Science Education, Center for Education. Division of Behavioral and Social Sciences and Education. The National Academies Press, Washington, DC (2009)
7. NRC: Education for life and work: developing transferable knowledge and skills in the 21st century. In: Pellegrino, J.W., Hilton, M.L. (eds.) Committee on Defining Deeper Learning and 21st Century Skills. Board on Testing and Assessment and Board on Science Education, Division of Behavioral and Social Sciences and Education. The National Academies Press, Washington, DC (2012)
8. NRC: STEM Integration in K-12 Education: Status, Prospects, and an Agenda for Research. The National Academies Press, Washington, DC (2014)
9. National Academy of Engineering (NAE): The Engineer of 2020: Visions of Engineering in the New Century. National Academies Press, Washington, DC (2004)

10. UNESCO: World Declaration on Higher Education for the Twenty-First Century. UNESCO, Paris (1998)
11. Herrera, A., Didriksson, A.: La construcción curricular: innovación, flexibilidad y competencias. Revista Educación Superior y Sociedad **10**(2), 29–52 (1999)
12. Díaz, C.R., López, O.: Estrategia de enseñanza-aprendizaje basado en problemas (ABP) para el logro de la competencia del proyecto Tuning: desarrollo de un planteamiento estratégico. Táctico y operativo en los estudiantes universitarios. Didáctica, innovación y multimedia, (34), 01–10 (2016)
13. NRC: Assessing 21st Century Skills: Summary of a Workshop. National Academies Press, Washington, DC (2011)
14. Centre for the Development of Vocational Training (Cedefop): Skill shortages and gaps in European enterprises: striking a balance between vocational education and training and the labour market. Cedefop reference series no. 102. Publications Office of the European Union, Luxembourg (2015)
15. Eurofound: Third European company survey (ECS) (2013). https://www.eurofound.europa.eu/data/european-company-survey. Accessed 22 Jan 2020
16. Rodríguez-Pose, A., Wilkie, C.: Innovation and competitiveness in the periphery of Europe. In: Huggins, R., Thompson, P. (eds.). Handbook of Regions and Competitiveness: Contemporary Theories and Perspectives on Economic Development, pp. 351–373. Edward Elgar Publishing, UK (2017)
17. Education Council (EC): Report from the Education Council to the European Council. "The concrete future objectives of education and training systems" Brussels, 14 February 2001. European Council, Brussels: (2001) https://goo.gl/b7AbMP. Accessed 22 Jan 2020
18. Council of the European Union (CEU): Recommendation of the European Parliament and of the Council, de 18 December 2006, on the key competences for lifelong learning (No. (2006/962/CE)). European Union, Brussels (2006)
19. Davies, A., Fidler, D., Gorbis, M.: Future work skills 2020. Institute for the Future for University of Phoenix Research Institute, 540 (2011)
20. Fisher, D.R., Bagiati, A., Sarma, S.: Developing professional skills in undergraduate engineering students through cocurricular involvement. J. Stud. Affairs Res. Pract. **54**(3), 286–302 (2017)
21. González, J., Wagenaar, R. (eds.): Tuning educational structures in Europe. University of Deusto, Bilbao (2003)
22. Curiel, M.L.M.: El proceso de Bolonia y las nuevas competencias. Tejuelo: Didáctica de la Lengua y la Literatura. Educación, (9), 19–37 (2010)
23. González, J., Wagenaar, R. (eds.): La contribución de las universidades al proceso de Bolonia. Deusto, Groningen (2006)
24. Bybee, R.W.: The BSCS 5E instructional model and 21st century skills. BSCS, Colorado Springs, CO (2009)
25. Tuning Latin America: Proyecto Tuning. Innovación Educativa y Social (2013). http://www.tuningal.org. Accessed 22 Jan 2020
26. Accreditation Board for Engineering and Technology (ABET): 2010–2011 Criteria for Accrediting Engineering Programs. Engineering Accreditation Commission. Effective for evaluations during the 2010–2011 accreditation cycle (2009)
27. ABET: Criteria for Accrediting Engineering Programs. ABET, Baltimore, MD: (2012)
28. Lattuca, L.R., Terenzini, P.T., Volkwein, J.F.: Engineering Change: A Study of the Impact of EC2000. ABET, Baltimore (2006)
29. Crawley, E.F.: The CDIO Syllabus. A Statement of Goals for Undergraduate Engineering Education (2001). http://www.cdio.org. Accessed 22 Jan 2020

30. Crawley, E.F., Malmqvist, J., Lucas, W.A., Brodeur, D.R.: The CDIO syllabus v2.0. An updated statement of goals for engineering education. In: Proceedings of 7th International CDIO Conference, Copenhagen, Denmark (2011)
31. International Project Management Association (IPMA): ICB-IPMA competence baseline version 3.0. IPMA, Nijkerk, the Netherlands (2006)
32. Nahod, M., Radujković, M.: The impact of ICB 3.0 competences on project management success. Procedia **74**, 244–254 (2013)
33. Palma, M., Miñán, E., de los Ríos, I.: Competencias genéricas en ingeniería: un estudio comparado en el contexto internacional. In: XV Congreso Internacional de Ingeniería de Proyectos. IPMA, Huesca, Spain (2011)
34. Fisher, D.R.: Educating engineers for the 21st century: a framework for skill development through co-curricular and extracurricular involvement (Doctoral dissertation). Massachusetts Institute of Technology (2013)
35. Fernández, T.E., Aubin, V.I., Guatelli, R., Bellani, M., Blautzik, L.: Influencia de la educación inversa en el aprendizaje y adquisición de competencias transversales. In: XIX Workshop de Investigadores en Ciencias de la Computación. ITBA, Buenos Aires (2017)
36. Córdoba, M.M.: Proyecto Integrador Corporación Universitaria Remington PICUR: Una estrategia innovadora para la formación de Ingenieros para el siglo XXI Piedad M. Metaute P. Giovanny A. Flórez O. 2. Desarrollo e innovación en ingeniería, 408 (2018)
37. Gómez, M.R.B., Esquiva, I.C.: Desarrollo de competencias mediante ABP y evaluación con rúbricas en el trabajo en grupo en Educación Superior. REDU: Revista de Docencia Universitaria **17**(2), 9 (2019)
38. Martínez, J.A., García, N.I.E., Hermosillo, C.A.O.: El proyecto como estrategia didáctica para desarrollar competencias en estudiantes de ingeniería. Pistas Educativas **39**(129) (2018)
39. Metaute, P.M., Flórez, G.A.: Diagnóstico sobre el desarrollo de competencias en los Ingenieros de Uniremington: Un aporte para la construcción de Estrategia Pedagógica, acorde a los retos de la ingeniería para el siglo XXI. Desarrollo e innovación en ingeniería, 304 (2017)
40. Queiruga, M.A., Diez, M., Saiz, M.C.: Análisis de la apreciación sobre la implicación y desempeño y las dificultades y aprendizajes en la realización de proyectos STE(A)M atendiendo al género en alumnos de secundaria. In: Pérez Fuentes, M.C. (ed.) Innovación Docente e Investigación en Ciencias de la Educación, vol. 68, pp. 989–998. Editorial Dykinson, Madrid, Spain (2019)
41. Larmer, J., Mergendoller, J., Boss, S.: Setting the Standard for Project Based Learning: A Proven Approach to Rigorous Classroom Instruction. ASCD, Alexandria, VA (2015)
42. Queiruga, M.A., Sáiz, M.C., Montero, E.: Problemas-Proyectos Adaptativos y Creativos en la enseñanza de las ciencias. Descripción de la metodología y apreciación de los estudiantes involucrados. Res. Educ. Learn. Innov. Arch. **23**, 1–23 (2019)
43. Sáiz, M.C., Escolar, M.C., Marticorena, R., García, C.I., Queiruga, M.A.: Aprendizaje basado en proyectos utilizando LMS: una experiencia en Ciencias de la Salud. In J.C Núñez., et al. Temas actuales de investigación en las áreas de la Salud y la Educación, pp. 739–746. SCINFOPER, Oviedo (2017)
44. Sáiz, M.C., Escolar, M.C., Marticorena, R., García, C.I., y Queiruga, M.A.: Formación del profesorado en Metodologías Activas desde Plataformas interactivas. In: Núñez, J.C., et al. (eds.) Temas actuales de investigación en las áreas de la Salud y la Educación, pp. 39–44. SCINFOPER, Oviedo (2017)
45. Sáiz, M.C., Montero, E.: Metodologías activas en docencia universitaria. Diseño de una asignatura de ciencias de la salud en la plataforma virtual. Universidad de Burgos, Burgos, Spain (2016)

46. Dede, C.: Comparing frameworks for 21st century skills. In: 21st Century Skills: Rethinking How Students Learn, vol. 20, pp. 51–76 (2010)
47. Voogt, J., Roblin, N.P.: 21st Century Skills Discussion Paper. Universiteit Twente, Netherlands (2010)
48. ENAEE: EUR-ACE Framewok Standards for the Accreditation of Engineering Programmes (2015). http://bit.ly/2OozZKS. Accessed 22 Jan 2020
49. Acevedo-Diez, J.A., Acevedo, P., Manassero, M. A., Oliva, J. M., Paixão, F., Vázquez, Á.: Naturaleza de la ciencia, didáctica de las ciencias, práctica docente y toma de decisiones tecnocientíficas. III Seminário Ibérico CTS no Ensino das Ciências, 23–30 (2004)
50. Avilés, R.M.H., Martínez, P.M.: La importancia de enseñar a pensar en el aprendizaje de la historia. Educar en el 2000 **9**, 34–40 (2006)
51. Capote, G., Rizo, N., Bravo, G.: La formación de ingenieros en la actualidad. Una explicación necesaria. Revista Universidad y Sociedad **8**(1), 21–28 (2016)
52. Castrillón, J.E.P.: Construcción de la competencia investigativa en ingeniería. Revista Educación en Ingeniería **13**(25), 12–19 (2018)
53. Consejo Federal de Decanos de Ingeniería (CONFEDI): Declaración Valparaiso" sobre Competencias Genéricas de Egreso del Ingeniero Iberoamericano. Universidad FASTA, Argentina (2014)
54. Murphy, M., Hawwash, K., Vigild, M., Fouge, X.: SEFI Position Paper on Engineering Skills (2016). https://www.sefi.be/publication/position-paper-on-engineering-skills. Accessed 22 Jan 2020
55. Nair, C.S., Patil, A., Mertova, P.: Re-engineering graduate skills–a case study. Eur. J. Eng. Educ. **34**(2), 131–139 (2009)
56. Scott, C.L.: The futures of learning 2: what kind of learning for the 21st century? In: Research and Prospective in Education. UNESCO, Paris (2015)
57. Suarez, L., Ramírez, J., Hurtado, S.M.: Enfoque Integral del Ingeniero del Siglo XXI: una Revisión de la Literatura. Revista Politécnica **14**(26), 9–18 (2018)
58. Sáiz, M.C., Queiruga, M.A.: Evaluación de estrategias metacognitivas: aplicación de métodos online. Revista de Psicología y Educación **13**(1), 23–35 (2016)
59. Sáiz, M.C., Queiruga, M.A., García, C.I., Montero, E., Rodríguez, J.: Observation of metacognitive skills in natural environments: a longitudinal study with mixed methods. Front. Psychol. **10**(2398), 1–13 (2019)
60. Carrillo, J., Arco, J.L., Fernández, F.D.: Investigando la mejora de la enseñanza universitaria a través del Aprendizaje-Servicio. Editorial Universidad de Granada, Granada (2017)
61. Hernández, A.M.: Aprendizaje-Servicio como elemento para el Fortalecimiento de la Calidad Académica. Universidad Valle del Momboy, Revista Semestral Dialogus (2009)

Towards an Innovative Learning of Chemical and Enzymatic Catalysis for Moroccan Secondary Students

Mohammed Amine Merzougui[1], Rachid Janati-Idrisi[1], Mohamed Dakkach[1,2], Mourad Madrane[1], Mohammed Laafou[1], and Fernando Aguilar[3](✉)

[1] Laboratoire Interdisciplinaire de Recherche en Ingénierie Pédagogique (LIRIP), ENS de Tétouan, Martil, Morocco
[2] Institut Supérieur des Professions Infirmières et Techniques de Santé (ISPITS), Tanger, Tétouan, Morocco
[3] Departamento de Ingeniería Electromecánica, Escuela Politécnica Superior, Universidad de Burgos, 09006 Burgos, Spain
Faguilar@ubu.es

Abstract. All information and communication technologies (ICT) bring innovative resources, not only for the transmission of information, teaching objects but also for the exploration of learning strategies for learners by promoting the development of skills. Among the most promising resources in the teaching of experimental sciences, we find interactive graphic animations intended for classroom or distance learning. These educational applications can create, model, animate and simulate real phenomena by creating models, diagrams or virtual experiences, allowing learners to understand the situation studied, formulate hypotheses, interpret data and facilitate the understanding of complex phenomena.

In order to improve the learning of chemistry catalysis course among Moroccan secondary students, in this work we have proposed a course in catalysis using different means of ICT, elucidated by an interactive graphic animation. The course is exported in different formats: Small Web Format (SWF) for the various materials that work with Windows and another version Android Package (APK) intended for the Android operating system. The results obtained, based on an anonymous questionnaire and a test, showed an improvement in learning among students who attended the course with ICT in comparison with other students who attended the same course, under the same conditions, but using the classic method of transmission pedagogy based solely on the school textbook and the workbook. We also think that the catalysis course that we have prepared could be used by secondary students from other countries, especially for Francophone education systems.

Keywords: Information and Communication Technologies · Learning of catalysis · Secondary students

Á. Herrero et al. (Eds.): ICEUTE 2020, AISC 1266, pp. 268–277, 2021.
https://doi.org/10.1007/978-3-030-57799-5_28

1 Introduction

Despite efforts in the field of science education, the impact of science education is limited compared to the educational intentions set by the official authorities [1–3]. To this end, developing countries have tried to correct their educational system through several solutions to achieve significant progress and to provide high-quality education; among the solutions offered are the integration and use of information and communication technologies (ICT).

Sugata (2013) conducted a 10-year study and demonstrates that learners can co-learn, alone, without teachers: that is, learning is built by itself with computer equipment and/or teaching materials [4]. He concludes that education is a self-organizing system, and where learning is the emerging phenomenon. This enables the use of ICT tools to facilitate the learning of complex phenomena [4].

Like all other countries Morocco has understood the importance of the use of ICT in its system, so it has created an educational committee, to find a regime or discipline to promote the integration of ICT into the Moroccan education system, which was crowned by the adoption of the "engineering" program in 2005 [5]. The objective is to find disciplines that put ICTs at the heart of any educational development, and that is committed to the design of an educational system that aims at evolution and innovation, by deploying new technologies for education and teaching.

In the same context, chemistry in the Moroccan education system is a field that has benefited too little from the pedagogical advances resulting from the use of ICT tools. There are many causes: lack of knowledge or skills in the use of these technologies, difficulty in accepting changes in working methods that have been proven for many years. All too often, courses are still presented by traditional teaching and learning methods, while there are many opportunities to present these courses in a captivating and attractive way through videos, animations, simulations, media and interactive documents. Indeed, innovative practices that focus on digital assets can partly address the quality and performance of digital education.

Due to its industrial and scientific importance, the development of catalysts is now one of the most studied topics in research. Catalysis, whether heterogeneous, homogeneous or microbiological (enzymatic), plays a key role in the development of chemistry [6]. Thus, a good understanding of the concept of catalysis is very interesting for learners for better use in practice.

In the course of our research in the field of education, we have discussed the problems faced by students in high school on the chemistry. We have found that there was some difficulties for high school students in studying digestion and the intervention of digestive enzymes in the simplification of foods because of a lack of understanding of the concept of catalysis. This is a real educational problem for students in learning different chemical and biological phenomena: digestion and absorption, kinetics, ATP synthesis, breathing, reproduction, muscular activity, fermentation…

In addition, our research in textbooks of chemistry, among Moroccan high school students, showed us that the concept catalysis is poorly formulated, because theses textbooks present, only in one page, some textual definitions of catalyst without any figure, image or practical work. Furthermore, we have observed the absence of use of the ICTs, which could influence negatively on the comprehension of the concept of catalysis, as we have showed previously in other study about the atom concept, in the textbook of secondary education in Morocco [7]. All of this led us to look for an alternative solution to help address this gap in teaching and learning of catalysis.

To help identify the true benefits of ICTs, and in particular, to assess the impact of ICT use on learners, our objective, in this work, is to implement online and remote training using ICTs and web tools 2.0, as well as to use all technological innovations for the education system to teach the concept of catalysis, taking as an example the use of graphic animations for easy and proactive access to learning.

2 Use of ICT in Chemistry Education

Different studies highlighted multiple interests in the use of ICT in science education, and in particular in chemistry education [8, 9]. Indeed, according to Beaufil [10] and Bryan [11], ICT can be teaching tools, laboratory tools (computer-aided experimentation (ExAO), data processing, etc.).

Computer instruments specific to chemistry education [8–11], those who logically come first, because they are experimental science, are linked to what is called ExAO. The latter involves the creation of computerized data acquisitions based on real-world experiences. Modeling is another form of specific use of ICT chemistry: this may involve dissecting as much as possible a chemical phenomenon to make it easier to disentangle concepts (conceptual model), to help students better understand and construct a mental representation [11]. The model can also be constructed in such a way as to be as close as possible to a phenomenological model, a process integrated into a scientific model on which the learner can conduct investigations; test his hypotheses, study relations and theories. Such simulation can be used in place of real-world experience and can be both conceptual and phenomenological.

The availability of current means, as well as their positive effects on chemical science learning, as reported by some research studies [8, 9] is expected to lead to significant use of ICT in the classroom, particularly in science education. However, international statistics and research by many authors have recently shown that the availability of ICT education resources for science, chemistry in particular, is far from rich [12, 13]. Thus, the use of ICT in education is not up to the expectations of developers, both quantity and quality.

According to Baek and Layne (1988) research on the place of ICTs in chemistry education is still a minority. Nevertheless, all stress the value of such means in helping students to understand complex phenomena, three-dimensional structures and spatial relationships between molecular objects [14]. As a result, the teacher may be led to propose computer models in the form of animations or simulations, thus giving the opportunity to see in a reduced or accelerated model, reactions that are not accessible to direct observation for reasons of time (e.g. very low speed of certain chemical reactions) or material (a lack of products that are mostly dangerous, prohibited to sell, or very expensive). The integration of ICTs into education is accompanied by a constant questioning of the expected and potential effects of these tools on learning [14].

Moroccan researchers have also studied the impact of ICT integration on teaching and learning of various scientific disciplines within the school in Morocco [15–23]. The results of their studies have shown that this integration can facilitate the learning of students in science. However, it faces several obstacles, difficulties, and constraints; like: Lack of high-performance development equipment and lack of connection among students to access online training.

On the other hand, the computer used to simulate or model a scientific phenomenon or experiment. It is an artificial reproduction of the phenomenon to be studied, it must not be ignored that the results obtained by the computer are programmed, and therefore do not take into consideration the complexity of reality and the difficulty of its perception. Animations are just models that simplify real-world experiences.

Thus, we will try through this study to answer the following question: Does the educational use of digital animation have a positive effect on the performance of high school students?

3 Case Study: Application of Active Learning in Chemical and Enzymatic Catalysis for Moroccan Secondary Students

Within the limited framework of this work, we have set two major goals:

- Assess the place and importance of using graphic animation plus the different means of ICT in understanding a concept related to chemistry science teaching, especially catalysis;
- Design, develop, and implement digital tools and materials in the teaching and learning process.

3.1 Preparation of an Active Catalysis Course

The catalysis course prepared, using ICTs, was intended for a sample of 60 students in second-year baccalaureate degree with an average age of 18 years-old, with learners representing classes in different school and institutions in the Fez-Meknes region (north central Morocco) during the 2018–2019 school year. The course is divided into three parts:

The first step, it's develop a catalysis course that meets pedagogical and didactic standards by adopting the investigative approach to manipulate practical work. Our pedagogical approach is described in a file 'Fiche pedagogique' in the following open-access link: https://drive.google.com/file/d/1JqLK4mqhQnOyTbTUgPtjyiebVccv0ZjE/view?usp=drivesdk.

The completion of a course in the format of a fun video, by the Wondershare Filmora softwarewith version 8.7.0, which contains the different parts of the catalysis course with attractive interfaces. For the implementation of this course we have chosen the well known E-Learning platform: MOODLE.

The third part is the graphical animation which contains three animations: Uncatalyzed reaction, chemically catalyzed reaction and enzyme catalyzed reaction.

The latter is exported in two SWF and APK formats, the first is for computers, and the second is for use by smartphones and android's users.

The animation we have done is a flash animation, created with Adobe Animate CC 2017 software. This animation is an ideal support allowing a good understanding of the phenomenon of catalysis, a good acquisition of skills which are sometimes difficult to acquire by the traditional course. Also, the animation makes the course more fun and easier.

All the material described above, it is developed by us, and to facilitate its consultation by all interested we give the following open-access link: https://drive.google.com/file/d/1JqLK4mqhQnOyTbTUgPtjyiebVccv0ZjE/view?usp=drivesdk.

3.2 Evaluation of an Active Catalysis Course

To provide answers and explanatory ideas related to the objectives assigned to this work (To help identify the true benefits of ICTs and particularly to assess the impact of the use of interactive graphics animation on learners), we have adopted two tools: a questionnaire and test. The questionnaire was intended for the sample of 60 high school students who attended our course using ICTs, while the test was done for these high school students and another sample of 60 high school students who took the same catalysis course, but with the traditional method; using only the textbook in the absence of ICTs. This test was carried out with the aim of comparing the effect of ICT on learning the concept of catalysis in these two types of samples. The test was done based on the language of teaching and the content of the proposed problem learning activity to compare the results of the samples. For a better comparison between the two samples (test group and control group), we have chosen the students for the two groups belonging to the same institution, the same course, the same teacher, the same study program, the same academic background and having the same grade in the three last physics-chemistry exams.

The construction of the questionnaire was carried out taking into account the scale of Likert. The questions asked were constructed in the way that we consider likely to achieve the objectives of our analysis. This is done by ensuring that each of the main axes is covered by a consistent set of questions:

- First axis: distance learning: importance and motivations;
- Second axis: quality of work: attractiveness, importance, and graphics;
- Third axis: improving work: Highlights and better.

The data collected is then analyzed with Sphinx (statistical analysis software).

The questionnaire and the test are available at the following address:

https://drive.google.com/file/d/1JqLK4mqhQnOyTbTUgPtjyiebVccv0ZjE/view?usp=drivesdk.

4 Results and Discussion

After reviewing the learners of the online course and collecting different responses from the learners, we achieved the following results (Fig. 1):

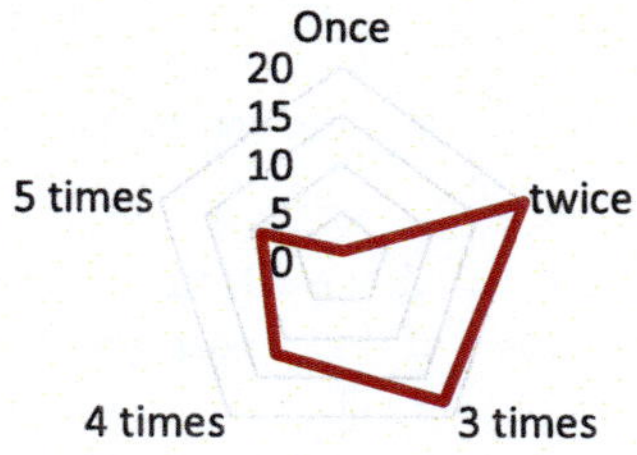

Fig. 1. Diagram showing the number of times the video is viewed.

The results about the many time consulting all content of the video, tell us that student can view the video multiple times without worrying about others' eyes to learn at their own pace. Research on ICT's contribution to learning finds that there are certain advantages to understanding in teaching ICTs: after a collective presentation that allows the progressive display of information, students can view these documents at their own pace from anywhere [24] (Fig. 2).

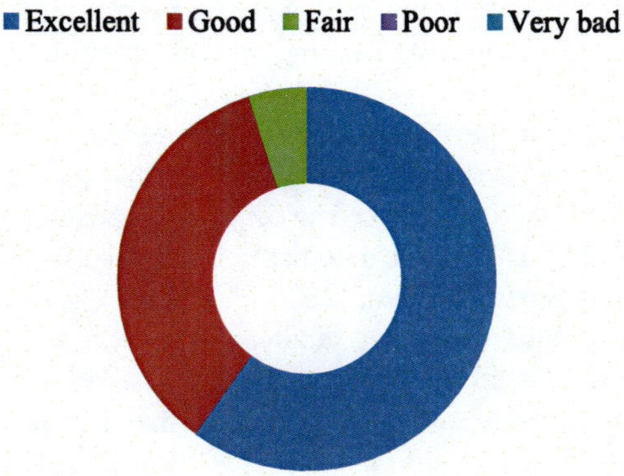

Fig. 2. Diagram showing learner satisfaction with the quality of the content of course with integrated ICTs.

Based on the results for the quality of the content, shown in Fig. 3, the majority of secondary students (36) who represent (60%) express their excellent satisfaction with the content presented using ICTs. This satisfaction mainly due to the use of technologies that facilitate the conceptualization of the concept such as catalysis. These results show the importance of the use of ICT in education, which has been approved by several previous studied [8–11].

Fig. 3. Chart showing balance between course and the graphical animation.

Based on the results shown in the above diagram, the majority of learners (51) who represent (85%) express satisfaction of this course and its equilibration.

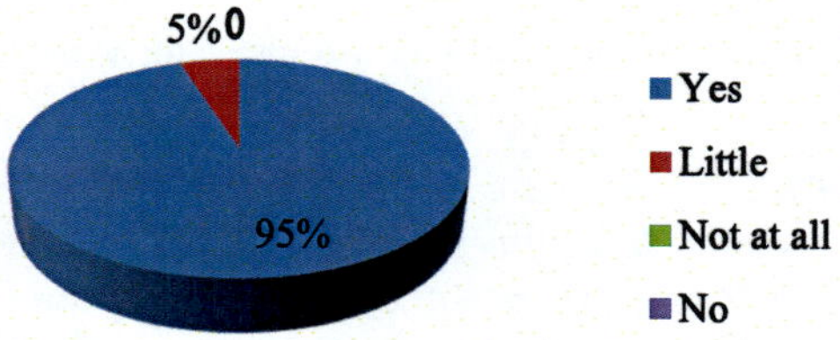

Fig. 4. Diagram showing the comprehension of the course content using ICTs.

Figure 4 shows that almost all of the students (95%) understood well the course content using ICTs. This result is in complete agreement with previous studies which affirm that the use of ICTs improves learning and the elucidation of various scientific phenomena [25].

Other results show that the majority of students (65%) are motivated to use the interfaces implemented in the course. And also tell us that (85%) of students consider the video capsule to be easy to understand so they have a positive impact to understand the phenomenon of catalysis by the excitement of their pleasures of knowledge using images, sound and animated interfaces.

According to the results of the test addressed to high school students after the presentation the course of catalysis with integrated ICTs. We have observed that 95% of the first sample (test group), students who studied the concept of catalysis using the various ICTs tools (animation and video capsule), have obtained a score of 15/20 or greater, which equals 75%, and which confirms that they well understood the course. By contrast, most second sample (control group) students who are limited with what he had studied in class with textbooks only (without ICT), do not have the means of the score, because 55% did not reach the qualifying score, which means that 33 students with a scoreless than 10/20.

Since the sample studied is small, we believe that our results cannot be generalized, however, these results can highlight the contribution of ICTs and especially interactive graphical animations in improving learners' learning quality, by overcoming some learning and understanding difficulties related to chemistry teaching and also improving learners' performance. The observations made on the performance of learners using the medium courses and interactive graphics animations are sufficiently decisive to draw a general conclusion. However, the present study revealed positive results in the Moroccan educational context that highlighted the positive impact of the pedagogical use of these educational devices on student learning in experimental disciplines.

Our study shows that equipment no longer poses a problem for this generation of young people, more than 95% of whom have at least one technology tool. The connection issue is also resolved as parents offer home-based connections. Their duration of use of various ICT tools is between 14 h and 20 h per week [26].

From the above, it is inferred that graphic animation has become a very interesting object in different disciplines, and several studies have been devoted to showing the importance of integrating animations into teaching and learning, and its impact on learners [27].

5 Conclusion

Catalysis is a very important concept in chemistry, it is an essential factor for understanding different chemical and biological phenomena, which is why the course has a heavy weight in chemistry books, but despite the efforts made by the Moroccan education system to give perfect conceptual design on this concept in the use of images and give examples. The results show the opposite because the courses are still presented to students by textual books only, while there are many opportunities to present these courses in a captivating and attractive way through videos, animations, simulations, and media and interactive documents. To make students very active by the excitement of their knowledge pleasures using images, sound and animated interfaces.

Graphical animation is a powerful tool for presenting the concept of catalysis, thanks to these educational applications that can create, model, animate and simulate real-world phenomena by creating models, diagrams or virtual experiences, allowing learners to understand the situation studied, formulate hypotheses, interpret data and facilitate understanding of this complex phenomenon.

By its specificities, the various ICTs tools complement the means available to teachers and students to implement these different aspects of real chemical activity. The use of ICTs helps to improve teaching practices in experimental science.

In short, the teaching and pedagogical process have gained added value through it.

References

1. Piaget, J.: Où va l'éducation. Gallimard, Paris (1972)
2. Giordan, A.: Une pédagogie pour les sciences expérimentales. Centurion, Paris (1978)
3. Astolfi, J.R., Cauziline-Marmèche, E., Giordan, A., Henriques-Christofidès, A., Mathieu, J., Weil-Barais, A.: Expérimenter: sur les chemins de l'explication scientifique. Privat, Toulouse (1984)
4. Sugata Mitra. https://www.ted.com/talks/sugata_mitra_the_child_driven_education. Accessed 23 Apr 2019

5. Mastafi, M.: Intégration et usages des TIC dans le système éducatif marocain : Attitudes des enseignants de l'enseignement primaire et secondaire. Adjectif.net (2013)
6. Ecole nationale supérieure de chimie de Rennes, https://www.ensc-rennes.fr/recherche/thematique-catalyse/. Accessed 26 December 2019
7. Ben mesaoud, M., Belfazga, M., Radid, M., Dakkach, M., Janati-Idrissi, R., Madrane, M.: Atome dans le manuel scolaire de l'enseignement secondaire au Maroc. Eur. Sci. J. **14**(28), 154–172 (2018)
8. Cox, M.J.: Information et apprentissage des sciences tendances dilemmes et conséquences pour l'avenir. Recherches en didactique des sciences et des technologies (6), 23–52 (2012)
9. Saltarelli, A., Saltarelli, W., Roseth, C., Grand, A.: Using multimedia and virtual simulations to enhance learning transfer in anatomy instruction. In: Koehler, M., Mishra, P. (eds.) Society for Information Technology and Teacher Education International Conference 2011, pp. 1663–1670. AACE, Chesapeake (2011)
10. Beaufils, D.: L'ordinateur outil de laboratoire en physique: quelles transpositions. INRP, Lyon (2005)
11. Bryan, J.: Technology for physics instruction. Sci. Contemp. Issues Technol. Teach. Educ. **6** (2), 230–245 (2006)
12. OECD Study Design The New Millennium Learners: ICT use in initial teacher training. http://www.oecd.org/docment/13/0,3746,en_21571361_49995565_41676365_1_1_1_1,00.html. Accessed 03 Jan 2020
13. Burgess, G., Young, B.A.: Technology walk into the 21st century: preparing future teachers in a changing landscape. In: Koehler, M., Mishra, P. (eds.) Society for Information Technology and Teacher Education International Conference 2011, pp. 912–918. AACE, Chesapeake (2011)
14. Baek, Y.K., Layne, B.H.: Color, graphics, and animation in a computer-assisted learning tutorial lesson. J. Comput.-Based Inst. **15**, 131–135 (1988)
15. Guennoun, B., Benjelloun, N.: Regards des étudiants sur l'intégration des TIC dans l'enseignement supérieur scientifique. Int. J. Technol. Higher Educ. **13**(1), 64–94 (2016)
16. Zerhane, R., Janati-Idirissi, R., Blaghen, M., Talbi, M.: Etude des besoins et des difficultés dans l'enseignement de l'immunologie, enquête dans quatre délégations du Nord - Ouest marocain. Prospective Pédagogique **24**, 38–39 (2002)
17. Janati-Idrissi, R., Zerhane, R., Khaldi, M., Aride, J., Blaghen, M.: L'environnement hypermédia au service de l'apprentissage et de l'enseignement de la Biochimie. Revue de l'EPI (2003)
18. Aboussaouira, T., Chehab, F.: Méthodes pédagogiques utilisées en auto apprentissage de biologie cellulaire. In: 17ème Journées Universitaires Francophones de Pédagogie des Sciences de la Santé, CIDMEF, Lille (2008)
19. Ahaji, K., El Hajjami, A., Ajana, L., El Mokri, A., Chikhaoui, A.: Analyse de l'effet d'intégration d'un logiciel d'optique géométrique sur l'apprentissage d'élèves de niveau baccalauréat sciences expérimentales. EpiNet (2008)
20. Berrada Fathi, W., Chraïbi, S.: Comparaison d'expériences d'insertion de dispositifs techno-pédagogiques dans le milieu universitaire marocain. In: 26ème Congrès International de l'Association Internationale de Pédagogie Universitaire, AIPU, Rabat (2010)
21. Pelgrum, W.-J., Law, N.: Les TIC et l'éducation dans le monde, tendances, enjeux et perspectives. International Institute for Educational Planning, Paris (2004)
22. OCDE. http://www.oecd.org/dataoecd/49/34/1911915.pdf. Accessed 29 Dec 2019
23. Karsenti, T., Tchameni, N.: Intégration pédagogique des TIC: stratégies d'action et pistes de réflexion. Centre de recherches pour le développement international, Ottawa (2009)

24. Bullat-koelliker, C.: Les apports des TIC.à l'apprentissage ce qu'en pensent les enseignants qui utilisent les ateliers d'informatique avec leurs élèves. Université de Genève, Genève (2003)
25. Huk, T.: Who benefits from learning with 3D models? The case of spatial ability. J. Comput. Assist. Learn. **22**, 392–404 (2006)
26. Brotocorne, P.: Les outils numériques au service d'une participation citoyenne et démocratique augmentée. http://www.ftu-namur.org/fichiers/TIC_participation_citoyenne.pdf. Accessed 17 Apr 2020
27. Janati-Idrissi, R., Zerhane, R., Khaldi, M., Madrane, M., Aride, J., Blaghen, M., Talbi, M.: Contribution des animations et des simulations informatiques dans l'enseignement et l'apprentissage de l'Enzymologie. In: Congrès de l'Association Internationale de la Pédagogie Universitaire, Université Cadi Ayad, Marrakech (2004)

The International Mobility Strategy of the ENSA El Jadida in Morocco: The Case Study of Student's Exchange with Spain

Mohamed Lifi[1(✉)], Sanae El Hassani[2], and Fernando Aguilar[1]

[1] Escuela Politécnica Superior, Universidad de Burgos, Avenida Cantabria s/n, 09006 Burgos, Spain
mlifi@beca.ubu.es

[2] Ecole Nationale des Sciences Appliquées d'El Jadida, Université Chouaib Doukkali, Route d'Azzemour, Nationale N°1, 1166 El Jadida, Morocco

Abstract. The call for more student and scientist exchanges, because of the globalization processes, is the present trend in science and research institutions worldwide. Nowadays, many universities and institutions aim offering an opportunity for transferring knowledge and experience to their students, and have launched international cooperation programs. Institutional programs are usually centered in collaborative projects amongst higher education and research organizations, with the aim of strengthening their performance capacity. This contribution presents the initiative of internationalization of the Ecole Nationale des Sciences Appliquées d'El Jadida at the Chouaïb Doukkali University in Morocco. The aim is to facilitate the professional integration of students in an internationalizing environment and to promote the international dimension of research activities. The cooperation for exchange of its undergraduate engineering students with the University of Burgos in Spain is described. For six years, eleven students of the engineering cycle Energy and Electrical Engineering have performed an internship in summer as part of a collaboration agreement with the Energy Engineering research group at the University of Burgos. The experience demonstrates the improvement of scientific, technical, operational, research and personal competences of those students participating in the experience. the institutional support of the respective universities and the need of leading persons within the respective universities able to carry out the experiences in the long term have been estimated as the key factor for success.

Keywords: International cooperation · Problem-based learning · Engineering education

1 Introduction

One of the key factors to boost socio-economic growth in many countries is the cooperation programs in the field of education. Mobility of people and ideas at different educational levels are the main characteristic of such programs. Cooperation programs use to focus mainly on providing continuing education and training for young academic

Á. Herrero et al. (Eds.): ICEUTE 2020, AISC 1266, pp. 278–287, 2021.
https://doi.org/10.1007/978-3-030-57799-5_29

staff. Frequently educational institutions report their experiences and results reached in the recent times [1]. The call for more student and scientist exchanges, because of the globalization processes, is the present trend. Then, universities aiming to become modern, top-quality research and innovative should integrate international exchange in their respective strategies. At present, many universities and institutions aim offering an opportunity for transferring knowledge and experience, and have launched international cooperation programs [2, 3]. Engineering education institutions are also frequently concerned by this approach [4, 5].

Both institutional and individual exchange programs could be possible. The fellowship/scholarship programs are addressed to help candidates from developing countries to obtain a diploma or degree abroad. While, institutional programs are usually centered in collaborative projects amongst higher education and research organizations, with the aim of strengthening their performance capacity. The higher education cooperation programs present as common objectives the capacity building (teaching, research), organizational and/or institutional development, networking between partners and research collaboration. Individual and institutional objectives are to be combined in many higher education cooperation programs.

2 The Ecole Nationale des Sciences Appliquées d'El Jadida and the Promotion of Students Mobility

The Ecole Nationale des Sciences Appliquées d'El Jadida (ENSAJ) [6] is a public institution dedicated to the training of state engineers, continuing education and scientific research.

Fig. 1. The main building of the Ecole Nationale des Sciences Appliquées d'El Jadida at the campus of the Chouaïb Doukkali University.

Its creation at the Chouaïb Doukkali University is part of a strategy to diversify the training provided in the university's institutions taking into account the new requirements of the national and international job market, to strengthen the gateway system by providing students with new opportunities for reorientation and to intensify interactions with socio-economic operators through technology transfer processes, scientific research and business creation. The main building is shown in Fig. 1.

The training course at ENSAJ is spread over five years of study. It is structured in two cycles: a cycle of two preparatory years of 2 years (or 4 semesters) and an engineering cycle of 3 years (or 6 semesters), following the picture of Fig. 2. The cycle of the two preparatory years, which covers the general scientific teachings, is structured according to an educational teacher's notebook designed and developed in close collaboration with the ENSA network of Morocco. The engineering cycle currently dispenses four courses in the following specialties:

- Information and Communication Systems Engineering
- Energy and Electrical Engineering
- Computer Engineering and Emerging Technologies
- Industrial Engineering

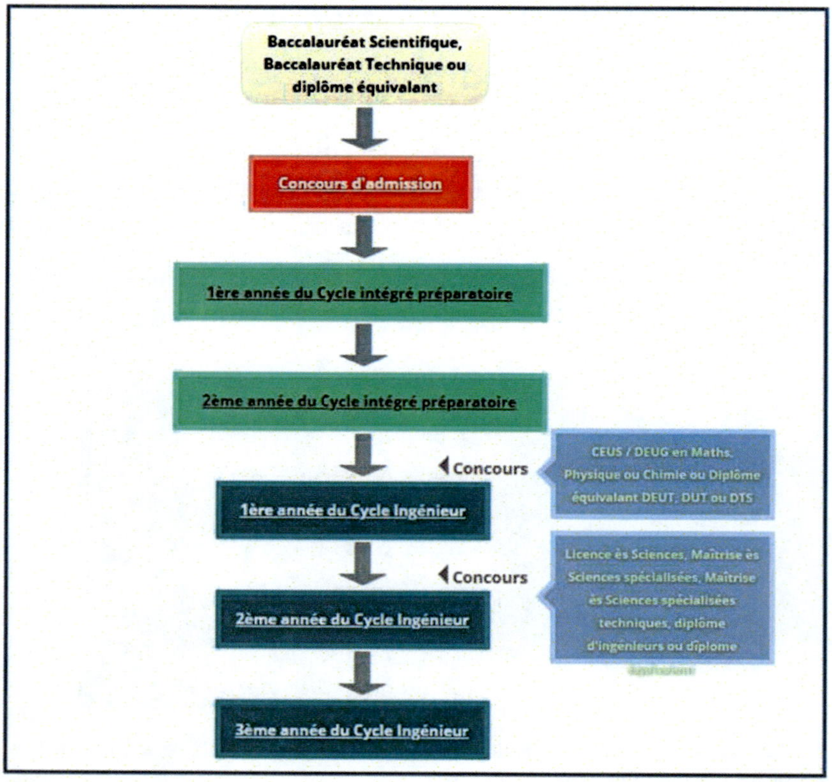

Fig. 2. The training course at the Ecole Nationale des Sciences Appliquées d'El Jadida [6].

The international strategy led by ENSAJ has two priority objectives: to facilitate the professional integration of students in an internationalizing environment and to promote the international dimension of research activities. It helps to strengthen ENSAJ's international positioning. To do this, the ENSAJ encourage the departure of engineering students abroad, for short periods of internships or training.

Each year the first 8 or 10 candidates of the first year of the engineering cycle Energy and Electrical Engineering will leave for Europe for an internship of 1 to 3 months in summer as part of a collaboration agreement with foreign universities.

The academic coordination of these international internships is supported by the lecturers of the Department of Sciences and INdustrial Technologies (STIN). It is a department of education and research, which provides many teaching activities in both the engineering curriculum and master's courses in the field of energy, renewable energy and its interfaces: industrial engineering environment, mechanics and logistics. The department includes the Laboratory of Engineering Sciences for Energy, very involved to promote research and development of energies in Morocco (production, new materials, storage…).

3 Case Study: The Exchange of Undergraduate Students Between the ENSAJ and the Escuela Politécnica Superior of the University of Burgos in Energy Engineering Topics

Morocco is now experiencing a remarkable development in all industrial sectors using electric energy. Engineers specializing in energy and electrical engineering are increasingly demanded in the areas of electrical systems and grid management, power generation and transport.

The launching of the Energy and Electrical Engineering degree at the ENSAJ is part of the actions undertaken to support the national strategy for the Morocco Energy Plan. It aims to train engineers who are able to control electrical installations in terms of production, optimization, recovery, transport, distribution and conversion of electrical energy and who are capable of designing and implementing sustainable energy systems.

The engineering students are trained in a cross-disciplinary approach that allows them to understand the complex problems that will constitute their future missions. The skills provided enable the production, conversion, distribution and use of electrical energy. These skills are addressed by confronting the proposed solutions with economic, societal and environmental impacts.

By 2014, the ENSAJ and the Escuela Politécnica Superior at the University of Burgos (UBU) started their academic collaboration. Every summer period, the Energy Engineering research group at the UBU received 2 students coming from the ENSAJ for a short training period of 1 month. The Energy Engineering research group at the University of Burgos (UBU) started its activity by 1997 [7]. Since then, 10 students from El Jadida have come to Burgos in the period 2014–2019 to develop experimental training at the laboratory of energy engineering shown in Fig. 3, dealing with heat transfer and energy storage devices.

Fig. 3. The laboratory of Energy Engineering of the Escuela Politécnica Superior at the University of Burgos.

These 10 students were enrolled in the first year of the engineering cycle, as described previously. Their laboratory works were developed under the problem-based learning approach commonly used by the lecturers of the Energy Engineering research group [8, 9]. The students are placed as if they were just graduated engineers recently employed by and industrial company where energy consumption is a critical cost factor. The teacher plays the role of the senior engineer of the company. Students are committed to prepare a report on a specific problem on energy analysis, to be ready after a short period of time.

The energy topics covered by the Moroccan students along their training stays in Burgos are mainly related to heat transfer in industrial process and energy storage. We present here a short summary of the most relevant.

The first one concerns the heat transfer in a jacketed vessel to control the temperature a liquid stored in a tank. This method for temperature control presents several advantages: (i) many fluids could be used for cooling or heating inside the jacketed coils; (ii) The temperature and speed of the liquids is easy to manage; (iii) the jacket is cheaper in comparison with the storage tank itself. The Fig. 4 shows the experimental prototype available at the laboratory. It consists of a stainless steel tank of 1000 L of volume, a set of four types of coils in the jacket, and the needed measurement instruments (temperature probes, pressure manometer, flowmeter, power-meter). The students conducted several experiences using water, varying the coil type and the mass flow of cooling water to keep the water inside the tank at a selected constant temperature. Classic combined conduction and convection analysis is performed to compare the heat exchange. The criteria used to propose the best option are the heat transfer global coefficient U (W/m^{2}K) and the power consumption (W) of the water pump.

(a) (b)

Fig. 4. The jacketed vessel. (a) general view of the vessel. (b) the four types of coils in the jacket: G1, half pipe coil 180°; G2, half pipe coil 120°; G3, plate-type coil; G4, dimple jacket.

The second experimental device served to study the energy performance of the air as refrigerating fluid using combined convection and radiation heat transfer, as shown in the Fig. 5(a). It includes a centrifugal fan with a vertical outlet duct at the top of which is mounted a heated, horizontal cylinder. The mounting arrangement for the cylinder is designed to minimize loss of heat by conduction to the wall of the duct allowing the combined effects of convection and radiation to be measured. A variable throttle plate at the inlet to the fan allows the speed of the air through the outlet duct to be varied by adjusting the screw. The power supplied to the heated cylinder can be varied and measured on the control box. Two thermocouples and an anemometer in the duct are the rest of measuring instruments. When operating the bench, the energy balance and the heat transfer convection and radiation transfer between the air flow and the surface of the cylinder can be calculated. Natural and forced convection by means of empirical equations could be studied. Convective heat transfer coefficients at different speeds of air flow can be determined. Domination of the radiation heat transfer coefficient depending on the low or high temperature surface of the cylinder can also be estimated.

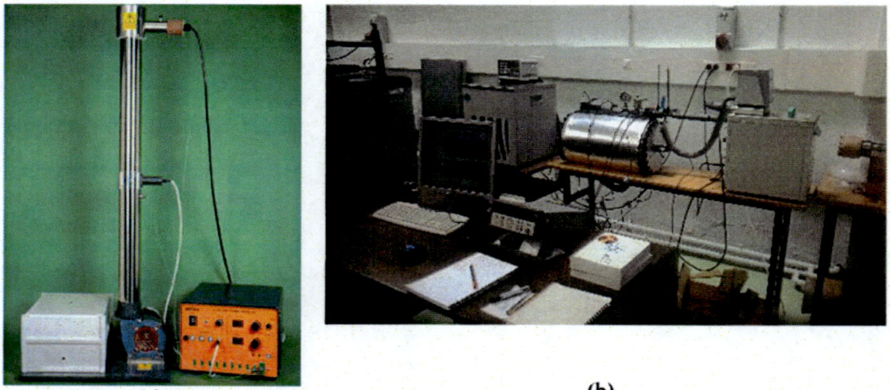

(a) (b)

Fig. 5. (a) The combined convection and radiation heat transfer experimental bench. (b) The combined convection and radiation heat transfer experimental bench.

The third experimental device used by some of the Moroccan students is an experimental bench for energy storage by using phase-change materials (PCM). Nowadays, the use of renewable energies is increasing considerably due to their low environmental impact, as well as in order to reduce the dependence on fossil fuels. One of their disadvantages is its variability; in some conditions the production of energy is too low to cover the demand, whereas in other moments energy production is rather higher than the demand, leading to an energy loss. Energy storage systems have been developed to supply this lack. These systems allow storing energy when its production is higher than the demand or when the energy demand is low, for example during the night.

Figure 5(b) presents the experimental bench, which is composed mainly of a thermal storage system, and two heat reservoirs, hot and cold. The generator reservoir corresponds to the thermal energy producer, that is, the hot source, for example a solar cell or a geothermal pump, whereas the consumer corresponds to the cold source, which is related to the final application of the system, the thermal energy consumer. The thermal energy storage system is the one that stores thermal energy during the charging phase, and that releases it during the discharging phase. It is similar to a mixing chamber, where the two fluxes coming from the hot and cold reservoirs are mixed inside it. The more important elements are the shell and caps, the baffles, and the tubes containing the PCM, all made of stainless steel.

The PCM used for the experiments were organic compounds coming from renewable sources, such as fatty acids. As hot source, a recirculating bath was used. This bath provides with the hot water necessary to melt the PCM contained in the tubes, in the same manner as a heat pump or a boiler would do. The consumer or cold source will use the energy stored in the thermal energy storage system for different purposes. In this case, a second recirculating bath was employed to extract thermal energy from the hot water. This bath is provided with a little pump that allows recirculating water coming from the system. Several temperature probes and flowmeter were also installed and connected to a computer to register continuous operation. The bench allows two operation modes, the charging and discharging of energy in the PCM. Evaluation of the energy balance and transient heat transfer to or form the PCM can be calculated, in order to estimate the energy efficiency of the thermal storage system.

Last but not least is the only stay in 2018 of one student of the third year of the engineering cycle, aimed to develop his final master thesis. This work was close to the frontier between professional training and research activity, which is also an objective for the ENSAJ. This experience was developed at the research laboratory of the Energy Engineering group in Burgos. In this work, new experimental data of density, sound velocity and refractive index and other thermodynamic properties such as excess volume, isentropic compressibility are reported for pure fluids and formed binary mixtures by different hydrocarbon groups (Methylcyclohexane, Cyclohexane, 1-Hexene, Heptane and 2.2.4 Trimethylpentane) for 11 mol fractions and at a temperature of 298.15 K and at atmospheric pressure. The measurements were made using an automatic densitometer and a refractometer. Two equations of polynomial and Redlich-Kister type were used to adjust the experimental data of the different thermodynamic properties, with small standard deviations. Experimental data of the density of the different systems were modeled by two state equations (Peng-Robinson and Soave-Redlich-Kwong) using a program on MATLAB.

Subsequently, experimental excess enthalpies of the pure and binary systems formed by complex fluids (Methyl-carbitol, Methylcyclohexane, Cyclohexane and 1-Hexene) at T = 313.15 K and at atmospheric pressure were measured using a quasi-isothermal flow calorimeter, shown in Fig. 6.

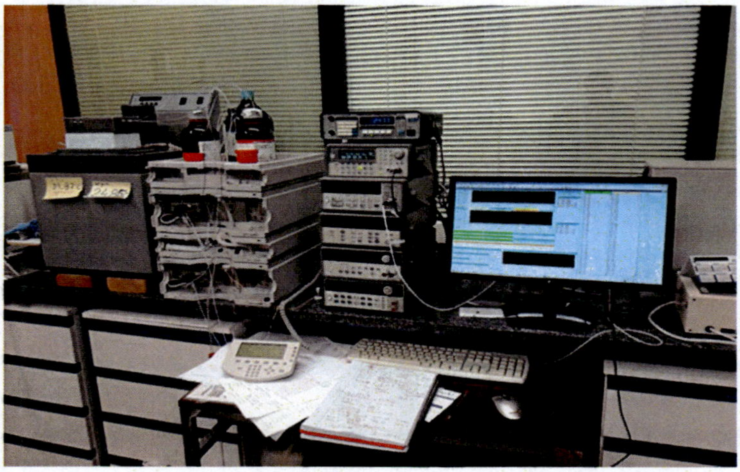

Fig. 6. The quasi-isothermal flow calorimeter used for the measurement of the excess enthalpy of mixtures.

All binary systems showed an endothermic behavior. Experimental data for the different pure and binary systems were adjusted using the Redlich-Kister and Margules empirical equations, and the NRTL and UNIQUAC predictive models. The standard deviation values showed a good agreement between the experimental results and those calculated from these equations.

The open approach taken for the laboratory experiences of the students lead to the acquisition of valuable competences:

- Scientific and technical competences: relationship between energy science and engineering problems, application of energy balance and heat transfer techniques, estimation of energy efficiency in steady-state and transient cases.
- Operational competences: use of measurement instruments, estimation of uncertainty of calculated results, planning of experiments to achieve an objective, writing technical reports based on data.
- Research competences: use of high precision research techniques, culture of rigorous methods to assure quality of research, use of scientific literature and databases for comparison and validation of results.
- Personal competences: working in a team in an international context of different culture and uses, facing up to open problems, self-organization of work, effective oral and written communication.

Two critical factors have contributed to the success of the described cooperation actions. First, the institutional support of the respective universities. As stated previously both universities promote, through their internationalization policies, the participation of staff and students in international cooperation activities. Even the budget available could be quite low, the initiative has boosted the number of activities performed abroad by students and teachers along the years. Once any initiative is known, the signature of framework agreements between universities are strongly recommended before any activity is started. These agreements give legal cover to any research or educational exchange, and allow the Head of Faculties and Departments to develop correctly the actions. The University of Burgos signed the agreement with the University Chouaïb Doukkali - El Jadida in 2014, once previous research relations were stablished.

The second factor of success is the need of leading persons in the respective universities. The mutual knowledge and confidence-building of the partners plays the critical role to boost cooperation for development actions.

4 Conclusions

This contribution presents a case study of the cooperation for education and research program developed between faculties of engineering at the University of Burgos, Spain, and at the University Chouaïb Doukkali - El Jadida, Morocco. The international strategy led by the Ecole Nationale des Sciences Appliquées d'El Jadida is aimed at facilitating the professional integration of its students in an internationalizing environment and promoting the international dimension of research activities, which reinforces its international positioning. This policy has found an echo in the University of Burgos, exchanging students of the engineering cycle along 6 years. The field of study was energy engineering, in correspondence with thee Energy and Electrical Engineering degree of the students. The experience demonstrates the improvement of scientific, technical, operational, research and personal competences of those students participating in the experience. Two factors could be pointed out as critical factors to the success of the described cooperation for development actions. First, the institutional support of the respective universities. Second, the need of leading persons within the respective universities able to carry out the experiences in the long term.

Acknowledgments. Mohamed Lifi wishes to acknowledge the Centre National pour la Recherche Scientifique et Technique CNRST (Morocco) for the funding of his doctoral scholarship (Pre-Doctoral Grants 2019).

References

1. Morgan, C., Volante, L.: A review of the organisation for economic cooperation and development's international education surveys: governance, human capital discourses, and policy debates. Policy Fut. Educ. **14**(6), 775–792 (2016)
2. Greenberg, J.K., Hefley, D., Brice, P., Hoti, A., Engel, A., Erkman, C., Yates, J., Webb, F., O'Rourke, K., Ademi, K.: The many educational facets of development cooperation between a kosovan village and earth scientists. J. Geosci. Educ. **60**, 201–211 (2012)
3. Eerdewijk, A., Westeneng, J., Hoop, T., Ruben, R.: Raising societal support for development cooperation: the role of students' exchange programs. Voluntas **20**, 351–368 (2009)
4. Lucena, J., Schneider, J.: Engineers, development, and engineering education: from national to sustainable community development. Eur. J. Eng. Educ. **33**, 247–257 (2008)
5. Lappalainen, P.: Development cooperation as methodology for teaching social responsibility to engineers. Eur. J. Eng. Educ. **36**(6), 513–519 (2011)
6. Ecole Nationale des Sciences Appliquées d'El Jadida. http://www.ensaj.ucd.ac.ma. Accessed 21 Jan 2020
7. University of Burgos, Energy Engineering Group. https://www.ubu.es/energy-engineering-research-group-ienergia. Accessed 21 Jan 2020
8. Montero, E., González, M.J.: Student engagement in a structured problem-based approach to learning: a first-year electronic engineering study module on heat transfer. IEEE Trans. Educ. **52**, 214–221 (2009)
9. Montero, E., Alaoui, F., González M.J., Aguilar, F.: Teaching thermodynamics to electronic engineers through active teaching strategies. In: Proceedings of IEEE Education Engineering-The Future of Global Learning Engineering Education EDUCON 2012, pp. 374–378. IEEE, Marrakesh (2012)

Check for updates

The Impact of International Mobility in Doctoral Training in Novel Research Groups: A Case Study

Gabriel Rubio-Pérez[1], Mohamed Lifi[1,2(✉)], Raúl Briones-Llorente[1], and Fernando Aguilar[1]

[1] Escuela Politécnica Superior, Universidad de Burgos, Avenida Cantabria s/n, 09006 Burgos, Spain
mlifi@beca.ubu.es

[2] Ecole Nationale des Sciences Appliquées d'El Jadida, Université Chouaib Doukkali, Route d'Azzemour, Nationale N°1, 1166 El Jadida, Morocco

Abstract. Mobility of doctoral students is considered a critical factor which contributes to the doctoral graduate research performance and the knowledge transfer. The research productivity of doctoral students is still more critical for those novel research groups which have to overcome several barriers before they reach a reasonable stability in research productivity. Getting started with a novel group could take several years until the group reaches a minimum productivity, depending on the field of study. Besides, internationalization of a novel group means its regular participation in the international research networks in the respective scientific domain. This contribution presents a case study of the impact of international mobility of doctoral students in the global research productivity of the hosting novel research group. The increase in research productivity due to mobility of doctoral students is shown analyzing the evolution of the number of articles published and comparing it to the critical decisions of internationalization taken by the group, suggesting a relationship between both arguments. Results show that the impact of the doctoral mobility reaches 45% of the total productivity in a period of 12 years.

Keywords: Doctoral mobility · Research productivity · Novel groups

1 Introduction

The doctorate is at the top of academic qualifications in most countries. It is a post-graduate research degree, and is the highest academic degree awarded to those students who have successfully completed a program of work in a particular field of study at the university. The doctorate is almost always based largely or exclusively on research, with the student as an apprenticeship under the guidance of a principal supervisor. Many new forces are nowadays promoting national and international debate on the nature of the doctorate [1], being internationalization one the most frequently cited. Internationalization deals with important issues including global competition for doctoral students, international research networks and the promotion of internationally competitive doctoral programs [2].

Á. Herrero et al. (Eds.): ICEUTE 2020, AISC 1266, pp. 288–294, 2021.
https://doi.org/10.1007/978-3-030-57799-5_30

Mobility of doctoral students is another issue concerning internationalization, as it is considered another critical factor which contributes to the doctoral graduate research performance and the knowledge transfer. For example, reference [3] highlights the impact and correlation between scientific migration and the transfer of knowledge within the European Union, while [4] describe how measures oriented to boost collaboration shape collaborative behavior and researcher productivity. The doctoral graduate research performance is traditionally evaluated by means of two alternative approaches [5]. The quantitative approach consists in analyzing the number of publications in selected repositories, such as academic journals, or calculating comprehensive indices from records of conference papers, journal publications, and books. Another frequent option used is to measure individual research performance by the number of citations and the impact factor of journals to show indirectly the quality of the research productivity. The qualitative approach involves the analysis of perceptions from expert-panel discussions, peer-reviews, and so on. The advantage of this perception analysis is the respondents' expertise and knowledge in related fields. However, it is a matter of fact that peers from different cognitive domains may evaluate a given scientific contribution quite differently, as their evaluations can be influenced by their own level of knowledge and research biases.

Additional studies present empirical evidence on the relationship between collaboration and researcher productivity, suggesting that the duration of collaboration is positively related to researcher productivity [6]. Later studies suggest that collaborations, while not effective in enhancing researcher productivity in the short run, may be an important promoter of effective collaborations in the longer run [4].

Some other studies focus on the controversial between academic inbreeding and international mobility of doctorates [7], the learning experiences of international doctoral students [8] or the supervisory feedback practice with international students and its impact on the thesis writing process [9].

The results of a Spanish survey on international mobility of scientists indicate that there is a strong influence of institutional forces constituting a scientific culture which promotes rules and certain habits about how scientists should progress in their careers [10]. The aim of this contribution is to show a case study of the impact of international mobility of doctoral students in the global research productivity of the hosting research group. The research productivity is still more critical for novel research groups, as it becomes the mean of obtaining research funding, doctoral grants and recognition in the respective cognitive domain.

2 Novel Research Groups and Doctorate Students

Novel research groups in universities and other research institutions have to overcome several barriers before they reach a reasonable stability in research productivity. In this paper, we refer to groups that initiate their activity in departments with no previous research activity. They develop a much more challenging task than those groups which are a branch of a previous experienced team or that are born in a department or faculty with a long research tradition.

Very frequently novel research groups grow around an initial researcher who has got his or her doctoral and research background in a university different of the one of the present position. Very often an empty or no laboratory or research infrastructure is available, and getting funding as a stand-alone researcher becomes a herculean task. This seminal researcher has to face up with two stressing responsibilities: to create a new competitive research team in a short period and to keep research productivity in a satisfactory level of quality as to get positive evaluations in competitive calls for projects, as projects mean funding. The most plausible option to cope with this last issue is to keep, along a certain period of time, the scientific collaboration with the previous research group where he or she got the Ph. D. degree, as a way of producing results (in terms of research articles or books, patents, etc.). Getting started with a novel group could take several years until the group reaches a minimum productivity, depending on the field of study. In general, those researches which need, for example, experimental equipment and installations will take more time to get started than those which require digital libraries or computing facilities, for the same level of complexity and deepness.

With the exception of a limited number of scientific fields, focused on local or regional research, the majority of research activities are developed in an international scenario, as science and knowledge have no frontiers. Then, internationalization of research becomes a key factor for the long term success of novel research groups. Internationalization of a novel group means its regular participation in the international research networks in the respective field of study. These networks present usually some formal and stable structures, such as scientific associations, editorial committees, international agencies, cluster, etc., and also non-formal or temporary ones, like international conferences, research projects, consortia, etc. For a novel research group, the access to the non-formal networks is the first step, as the access to the formal ones requires demonstrating more experience and competence in research, which sometimes is reserved for excellence groups.

Collaboration with others is then the first step for internationalization. Fruitful collaborations between research groups are based mainly in mutual confidence and compromise between individuals. Exchange of researchers is one of the most proper ways to generate mutual benefits for both the hosting and the invited groups. Sometimes senior or post-doctoral researcher's mobility serves to consolidate strong links between groups, but the most frequent is the exchange of young researchers, many of them pre-doctoral students. Short or long research stays of pre-doctoral students in a foreign institution use to yield shared results contributing to an increase of the productivity of both groups. From the point of view of the student, the life experience in a different environment and culture will play a positive part in his/her maturity as scientist.

3 Case Study: A Novel Research Group in Energy Engineering

The Energy Engineering research group at the University of Burgos (UBU) started its activity by 1997 [11]. The first researcher and founder of the group hold the position of lecturer at the UBU School of Engineering since 1987, at that time a vocational school of engineering, with no research activity at all. In the same period, taking in consideration the non-existent offer of a doctoral program in engineering at the UBU, he started his doctoral studies at the University of Valladolid (UVA) and finally got his Ph. D. in 1996. Then he became the first doctor of his department of engineering and cope with the challenge of creating a novel research group. That meant looking for funding, installation of an experimental laboratory devoted to the measurement of thermodynamic properties of industrial liquids, promoting the offer of a doctoral program and looking for additional members of the team. And the most important, developing this pioneer work in a faculty with not only a total lack of tradition and culture of research but many times opposed to this activity. Later changes at the University policy enabled the culture change at a very slow pace.

It took 10 years to become an autonomous research group, it is to say, to be a group able to produce scientific results at its own laboratory. Along this time, the research productivity of the group was linked to the mother group Termocal in the UVA [12]. The average productivity of this period was, measured in terms of research articles in international journals, of 1 article per year. Though the quality of research was quite good (all of them present an impact factor placed in the first quarter of his category following the Journal of Citation Reports [13]), the amount of articles was quite poor, 10 articles in 10 years. Nevertheless, even the research productivity was so low, it allowed the novel group to obtain resources and funding as to be an autonomous group, and then the mother group at the UVA became a partner.

2007 meant a critical date of change as the novel group received its first doctoral student from abroad. In addition, the group took the decision of opening international collaborations with other research groups. After some inquiries with senior researchers of the UVA, the group decided to contact the "Laboratoire des Fluides Complexes et leur Réservoirs" (LFCR) at the University of Pau in France [15]. Fortunately, they opened their doors and trusted in the novel group, and they are very grateful for that. As a result, the group started this new collaboration with the University of Pau, exchanging pre-doctoral students since then for research stays ranging from 4 to 6 months, with a total number of 3 students in the period 2008–2020.

The Fig. 1 shows the evolution of the research productivity of the group in terms of articles published in journals of high impact factor in the period 1995–2019 (the year 2020 is not significant at the moment of writing this report). It can be observed that there is a significant increase in the number of articles since 2009, with a positive slope in the next period 2009–2019. In correspondence, the number of citations presents an increasing average profile in the same period, as an indicator of the relevance of the research.

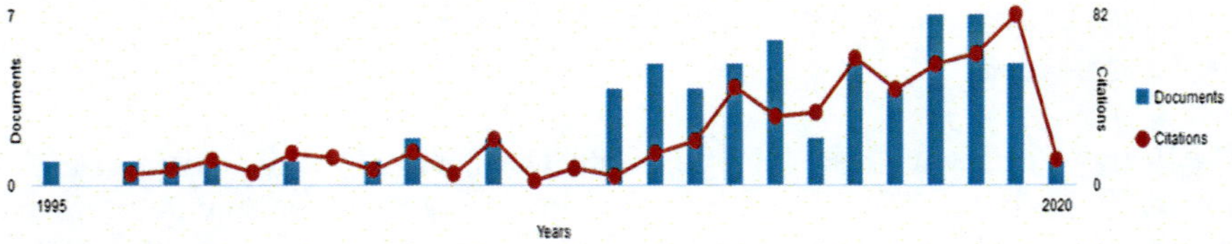

Fig. 1. Evolution of the research productivity of the Energy Engineering research Group in the period 1995–2019. (The year 2020 is not significant at the moment of writing this report). Source: Scopus ® [14]

Table 1 reflects the main impact of the international mobility of doctoral students, incoming or outgoing, in the research productivity of the novel group. The observed increase in the average of yearly articles published, from 1 art/year to 4.8 art/year, could be explained in a simple way due to the natural growing maturity end experience of the team, as well as the total number of doctoral students along time. But when looking at the influence of the doctoral mobility in the research productivity, it can be stated that 45% of the total productivity corresponds to this factor.

Table 1. Evolution of the research productivity of the Energy Engineering research Group in the period 1996–2019. % mobility stands for the percentage or articles coming from mobility of doctoral students, incoming or outgoing, at the UBU.

Energy Engineering Research Group: productivity		
Period	1996–2007	2008–2019
Average	1 article/year	4.8 articles/year
% mobility	0%	45%
Dependence	UBV + UVA	UBU (autonomous)

Another interesting issue is the relationship between international collaboration and young researcher's productivity. Figure 2 shows the evolution of the research productivity of a young member of the Energy Engineering research Group after her doctoral studies in 2017 and her inclusion as lecturer in 2018 (again the year 2020 is not significant at the moment of writing this report).

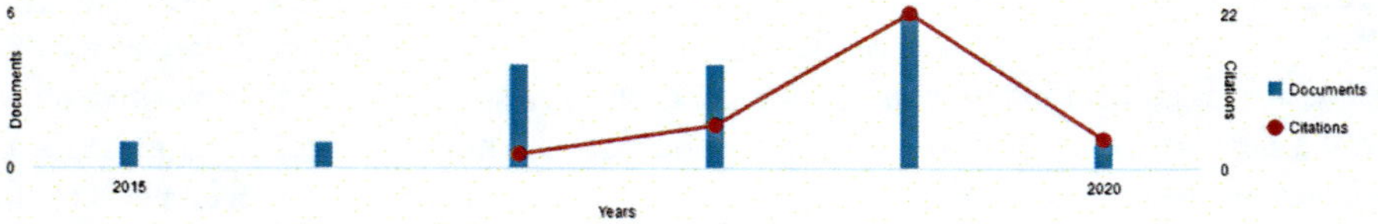

Fig. 2. Evolution of the research productivity of a young member of the Energy Engineering research Group after her doctoral studies in 2017 and her inclusion as lecturer in 2018 (The year 2020 is not significant at the moment of writing this report). Source: Scopus ® [14]

This member of the novel group carried out an outgoing research stay of 5 months during her doctoral period. The productivity data of the years following the doctoral studies presents an increasing behavior in the next 3 years, suggesting that the experience of international collaboration is positively related to researcher productivity.

After the key starting experience with the University of Pau, the group enhanced its international collaborations along the years, and nowadays has stable international exchanges of students and researchers with the Université Chouaïb Doukkali (El Jadida) and the Université Abdelmalek Essaâdi (Tetouan) in Morocco, the Cooper Union in United States and, more recently, the Universidade do Minho in Portugal and the Lucerne University of Applied Sciences and Arts in Switzerland. Because of the expansion of its international doctoral mobility, the group expects a huge increase in its future research productivity, keeping the influence of this factor in the range 45–55%.

4 Conclusions

This contribution presents a case study of the impact of international mobility of doctoral students in the global research productivity of a novel research group in energy engineering. The research productivity is measured in terms of research articles published in high impact factor journals.

The novel group developed its activity in two temporary steps: (i) a first step where the link with an external research group provided the needed productivity to overcome the initial barriers to start its own research line; (ii) a second step of autonomous research productivity. The increase in research productivity due to mobility of doctoral students is shown analyzing the evolution of the number of articles published and comparing it to the critical decisions of internationalization taken by the group, suggesting a relationship between both arguments. The impact of the doctoral mobility reaches 45% of the total productivity in a period of 12 years. The contribution could serve as an example for other novel groups.

Acknowledgments. Gabriel Rubio-Pérez wishes to acknowledge the Ministerio de Ciencia, Innovación y Universidades (Spain) for the funding of his doctoral scholarship.

Mohamed Lifi wishes to acknowledge the Centre National pour la Recherche Scientifique et Technique CNRST (Morocco) for the funding of his doctoral scholarship (Pre-Doctoral Grants 2019).

Raúl Briones-Llorente wishes to acknowledge the Consejería de Educación de la Junta de Castilla y León (Spain) and the Operational Programme European Social Fund for the funding of his doctoral scholarship.

References

1. Park, C.: Redefining the Doctorate. The Higher Education Academy, York (2007)
2. De Rosa, A.S.: New forms of international cooperation in doctoral training: internationalisation and the international doctorate – one goal, two distinct models. Higher Educ. Eur. **33**(1), 3–25 (2008)

3. Ackers, L.: Moving people and knowledge: scientific mobility in the European Union. Int. Migr. **43**(5), 99–131 (2005)
4. Defazio, D., Lockett, A., Wright, M.: Funding incentives, collaborative dynamics and scientific productivity: evidence from the EU framework program. Res. Policy **38**(2), 293–305 (2009)
5. Gu, J., Lin, Y., Vogel, D., Tian, W.: What are the major impact factors on research performance of young doctorate holders in science in China: a USTC survey. High. Educ. **62**, 483–502 (2011)
6. Porac, J.F., Wade, J.B., Fischer, H.M., Brown, J., Kanfer, A., Bowker, G.: Human capital heterogeneity, collaborative relationship, and publication patterns in a multidisciplinary scientific alliance: a comparative case study of two scientific teams. Res. Policy **33**(4), 661–678 (2004)
7. Horta, H.: Deepening our understanding of academic inbreeding effects on research information exchange and scientific output: new insights for academic based research. High. Educ. **65**, 487–510 (2013)
8. Evans, C., Stevenson, K.: The learning experiences of international doctoral students with particular reference to nursing students: a literature review. Int. J. Nurs. Stud. **47**, 239–250 (2010)
9. Wang, T., Li, L.Y.: 'Tell me what to do' vs. 'guide me through it': feedback experiences of international doctoral students. Act. Learn. High Educ. **12**(2), 101–112 (2011)
10. González Ramos, A.M., Malpica Lander, C.: International mobility of spanish men and women doctorate holders. Sociología y Tecnociencia/Sociol. Technosci. **3**(3), 31–54 (2013)
11. University of Burgos, Energy Engineering Group. https://www.ubu.es/energy-engineering-research-group-ienergia. Accessed 21 Jan 2020
12. TERMOCAL Laboratory. http://www.eis.uva.es/termocal/index.htm. Accessed 21 Jan 2020
13. Clarivate Analytics, Web of Science Group. Journal of Citation Reports. https://clarivate.com/webofsciencegroup/solutions/journal-citation-reports/. Accessed 21 Jan 2020
14. Elsevier, Scopus database. https://www.elsevier.com/solutions/scopus. Accessed 21 Jan 2020
15. University of Pau, Laboratoire des Fluides Complexes et leur Réservoirs (LFCR). https://lfc.univ-pau.fr/fr/index.html. Accessed 21 Jan 2020

Enhancing Interactive Teaching of Engineering Topics Using Digital Materials of the MERLOT Database

Natalia Muñoz-Rujas[1], Jennifer Baptiste[2], Ana Pavani[3], and Eduardo Montero[1(✉)]

[1] Escuela Politécnica Superior, Universidad de Burgos, Avenida Cantabria s/n, 09006 Burgos, Spain
emontero@ubu.es
[2] University of Trinidad and Tobago, Port of Spain, Trinidad and Tobago
[3] Pontifícia Universidade Católica do Rio de Janeiro, Rio de Janeiro, Brazil

Abstract. Open Educational Resources are freely accessible, openly licensed text, media, and other digital assets that are useful for teaching, learning, and assessing as well as for research purposes. They include free use for educational purposes by teachers and learners, and encompass all types of digital media. They are frequently used as support of teaching and learning activities, using course content available in repositories under open licenses. Availability of well-structured and peer-reviewed online repositories covering a broad range of topics are highly appreciated. MERLOT (Multimedia Education Resource for Learning and Online Teaching) is an online repository where its partners and members are devoted to identifying, peer reviewing, organizing and making available existing online learning resources in a range of academic disciplines for its use by higher education faculty and students. This article concerns the usefulness of MERLOT in engineering education. At the Engineering collection of MERLOT, lots of resources can be found on a wide variety of topics ranging from aerospace engineering to petroleum engineering to help teachers with their teaching and research. The MERLOT Engineering Community and Materials are described, highlighting the availability of learning resources in nineteen engineering disciplines and twenty one types of material, including animations, online courses, simulations and reference materials. The Engineering Board is in charge of expanding and managing the collection, as well as curating the same applying the rigorous peer review processes of MERLOT to those online learning materials of valuable high quality.

Keywords: Open educational resources · Engineering education · MERLOT

1 Open Educational Resources and the MERLOT Digital Library

There exist many definitions, both at institutional level [1] or in wikis, of what Open Educational Resources (OER) are. We can summarize that OER are freely accessible, openly licensed text, media, and other digital assets that are useful for teaching,

Á. Herrero et al. (Eds.): ICEUTE 2020, AISC 1266, pp. 295–306, 2021.
https://doi.org/10.1007/978-3-030-57799-5_31

learning, and assessing as well as for research purposes. They include free use for educational purposes by teachers and learners, and encompass all types of digital media. OER are frequently used in support of teaching and learning activities, using course content available in repositories under open licenses. Sometimes it is difficult to find OER's due to several reasons: (i) because of technical issues of search engines and repositories; (ii) due to a lack of user's search skills; (iii) the high amount of resources available in different subject areas [2]. To overcome these barriers, when available, well-structured and peer-reviewed online repositories covering a broad range of topics are highly appreciated.

In relation with the utility of OER, many experiences of using OER found in the literature refer to the use of OER in support of teaching and learning activities. Some others provide access to learning materials available in online repositories under open licenses. Less frequent is the description of experiences where the OER is integrated into the curriculum, that is to say that students are committed to use OER as part of their course, being evaluated through tests and exams [3, 4].

MERLOT (Multimedia Education Resource for Learning and Online Teaching) [5] is an online repository where its partners and members are devoted to identifying, peer reviewing, organizing and making available existing online learning resources in a range of academic disciplines for its use by higher education faculty and students. Promoted and supported by the California State University Center for Distributed Learning since 1997. It is an international consortium of institutions (and systems) of higher education, industry partners, professional organizations and individuals. MERLOT is a community of staff, volunteers, and members who work together in various ways to provide users of OER teaching and learning materials with a wealth of services and functions that can enhance their instructional experience.

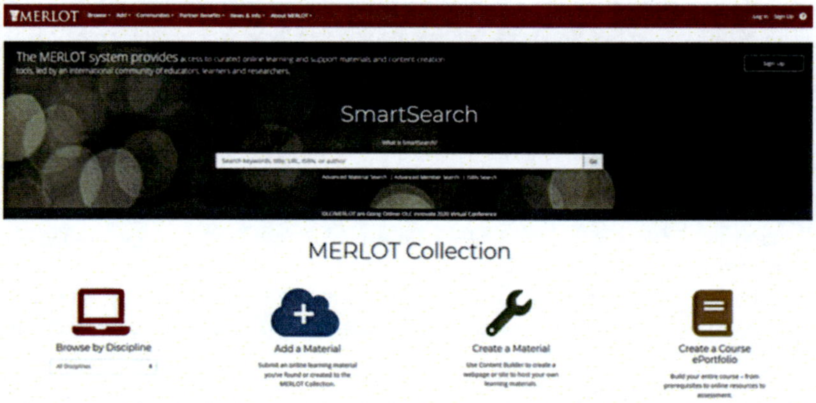

Fig. 1. MERLOT facilities website at https://www.merlot.org/merlot/

The MERLOT collection of OER is noteworthy among others because of the quality of the learning materials in the collection. These materials are catalogued in a variety of ways, one of the most important being the academic discipline in which the learning material might be best used for teaching and learning (Fig. 1). The collections of online peer reviewed higher education learning materials in MERLOT are catalogued by registered members (91,538 materials and 166,600 registered users, as of February 3, 2020) and a set of faculty development support services. MERLOT's strategic goal is to improve the effectiveness of teaching and learning by increasing the quantity and quality of peer reviewed online learning materials that can be easily incorporated into faculty-designed courses. MERLOT's activities are based on the creative collaboration and support of its individual Members, Institutional Partners, Corporate Partners and Editorial Boards, which are updated regularly [6].

This article concerns the use of OER in engineering education, which is an interesting topic of research during the last decade [7]. Some criteria dealing with advantages and drawbacks of using OER in engineering education could be found in [8]. In the Engineering collection, lots of resources can be found on a wide variety of topics ranging from aerospace engineering to petroleum engineering to help teachers with their teaching and research.

2 The MERLOT Engineering Community

Members of MERLOT communities are vital to its success. Community Members help grow the MERLOT collection by contributing materials, and adding assignments and comments. Additionally, they introduce others in the discipline to MERLOT by presenting at conferences, authoring articles about MERLOT, and encouraging others to use and contribute to MERLOT. The Engineering Portal on MERLOT has its own website at MERLOT [9]. Here, one will find lots of resources on a wide variety of topics ranging from aerospace engineering to petroleum engineering to help with teaching and research. The page, contains many Engineering resources. This includes the most recently added Engineering material and members, journals and publications and Engineering education alerts and twitter feeds.

At present, the Engineering Community of MERLOT involves 4,640 members, as shown in Table 1. It can be observed that most of the members are students, which reflects that looking for useful open educational resources is an interesting issue in engineering education. Moreover, as registration in MERLOT is voluntarily, that means an aim of continuity in the search of updated and new learning materials.

Table 1. Engineering Community of MERLOT: members by member type.

	Member type	Members
1	Faculty	1,623
2	Staff	274
3	Student	2,000
4	Administrator	95
5	Librarian	24
6	Health Care Professional	10
7	Consultant	110
8	Teacher (K-12)	208
9	Content Developer	76
10	Other	220
	Total	4,640

As expected, faculty members are the second largest group, over 1,620 members. The use of digital materials in higher education institutions is a trend worldwide, and engineering educators follow the same pattern. In contrast, the presence of teachers (K-12) is quite low in comparison with faculty, probably due to the origin of MERLOT, which was supported since its birthday by a university.

Other membership such as staff reveals the interest of higher education managers to be involved in open, online approaches to learning. The membership of consultants, on the other hand, shows the interest of professional partners and individuals in long-life learning opportunities.

We could complete this information by mentioning the affiliation of the engineering community members: 2,966 of them come from the education sector, 167 from corporation companies, 119 from non-profit organizations and 76 from government sector, as the most significant categories.

Table 2 present the distribution of the engineering members by the nineteen disciplines of engineering selected by MERLOT. This distribution corresponds, more or less, to the presence of these disciplines in the professional job market. The most populated ones are mechanical (981) and electrical engineering (866), followed by computer (501) and civil engineering (454).

On the other side, disciplines such as audio, geological, ocean and nuclear engineering are under 25 members. The total number of members reflected in Table 2 does not match the same in Table 1 because some registered members do not declare the respective discipline.

Table 2. Engineering Community of MERLOT: members by discipline.

	Discipline	Members
1	Aerospace and Aeronautical Engineering	94
2	Agricultural and Biological Engineering	40
3	Audio Engineering	14
4	Biomedical Engineering	96
5	Chemical Engineering	194
6	Civil Engineering	454
7	Computer Engineering	501
8	Electrical Engineering	866
9	Engineering Science	130
10	Environmental Engineering	154
11	Geological Engineering	20
12	Industrial and Systems	266
13	Manufacturing Engineering	164
14	Materials Science and Engineering	127
15	Mechanical Engineering	981
16	Mining Engineering	54
17	Nuclear Engineering	21
18	Ocean Engineering	20
19	Petroleum Engineering	33
	Total	4,229

3 The MERLOT Engineering Materials

Volunteer community Members work with MERLOT staff to develop policies and practices that govern MERLOT's operations. These volunteers are the Editorial Boards and Project Directors. Editorial boards are the cornerstones of MERLOT. They lead the development of MERLOT's discipline communities - one Editorial Board for each community - and are comprised of an Editor, Associate Editors and Peer reviewers. Project Directors have critical responsibilities in managing their institution's participation in MERLOT and MERLOT's connection to their institution's academic technology initiatives.

Each editorial board is responsible for:

- Expanding and managing the collection of online learning materials in its discipline.
- Educating and reaching out to the community of educators in its discipline.
- Implementing and managing the MERLOT peer review process for materials in its discipline.
- Recruiting and training peer reviewers.

Only MERLOT members can add learning materials to the MERLOT collection. This means adding the metadata for a material to describe the item and provide a URL.

If there isn't a URL for the item, it cannot be added to MERLOT. Only materials for which there is an Editorial Board will be considered for review.

MERLOT does not store learning materials. Only the metadata (descriptive items) are stored on the MERLOT site. To maintain an accurate site, MERLOT runs a link checker on a monthly basis to ensure the validity of the links.

Each Editorial Board has an Editor that triages each material in their discipline and assigns it to a reviewer in their discipline. The MERLOT editorial boards provide leadership, tools, and training in developing evaluation standards and processes for peer review which include quality of content, potential effectiveness as a teaching tool and ease of use.

Editorial boards decide on the process for selecting materials to be peer reviewed. First, the editorial board reviews its collection and "triages" materials with a rating to indicate order of importance. Note that it may not be practical for our reviewers to review all contributed materials. Editorial board members complete the triage and can, if they wish, post member comments based on their cursory evaluations. This results in an Editor Review which users see on the material detail page. Finally, only for those materials highly rated by the editorial board, the Editor assigns two peer reviewers to each material. They use their editorial board's review procedures, forms, and evaluation standards to independently review the material. Reviewers write peer review reports using the evaluation criteria for MERLOT learning materials as a guideline.

The engineering community of MERLOT presents 4,522 learning materials involving the nineteen disciplines, as shown in Table 3. The most part of materials are devoted to electrical (1,094) and mechanical engineering (864), followed by civil engineering (608), showing only a certain correspondence with the number of the respective memberships. Under 25 materials, we could find the audio, mining, petroleum and engineering science disciplines.

Table 3. Engineering Community of MERLOT: materials by discipline.

	Discipline	Members
1	Aerospace and Aeronautical Engineering	274
2	Agricultural and Biological Engineering	57
3	Audio Engineering	5
4	Biomedical Engineering	61
5	Chemical Engineering	223
6	Civil Engineering	608
7	Computer Engineering	195
8	Electrical Engineering	1,094
9	Engineering Science	25
10	Environmental Engineering	160
11	Geological Engineering	254
12	Industrial and Systems	121
13	Manufacturing Engineering	97

(*continued*)

Table 3. (*continued*)

	Discipline	Members
14	Materials Science and Engineering	376
15	Mechanical Engineering	864
16	Mining Engineering	6
17	Nuclear Engineering	67
18	Ocean Engineering	14
19	Petroleum Engineering	21
	Total	4,522

Table 4 exhibits the MERLOT 7,674 available learning materials in engineering disciplines classified by material type. At the MERLOT website [5], you can refine the material list by selecting a specific category, material type, mobile filters or other filters to narrow a search. The largest groups of materials (over 1,000 items) are animations, online courses, simulations and reference materials. While animations frequently show mostly qualitative effects of science and technology phenomena, simulations often include some degrees of freedom to the user in order to manipulate the app or computer package, allowing quantitative analysis of phenomena.

Table 4. Engineering Community of MERLOT: materials by material type.

	Discipline	Members
1	Animation	1,847
2	Assessment Tool	132
3	Assignment	538
4	Case Study	25
5	Collection	282
6	Development Tool	27
7	Drill and Practice	19
8	ePortfolio	10
9	Learning Object Repository	71
10	Online Course	1,235
11	Online Course Module	39
12	Open (Access) Journal-Article	75
13	Open (Access) Textbook	405
14	Presentation	728
15	Quiz/Test	2
16	Reference Material	1,035
17	Simulation	1,130
18	Social Networking Tool	4
19	Syllabus	7
20	Tutorial	49
21	Workshop and Training Material	14
	Total	7,674

Online Courses have come to be very popular today as the information technologies have been extremely widespread. Many massive open online courses (MOOC) are offered not only by some virtual platforms like Coursera, EdX, MiriadaX but also by many universities and higher education institutions. Reference Materials are frequently websites or wikis containing a comprehensive part of a scientific or technological domain in a discipline.

With the aim of distinguish those materials with an upper level of quality accordingly to MERLOT standards, some of them are rated in several ways. 709 of them are rated by users, which tell us the point of view of the opinion of other teachers or students with respect to the applicability of the material into a particular context. Also 638 have Editor reviews and 71 peer reviews, following the protocol previously described in this section.

The managers of MERLOT have some statistical tools which allow them recording the visits to the website, offering relevant information to the editorial boards on the use of the respective materials. For example, Table 5 presents the number of times somebody has found a certain engineering material of interest and has clicked on the "Go To Material" link placed at the MERLOT information sheet of the material, during the second half of 2019.

Table 5. Engineering Materials of MERLOT: "Go To" on the second half of 2019. Only materials with more than 10 "Go To" are reported.

#	Title	Material "Go Tos"
1	Games for Basic Electronics Skills	185
2	MecMovies	131
3	Structural Analysis Solved Problems	88
4	Shock Tube Calculator	66
5	Artificial Intelligence: Machine Learning	46
6	Fundamentals of Electrical Engineering I	44
7	The Distribution Game	36
8	Fabrication Steps of an FET and a BJT	33
9	E-Books Directory	33
10	Using the Oscilloscope to Measure the Time Constant	31
11	All About Circuits - Volume I - DC	31
12	Internal Combustion Engine	29
13	Electric Circuit Symbol Flashcards	29
14	Manufacturing Processes and Materials: Exercises	26
15	French Open Educational Resources (OER) Portal	26
16	The Semiconductor Applet Service	24
17	Engineering Graphics	24
18	Basic Electronics	24
19	Introduction to 3D Printing	23
20	JAVA Optics Course	23

(continued)

Table 5. *(continued)*

#	Title	Material "Go Tos"
21	PhET Teacher Activities: Hooke's Law	23
22	Thermal Power Plant Micro Module	22
23	Technical Drawing Animations	22
24	B.C. Campus Open Education: Open Textbook Self-Publishing Guide	21
25	Cosmo Learning	20
26	Strength of Materials	19
27	Workbook for AC Electrical Circuits	19
28	SteamTables Calculator	19
29	Civil Engineering and Engineering Mechanics Software	19
30	E6B Emulator	18
31	Technical Communication (Engineering)	18
32	Virtual Reality And It's Potential For Europe	18
33	15.053 Optimization Methods in Management Science (MIT)	17
34	SCHOLAR -CNKI (China Integrated Knowledge Resource Database)	17
35	Fostering Research Integrity: The Hong Kong Manifesto for Assessing Researchers	17
36	The Reactor Lab, chemical reactor simulations	16
37	Engineering Statics	16
38	GasTurb Program	16
39	Chemical Engineering Process Dynamics and Controls	16
40	Converging-Diverging Nozzle Applet	15
41	Use of the vernier caliper in millimeter with resolution of 0.05 mm (metrical sistem) - Simulator	15
42	A First Course in Electrical and Computer Engineering	14
43	Get The Research: Open Access Research Search Engine/Tool for General Public	14
44	Occupational Safety and Health: Free, Open-Access Resources	14
45	Lab Manual For AC Electrical Circuits	14
46	The Best Dam Simulation Ever	13
47	pH Scale	13
48	The Semiconductor Applet Service	13
49	Introduction to Robotics	13
50	Laboratory Manual for Computer Programming with Python and Multisim (tm)	13
51	Engineering Graphics & CADD - Yong Suh	13
52	Basic Electronics Tutorials	13
53	InTech: Open Access Publisher of eBooks and eJournals	12
54	Engineering Statics	12
55	Base 2 binary multiplication - basic	12

(continued)

Table 5. (*continued*)

#	Title	Material "Go Tos"
56	Engineering Design Process	12
57	Compressible Aerodynamics Calculator	12
58	All About Circuits - Volume II - AC	11
59	Physics for Scientists & Engineers	11
60	6.254 Game Theory with Engineering Applications (MIT)	11
61	Artificial Intelligence	11
62	ChemRxiv: Open Access Preprints Repository Archive for Chemistry	11
63	First and Second Law of Thermo Visual Calculator	11
64	The Supply Chain Game	11
65	Challenges for Electronics	11
66	Salts & Solubility	11
67	MAE 91: Introduction to Thermodynamics	11
68	ESD.273 J Logistics and Supply Chain Management (MIT)	11
69	Computer Hardware Course: Basic Digital Circuits and Introduction to Computer Processors	11
70	Classroom Presenter	11

This information is reported every six months to the respective Editorial Board, and it serves as a guide for curating the collection, taking decisions on triages and peer reviews.

Finally, MERLOT has an Awards program which recognizes and promotes outstanding online resources designed to enhance teaching and learning and to honor the authors and developers of these resources for their contributions to the academic community. Each MERLOT Editorial Board selects an outstanding resource from its discipline to receive the MERLOT Classics award.

Authors wishing to have their learning material considered to receive the Classics Award must first contribute the material to MERLOT. Once the material is submitted it needs to be peer-reviewed by the Editorial Board, which may take several months. Materials are only considered for the award when they are catalogued in the disciplines for which there are Editorial Boards. For example, last 2019, the website "Thermodynamics" [10], addressed to chemical engineering undergraduate students, received the Classics award in Engineering. This website covers difficult topics in thermodynamics including reversibility, entropy and the second law, equilibrium and steady state, reaction equilibrium and rate, internal energy and enthalpy, vapor-liquid equilibrium, and systems and surroundings. Each topic contains student made videos explaining various aspects of the topic and simulations to further solidify the material. Practice problems and outside resources are also available.

4 Conclusions

This contribution presents the MERLOT digital library and its Engineering collection. The Engineering Community of MERLOT involves 4,640 members. Most of the members are students and faculty, which reflects the fact that looking for useful open educational resources is an interesting issue in engineering education. Moreover, as registration in MERLOT is voluntarily, that means an aim of continuity is in the search of updated and new learning materials. The distribution of the engineering members by the nineteen disciplines of engineering selected by MERLOT corresponds to the presence of these disciplines in the professional job market.

The engineering community of MERLOT presents 7,674 available learning materials in engineering disciplines distributed into 21 categories of material type. At the MERLOT website one can refine the material list by selecting a specific category, material type, mobile filters or other filters to narrow the search. This allows the individual teacher to find the most suitable online learning resource for the respective interest or course. The largest groups of materials (over 1,000 items) are animations, online courses, simulations and reference materials.

MERLOT, with the aim of distinguishing those materials with a higher level of quality, presents some tools and protocols to rate the materials in several ways: rating by users, Editor's reviews and peer reviews. The MERLOT Engineering board provides evaluation standards and processes for peer review which include quality of content, potential effectiveness as a teaching tool and ease of use. This is why MERLOT is recognized worldwide as an organization boosting outstanding online resources designed to enhance teaching and learning.

Acknowledgements. The authors want to thank Aly El-Iraki, co-editor of the Engineering Board of MERLOT, for his revision and improvement of the present contribution.

References

1. Centre for Educational Research and Innovation: Organization for Economic Co-operation and Development (OECD): Giving Knowledge for Free: The Emergence of Open Educational Resources. OECD, Paris (2007)
2. White, D., Manton, M.: Open Educational Resources: The Value of Reuse in Higher Education. University of Oxford, Oxford, U.K., JISC-funded OER impact study (2011)
3. Hilton, J.: Open educational resources and college textbook choices: a review of research on efficacy and perceptions. Educ. Tech. Res. Dev. **64**(4), 573–590 (2016). https://doi.org/10.1007/s11423-016-9434-9
4. Wiley, D., Bliss, T.J., McEwen, M.: Open educational resources: a review of the literature. In: Handbook of Research on Educational Communications and Technology, pp. 781–789, Springer, New York (2014)
5. MERLOT (Multimedia Education Resource for Learning and Online Teaching). www.merlot.org. Accessed 20 Jan 2020
6. Tovar, E., Chan, H., Reisman, S.: Promoting MERLOT communities based on OERs in computer science and information Systems. In: IEEE 41st Annual Computer Software and Applications Conference (COMPSAC) (2017)

7. Duse, C.S., Duse, D.M., Bonnano, P.: The importance of OER in engineering education. MATEC Web Conf. **121**, 12006 (2017)
8. Tovar, E., Piedra, N.: Guest editorial: open educational resources in engineering education: various perspectives opening the education of engineers. IEEE Trans. Educ. **57**(4), 213–219 (2014)
9. MERLOT Engineering Board. https://www.merlot.org/merlot/Engineering.htm. Accessed 20 Jan 2020
10. MERLOT Engineering Board. https://www.merlot.org/merlot/viewMaterial.htm?id=1052199. Accessed 20 Jan 2020

Technologies Applied to the Improvement of Academic Performance in the Teaching-Learning Process in Secondary Students

Miguel Ángel Queiruga-Dios[1], Emilia López-Iñesta[2], María Diez-Ojeda[1(✉)], and José Benito Vázquez-Dorrío[3]

[1] Facultad de Educación, University of Burgos, c/Villadiego, 1, 09001 Burgos, Spain
maqueiruga@ubu.es
[2] Facultat de Magisteri, University of Valencia, Av. Tarongers, 4, 46022 València, Spain
[3] Escuela de Ingeniería Industrial, University of Vigo, Lagoas-Marcosende, 36310 Vigo, Spain

Abstract. The use of active methodologies and educational technologies can improve the academic performance of Secondary students in Physics, and it is one of the most predominant subjects in the field of engineering. This article describes a proposal of Physics activities that use various digital and technological tools incorporated into problem solving. This proposal was carried out in the subject of Physics and Chemistry of 3° of Compulsory Secondary Education (14–16 years) with 35 students (12 boys and 23 girls), forming two groups. The result of the implementation in the classroom was an improvement in the academic performance of the entire student body, being more appreciable in girls and in general, in students with a lower academic level.

Keywords: Physics · Secondary · Educational technologies · Academic performance

1 Introduction

There are many students who become disengaged from science from secondary education, and in many countries, the disengagement from scientific-technological education increases in the transition from secondary education to tertiary education; and to increase the number of graduates in scientific-technological careers it is important to increase the number of entrants [1, 2]. In addition, a third of higher education students drop out of science and technology studies, being the previous school performance one of the factors affecting retention [3]. These issues are not new and are a matter of concern that have a deep impact in both national and transnational Higher Education programs [4].

Physics is one of the scientific disciplines with the lowest rate of academic performance in Secondary Education [5], and, although it is one of the most prevailing subjects in the field of engineering, it is considered the most problematic and

Á. Herrero et al. (Eds.): ICEUTE 2020, AISC 1266, pp. 307–316, 2021.
https://doi.org/10.1007/978-3-030-57799-5_32

traditionally less attractive to students than other scientific disciplines [6]. However, the use of digital and technological tools can favor the Physics teaching-learning processes [7–9] facilitating the construction of concepts by the student. In meaningful learning, the student relates the new information with what he already has, composing meanings, integrating or assimilating the new information to the schemes he already possesses [10]. For this to occur, the student must provide the information received from meaning, so that, with the modification of previous knowledge arises new learning [11].

In the transmissive teaching of concepts, many of the students limit themselves to memorize and subsequently try to reproduce the teacher's words as faithfully, without understanding the physical concepts or phenomena [12]. On the other hand, active methodologies, such as problem-based teaching, directed research or inquiry, have some common characteristics: they intervene in the student's near development zone [13], being, the student, the protagonist of the teaching-learning process and the person responsible for their learning, building science and applying it, in most cases, to real-life situations; and in these methodologies, the teacher guides the student in the learning process by guiding him through continuous feedback [14]. Thus, using a variety of methodologies and technologies, the teacher can direct their interventions towards all students, favoring inclusion in the classroom, increasing their participation and "reducing exclusion in and from education" [15 p. 13]. This inclusion implies access to quality education without any discrimination, and along this line the transformation of the educational system must proceed [11].

2 Research Objectives

The objectives proposed in this research are as follows:

- Analyze student academic performance when active methodologies are incorporated using mobile devices and technologies.
- Determine if there are differences in student academic performance between both groups.

3 Methodology

3.1 Participants

The work sample is made up of 35 students (23 girls and 12 boys) from Compulsory Secondary School (14–16 years), enrolled in an educational center in the downtown area of a city in northern Spain of medium-high socio-economic level. These students are separated into two classrooms or groups: group A, consists of 17 students (6 boys and 11 girls), and group B is also composed of 18 students (6 boys and 12 girls). Group A usually has better academic results and a better attitude during classes than group B. Educational interventions were conducted in six sessions with each group.

3.2 Instruments

For the instruction, a program of Physics activities with incorporated technologies was used. This program is designed to work the following Physics contents: Difference between position, trajectory and displacement of a body. Average speed and instantaneous speed. Equations and graphs of rectilinear movements. Relationship between forces and movement. A brief description of the activities carried out and their purpose is given in the following lines:

1. Kahoot! Questionnaire: it is a questionnaire of previous knowledge about kinematics content to know the students' previous ideas.
2. Troubleshooting with the Smartphone: this activity is carried out by students in groups of 4 that have to use a physics tool Box app (https://bit.ly/3cXGhuY) and video Tracker (https://physlets.org/tracker/).
   - Student 1: a) travels a distance of 40 m in the yard (at constant speed), in which there are marks every 5 m. The experience is repeated three times and promised; b) try to travel the same distance by increasing its speed evenly.
   - Student 2 and 3: time the time it takes for student 1 to travel 40 m, and the times in which he passes through the 5-m marks. With this data they will calculate the average speeds in each interval and the average speed of the entire route.
   - Student 4: videotape the movement made by component 1 in each of the experiences, always from the same position without moving the phone.
3. Virtual Laboratory of *educastur* [16]: one of the objectives of this activity is that students can contrast the difference between the actual data collection (see activity 2) and the use of data obtained in a simulation. Students take data and use Excel to analyze them and find the differences between uniform movements and accelerated movements. This virtual laboratory can be visited at: https://fisquiweb.es/Laboratorio/Cinematica/LabCinematica.htm.
4. Oral presentation using PowerPoint: students expose the activities they have done.
5. QR codes: reinforcement, extension and other activities to promote reading and values were created. Students could access them using their mobile device. These are exercises that students can do and then ask the teacher. Some of these exercises contain texts that contextualize the concepts of kinematics in the real world. It can be downloaded at the following link: http://bit.ly/2ROGo41

For the evaluation of the academic performance of the student body, a questionnaire was prepared consisting of 20 test questions on the contents of Physics. A quasi-experimental design was used with pre-test/post-test measures in the group [17]. Figure 1 and Fig. 2 shows a sample of the 20 questions in the questionnaire.

1. Kinematics is the branch of Physics that studies...:

a) Forces;	**b)** Movement;
c) Energy;	**d)** Cinema

Fig. 1. A sample of the questions in the questionnaire for the evaluation of the academic performance of students.

4. We call the way that travels the mobile...:
a) Trajectory; **b)** Space traveled; **c)** Position

5. The path length between two given positions is...:
a) Trajectory; **b)** Space traveled; **c)** Position

6. The branch of Physics that studies movement is called...:
a) Mechanics; **b)** Kinematics;
c) Speed; **d)** Acceleration

7. The relation between the space traveled and the time taken to travel it is...:
a) Speed; **b)** Acceleration;
c) Time; **d)** The space traveled

10. The variation of the speed in the unit of time is called:
a) Average speed; **b)** Acceleration;
c) Space traveled; **d)** Time spent

14. In the International System, acceleration is measured in m/s^2.
a) True; **b)** False

17. The statement: "Space travel and movement of a mobile are two equal concepts," is:
a) Correct; **b)** Wrong

18. According to the graph:
a) The mobile reaches a speed of 30 m/s;
b) The mobile travels in total 30 m;
c) In the third section, it returns to the slowest starting point;
d) The mobile is stopped 10 s.

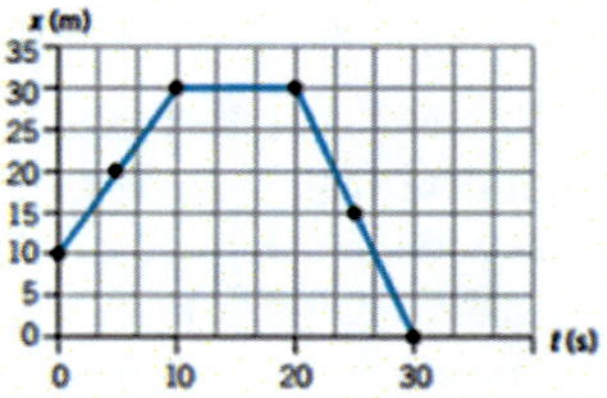

19. The following graph refers to a uniformly accelerated rectilinear motion:
a) True;
b) False

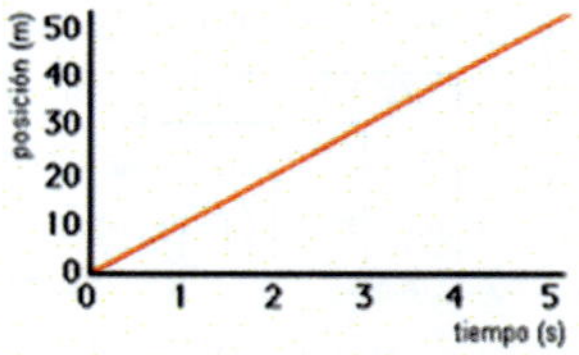

20. The following graph refers to a uniform rectilinear movement:
a) True;
b) False

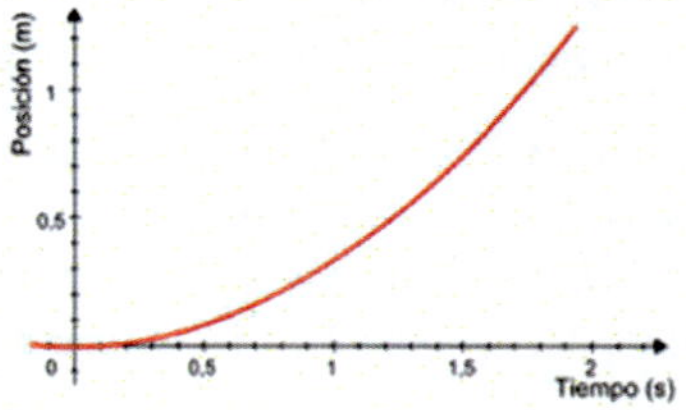

Fig. 2. An extract from the questionnaire for the evaluation of the academic performance of the students.

The data analysis as well as the graphical representation of the data was performed using the statistical software R (http://www.R-project.org).

3.3 Procedure

The consent was obtained from the Directorate of the Center where the study was carried out. Likewise, all participants and their families were informed of the objectives of said study and their consent was requested. The assignment of the students to each of the groups is not random, since in order to maintain the normal functioning of the classes in the center, it was necessary to work with each group-class configured in a natural way. Therefore, the study is carried out through a quasi-experimental type design with pre-test and post-test measures in the two groups in order to establish the necessary comparisons. The contents and activities of Physics teaching were structured to be addressed over six sessions. Some of these activities are designed for group work and others for individual work. In the first and last of the sessions, the student completed the pre-test/post-test evaluation questionnaire.

The students used the following technologies integrated in the teaching-learning activities: Smartphone, take videos of the movement, take photographs for the preparation of final reports, use a stopwatch, read QR codes; Physics Toolbox Suite, smartphone application that allows to determine the acceleration and speed to which the mobile is subjected; Video Tracker, program that allows to obtain data of the movement of an object from a video recording; Virtual Laboratory for the study of movement [16]; other elements such as: PC, Excel, QR codes.

4 Results

The comparison of the results of the pre-test and post-test allows to analyze the acquisition of mathematical competence and science and technology competences. Table 1 shows a brief descriptive analysis of the data of the pre-test, post-test and the difference between them.

Table 1. Average and standard deviations of the questionnaire score by group and gender.

	n	Pre-test		Post-test		Difference	
		M	SD	M	SD	M	SD
A Group	17	10.59	4.66	12.64	4.33	2.06	2.93
Girls	11	9.36	4.97	11.64	4.57	2.27	3.20
Boys	6	12.83	3.31	14.50	3.45	1.67	2.58
B Group	18	9.17	2.87	12.77	3.52	3.61	3.52
Girls	12	9.0	3.25	12.50	2.35	3.50	2.88
Boys	6	9.5	2.17	13.33	5.43	3.83	4.88

It can be seen that the starting situations of the groups in the pre-test differ by more than one point and that the score is higher in group A than in group B. Post-test results indicate that both groups have improved their performance and there is a slight improvement in group B versus A. However, it can also be seen that group B has increased its performance in the questionnaire by an average of more than 3 points, being able to intuit that the methodology has had a positive effect in group B that started from a more disadvantageous situation as to performance. Figure 3 graphically shows some of the patterns detected in Table 1: it can be seen how in group B boys and girls start from a similar performance in the pre-test, while in group A it is observed that boys have a superior performance compared to girls. Regarding the girls, between group A and group B, greater improvement is observed in group B.

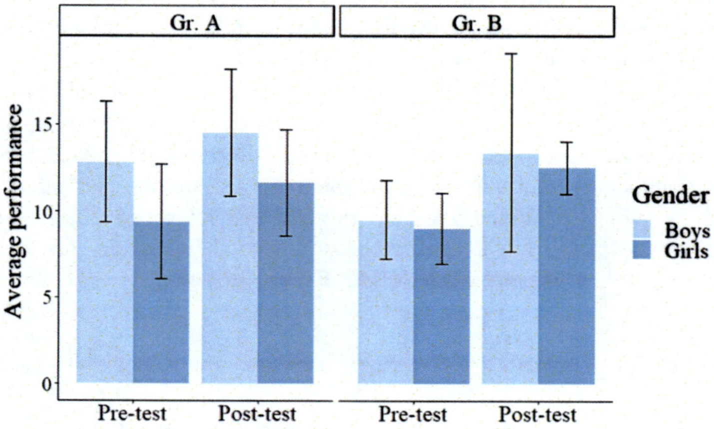

Fig. 3. Average performance in the questionnaire in pre-test and post-test by gender and Group.

An important aspect to check is whether the initial levels of the participants in terms of problem solving are really comparable. For this, a contrast of means of independent groups was made on the result of the pre-test, taking into account the gender and student group of belonging (A, B). Given the sample size, proceed with a non-parametric Wilkoxon test. The statistical analysis showed no significant initial differences in the performance of the participants depending on the group ($W = 182$, $p = 0.3447$) and no statistically significant differences were found in gender ($W = 180.5$, $p = 0.1426$).

Once the starting situation is comparable, it was studied whether there has been a gain between the post-test and the pre-test.

The analyzes establish that the gain scores in group A were not significantly different from those in group B ($W = 119.5$, $p = 0.272$), therefore, it is not possible to reject the null hypothesis that states that training using technologies does not encourage higher yields between groups.

With respect to the results of the gain in terms of the gender variable, non-statistically significant results are obtained (W = 142.5, p = 0.8885).

If the gain is observed for girls between group A (M = 2.27, SD = 3.20) and group B (M = 3.50, SD = 2.88), the results are not statistically significant (W = 56, p = 0.5551). For boys, in group A (M = 1.67, SD = 2.58) and group B (M = 3.83, SD = 4.88), the results are not statistically significant (W = 12, p = 0.3717).

To complete the previous results, we analyzed the students' responses to the questionnaire and based on the number of correct or failed answers of the pre-test and the post-test, a new classification of student performance was established, shown in Fig. 4 where it indicates if the performance improves (80%), worsens (14.29%) or does not vary (5.71%) after the educational intervention.

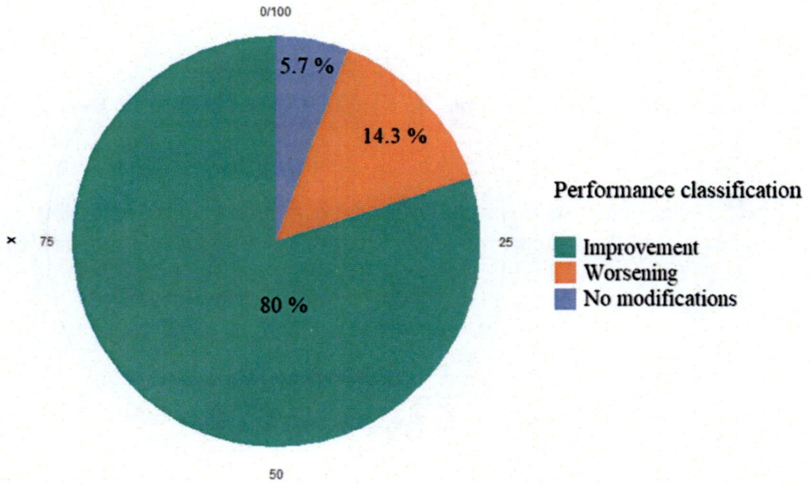

Fig. 4. Classification of student performance based on the number of correct or failed answers.

Making this new classification in each of the groups, it is obtained that although both groups achieve an appreciable improvement, a performance difference of 18 points is observed, better in group B with respect to group A (see Fig. 5).

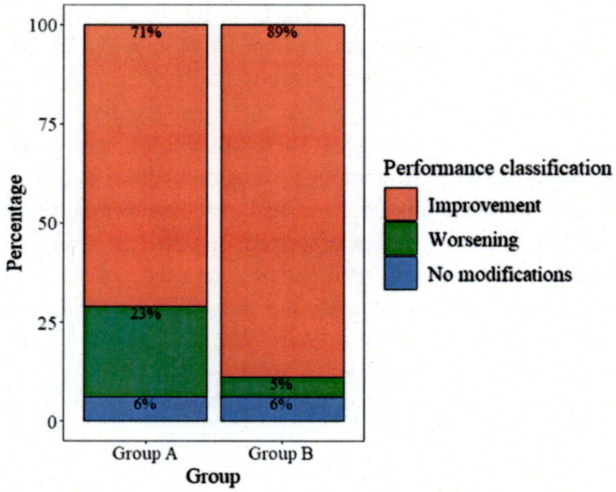

Fig. 5. Performance classification according to number of correct or failed answers (left group A, right Group B).

The performance analysis according to the gender variable is shown in Fig. 6. The results indicate a significant improvement in academic performance in girls (83%) with respect to boys (75%).

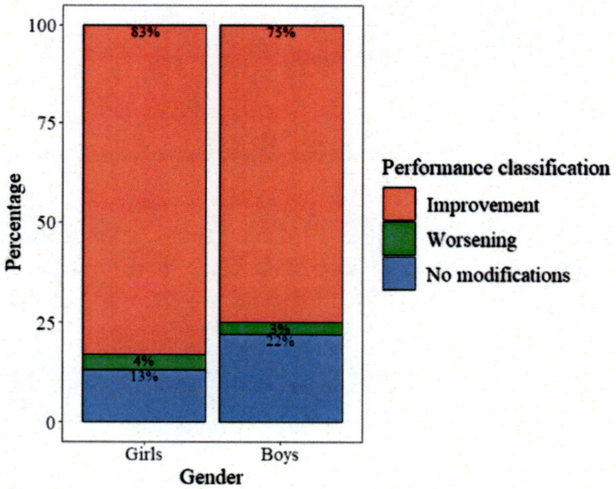

Fig. 6. Student performance by gender.

5 Analysis and Discussion of Results

Although the sample size forces the results to be taken with caution, an analysis of academic performance with the use of active methodologies and technological tools, shows an appreciably greater improvement for girls compared to boys. On the other hand, as indicated above, the class with the lowest academic performance was that of group B; however, the results show a higher academic performance than the other group (group A). That is, the use of active methodologies and teaching-learning technologies applied to the teaching of Physics produces an improvement in the academic performance of all students, but this performance is more noticeable in girls and in students with a lower level academic, despite the fact that the gender variable is usually a significant predictor of the scientific-technological performance of the student, with girls usually getting a lower average [18], although there is no difference between girls and boys in terms of capacity [19].

In subsequent investigations, the sample size should be increased and, in addition, generated content adapted to the Physics curriculum, so that they can be systematically implemented along the different levels of education, which, presumably, would improve performance academic in the Physics assignment.

6 Conclusions

The incorporation of active methodologies and technologies in the teaching-learning of Physics in Secondary improves the academic performance of students. As a consequence, presumably, there will be more students interested in science-technology careers [1, 2] and fewer students who drop out of these studies areas [3], in addition to contributing to transnational education due to problem-solving method makes students develop some skills like research, discovery, scientific thinking [20].

On the other hand, as has been shown in this article, the use of a variety of methodologies and technologies makes it possible to direct teacher interventions to the diversity of students [11, 15], among others, gender diversity [19, 21].

References

1. Boiko, A., Nistor A., Kudenko I., Gras-Velazquez, A.: The attractiveness of Science, Technology, Engineering and Mathematics subjects. Results from five countries. European Schoolnet, Brussels (2019)
2. Henriksen, E.K.: Introduction: participation in science, technology, engineering and mathematics (STEM) education: presenting the challenge and introducing Project IRIS. In: Understanding Student Participation and Choice in Science and Technology Education, pp. 1–14. Springer, Dordrecht (2015)
3. Ulriksen, L., Madsen, L.M., Holmegaard, H.T.: Why do students in stem higher education programmes drop/opt out?–explanations offered from research. In: Understanding Student Participation and Choice in Science and Technology Education, pp. 203–217. Springer, Dordrecht (2015)

4. Larsen, M.R., Sommersel, H.B., Larsen, M.S.: Evidence on Dropout Phenomena at Universities. Danish Clearinghouse for Educational Research, Copenhagen (2013)
5. Sáiz-Manzanares, M.C., Bol, A.: Cómo enseñar y cómo evaluar la resolución de problemas en física: una reflexión sobre la propia práctica. In: Queiruga Dios, M.A. (ed.) Innovación en la enseñanza de las Ciencias: reflexiones, experiencias y buenas prácticas, pp. 129–146. Editorial Q, A Coruña (2015)
6. Guido, R.M.: Attitude and motivation towards learning physics. Int. J. Eng. Res. Technol. **2**, 2087–2094 (2013)
7. Queiruga-Dios, M.A., Diez-Ojeda, M., Velasco-Pérez, N.: Utilización de las TIC en la construcción de la física: análisis de una propuesta didáctica. In: La educación ante el nuevo entorno digital. Comunicación presentada en Congreso Iberoamericano (2019)
8. Queiruga-Dios, M.A., Juez, S., Sáiz-Manzanares, M.C., Collado, S.: Mobile Learning: análisis y reflexión. Una propuesta de implementación en el aula. In: Membiela, P., Cebreiros, M.I., Vidal, M. (eds.) Nuevos retos en la enseñanza de las ciencias, pp. 517–521. Educación Editora, Ourense (2018)
9. Viloria, R., Tricio, V., Collado, S.: Los teléfonos móviles como herramientas TIC para la enseñanza de la Física. In: Congreso Iberoamericano de Docentes, Algeciras (2018)
10. Ausubel, D.P., Novak, J.D., Hanesian, H.: Psicología educativa: un punto de vista cognoscitivo, vol. 3. Trillas, México (1976/1983)
11. Queiruga-Dios, M.Á, Sáiz-Manzanares, M.C., Montero-García, E.: Análisis de Protocolos en Alumnos de Educación Secundaria Obligatoria. Doctoral dissertation, University of Burgos, Burgos (2016)
12. Queiruga-Dios, M.A.: Indagación, trabajo cooperativo y método científico en la enseñanza-aprendizaje de la física en Secundaria Obligatorio. Propuesta y reflexión. In: Gómez-Galán, J., López-Meneses, E., García, L.M. (eds.) Instructional Strategies in Teacher Training 2016, pp. 317–329. UMET Press, San Juan (2016)
13. Vygotsky, L.S.: El desarrollo de los procesos psicológicos superiores. Crítica, Barcelona (1979)
14. Queiruga-Dios, M.A., Sáiz-Manzanares, M.C., Montero-García, E.: Problemas-Proyectos Adaptativos y Creativos en la enseñanza de las ciencias. Descripción de la metodología y apreciación de los estudiantes involucrados. Res. Educ. Learn. Innovat. Arch. **23**, 1–23 (2019)
15. UNESCO. Guidelines for inclusion Ensuring access to education for all. UNESCO, París (2005). https://unesdoc.unesco.org/ark:/48223/pf0000140224. Accessed 19 May 2020
16. García-González, L.I.: Laboratorio virtual de cinemática. FisQuiWeb (2020). https://fisquiweb.es/Laboratorio/Cinematica/LabCinematica.htm. Accessed 22 Jan 2020 (16)
17. Campbell, D., Stanley, J.: Diseños experimentales y cuasiexperimentales en la investigación social. Amorrortu editores, Buenos Aires (1995)
18. Rodríguez-Mantilla, J.M., Fernández-Díaz, M.J., Olmeda, G.J.: PISA 2015: Predictores del rendimiento en Ciencias en España PISA 2015: Predictors of Science Performance in Spain. Revista de Educación **380**, 75–102 (2018)
19. Botella, C., Rueda, S., López-Iñesta, E., Marzal, P.: Gender diversity in STEM disciplines: a multiple factor problem. Entropy **21**(1), 30 (2019)
20. Aka, I.E., Guven, E., Aydogdu, M.: Effect of problem solving method on science process skills and academic achievement. J. Turkish Sci. Educ. **7**(4), 13–25 (2010)
21. López-Iñesta, E., Botella, C., Rueda, S., Forte, A., Marzal, P.: Towards breaking the gender gap in Science, Technology, Engineering and Mathematics (in press)

Data Collection Description for Evaluation and Analysis of Engineering Students Academic Performance

José Antonio López Vázquez, José-Luis Casteleiro-Roca(✉), Esteban Jove, Francisco Zayas-Gato, Héctor Quintián, and José Luis Calvo-Rolle

Department of Industrial Engineering, University of A Coruña, CTC, CITIC, Avda. 19 de febrero s/n, 15405 Ferrol, A Coruña, Spain
{jose.lopez,jose.luis.casteleiro,esteban.jove, hector.quintian,jlcavo}@udc.es

Abstract. This paper addresses the problem of low academic performance in engineering studies. The data collection is carried out from the databases of the University of A Coruña, corresponding to the students who have completed the degrees of Industrial Engineering. The objective is checking if this data is sufficient to perform an evaluation and analysis of the student academic performance, which allow to know the possible causes of the problem addressed.

Once the analysis and evaluation of the data records has been carried out, it is found that they are insufficient, it is tried with the individual academic performance of each student. Therefore, it is decided to expand the data collection, which will consist in individually consulting of the academic record from each student. The aim is to obtain the attempts and marks obtained in each of the subjects, in order to be able to know the academic performance of the students.

Keywords: Innovation in education · Academic performance

1 Introduction

The academic performance of university students is one of the most important issues of quality in Higher Education in Spain. According to [16], it can be affirmed that the low average marks, the high percentages of not presented to exams or fails, and also the high repetition rate, are not a recent phenomenon and is maintained over the years.

As shown in various studies on Technical Engineering [1,10–12], the percentage of students who finish their studies in the planned time (3 years) is 11% of the total number of graduates, students take an average 5.41 years to graduate, the average dropout rate is around 70% and the return average rate of return is 56%.

Á. Herrero et al. (Eds.): ICEUTE 2020, AISC 1266, pp. 317–328, 2021.
https://doi.org/10.1007/978-3-030-57799-5_33

Taking into account more recent works [6–8] carried out on the Spanish University System (SUE), the performance rates of Engineering and Architecture degrees are around 60% and they are the lowest of the SUE. Comparing them with the best performing degrees, Health Sciences, are 20% lower.

Currently, similar data are obtained to the previous ones, even lower, in the Degree in Electrical Engineering (performance rate of 51.90%) and Degree in Industrial Electronic and Automatic Engineering (performance rate of 50.03%), belonging to the University of A Coruña (UDC) and taught at the Polytechnic University School (EUP, *Escuela Universitaria Politécnica*).

From the UDC it is crucial to understand how to address the problem of low academic performance in engineering students, with the aim of trying to improve it [4,9,12,14]. For this, the first thing that must be done is the collection of existing data in the UDC to verify if it serves as a basis to analyze and evaluate the students academic performance.

The rest of the document is structured as follows. Sect. 2 describes briefly the case of study. After Sect. 2, it is explained the way to obtain the dataset. Sect. 4 shows the methods used to analyze the data, and in the next section, the results are shown. The paper ends with the conclusions and future works in Sect. 6.

2 Case Study

The dataset used in this research is collected from the degrees taught at the EUP of the UDC. The degrees lasts three academic courses and have a total of 25 subjects divided in nine, nine and seven subjects each year, with 236 academic credits.

The academic data of students enrolled in the EUP from 1996/1997 to 2009/2010 were considered. This dataset consisted of 2736 students, that studied the Electrical and Electronic Degree between these courses.

3 Obtaining Dataset

The procedure to obtaining the final dataset used in this research can be divide in two different stage. The process is described deeply in the next sub-sections.

3.1 First Stage

This first stage started when the UDC is requested to allow the used of the students academic records. This data includes the students enrolled in the Electrical and Electronic Engineering Degree. The request included the next data for each student, from the beginning until the end of the degree:

- The admission mark: the mark obtained by the students in their previous studies.

- The access method: the method used by the student to enroll the University. Two main cases are considered: high school or vocational training.
- The number of attempts required to pass each subject, and the mark obtained.
- Optional subjects for each student.
- Time spent to finish the degree.

Tables 1 and 2 shows an example of this records. 2001 students are included in this first dataset. Where:

- 1M/CN: 1st registration/Continuing studies.
- OB/TR/OP: Obligatory/Troncal/Optional.
- F/X: Final/June attempt.
- DEC: December attempt.
- SET: September attempt.
- NP: *No presentado.*
- AP: *Aprobado.*
- NT: *Notable.*
- MH: *Matrícula de honor.*

Table 1. Students academic records, 1st stage

Course	Student code	Access method	Subject code	Subject	Type	Credits	Attempt	Mark
2000/2001	162604	1M	770511107	*Estadística*	TR	6	ST	NP
2000/2001	162604	1M	770511108	*Química*	OB	6	F/X	MH
2002/2003	162604	CN	770511102	*Matemáticas I*	TR	12	DEC	NT
2001/2002	162604	CN	770511201	*Electrometría*	TR	3	F/X	NT
2001/2002	162604	CN	770511503	*Ampliación de Química*	OP	4.5	F/X	NT
2001/2002	162604	CN	770511204	*Regulación Automática*	TR	7.5	F/X	NP
2002/2003	162604	CN	770511304	*Oficina Técnica*	TR	9	F/X	AP

As the original data are confidential, the students are identify by a numerical code to preserve its privacy according to *Ley Orgánica 15/1999 de Protección de Datos de Carácter Personal*, as it is shown in Tables 1 and 2. Where:

- PAU: University access exam.
- LOGSE: *Ley Orgánica 1/1990 de Ordenaciń General del Sistema Educativo.*

3.2 Second Stage

After analysing the results of the information obtained in the first stage, and after several conversations with the Titles, Quality and New Technologies Vice-Rectorate, it was analyzed the possibility of complementing the dataset, accessing to the academic data of 2736 students with the following information:

Table 2. University access method for each student, 1st stage

Student code	Degree code	Access method	Access mark
162561	770511V01	*PAU*	5.57
164206	770511V01	*Bachillerato LOGSE y Selectividad*	Pass
164214	770511V01	*Bachillerato LOGSE y Selectividad*	Pass
164223	770511V01	*Bachillerato LOGSE y Selectividad*	Pass
164237	770511V01	*Bachillerato LOGSE y Selectividad*	Pass
164251	770611V01	*Bachillerato LOGSE y Selectividad*	Pass
164288	770611V01	*Bachillerato LOGSE y Selectividad*	Pass
164304	770511V01	*Bachillerato LOGSE y Selectividad*	Pass
164322	770611V01	*Bachillerato LOGSE y Selectividad*	Pass

- Gender.
- Access method and mark.
- Enrolling University age.
- Finishing degree age.
- Number of courses used to finish the degree.
- Average mark at the end of the degree.
- Geographical origin.

In Tables 3 and 4, an example of this information is shown in the original format.

Table 3. Students academic records, 2nd stage

Student code	Gender	Access method	Mark	Initial course	Initial age	Ending course
812570	Woman	PAU	5.85	1997/1998	20	2003/2004
812805	Man	PAU		2001/2002	19	2005/2006
813020	Man			2000/2001	15	2004/2005
813465	Man	PAU		2006/2007	19	
813700	Man	PAU	5.87	1997/1998	22	2005/2006
813815	Man	Vocational training	8.18	2002/2003	21	2012/2013
814175	Man	PAU		2006/2007	56	
821030	Man	PAU	6.98	2007/2008	19	2012/2013
821070	Man	PAU	5.64	2007/2008	22	

4 Methods

This section indicates the methods and tools used for the descriptive analysis of the data.

Table 4. Students academic records, 2nd stage

Student code	Finished?	Ending age	Penalized mean	Geographical origin
812570	Y	26	1.0572	Ferrol
812805	Y	23	1.3398	A Baña
813020	N			Narón
813465	Y	30	1.0981	Betanzos
813700	Y	31	1.5848	Aranga
813815	N			Fene
814175	Y	24	1.2024	Laracha
821030	N			A Coruña
821070	Y	25	1.3742	Lugo
821115	N			Ferrol

4.1 Organization of the Data. Frequency Distribution

The frequency tables and histograms are used to extract and interpret the information related to the dataset obtained in the second stage.

Frequency Tables. The frequency tables are used to analyze the data related to the age at the beginning of the studies, the age at the end of the degree and number of courses used.

Once the information has been collected, the frequency tables are elaborated. The following general concepts [3,13] are used:

- Absolute frequency n_i: the absolute frequency of the value x_i from the variable X, is the number of times n_i that the value is repeated.
- Cumulative absolute frequency N_i: the cumulative absolute frequency of the value x_i is called to the sum of the absolute frequencies of the values of the variable X less or equal to x_i.
- Relative frequency f_i: the relative frequency of the x_i value from the X variable is the ratio between the number of times that appear x_i and the total number of values of the variable (N). $f_i = \frac{n_i}{N}$.
- Cumulative relative frequency F_i: is the absolute cumulative frequency divided by the total number of values of the variable. $F_i = \frac{N_i}{N}$.

Table 5 shows a general frequency table.

Histograms. A histogram is a graphical representation of a set of observed values of a continuous quantitative variable [15]. To perform a histogram, the abscissa axis represents the values taken from the studied variable grouped in sections. A rectangle of height proportional to the absolute frequency of the observed values is raised on each section.

As an example, Fig. 1 shows the histogram with the students age at the beginning of the degree.

Table 5. Frequency table

X_i	n_i	N	f_i	F_i
x_1	n_1	$N_1 = n_1$	$f_1 = \frac{n_1}{N}$	$F_1 = \frac{N_1}{N}$
x_2	n_2	$N_2 = n_1 + n_2$	$f_2 = \frac{n_2}{N}$	$F_2 = \frac{N_2}{N}$
x_3	n_3	$N_3 = n_1 + n_2 + n_3$	$f_3 = \frac{n_3}{N}$	$F_3 = \frac{N_3}{N}$
$\vdots$	$\vdots$	$\vdots$	$\vdots$	$\vdots$
x_k	n_k	$N_k = n_1 + \cdots + n_k = N$	$f_k = \frac{n_k}{N}$	$F_k = \frac{N_k}{N} = 1$
	$\sum n_i = N$		$\sum f_i = 1$	

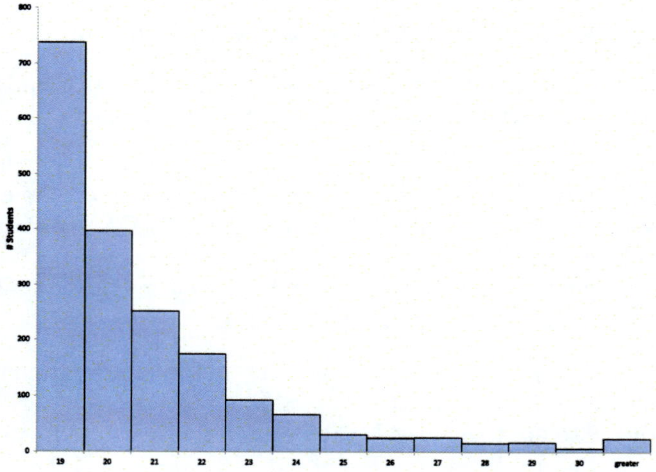

Fig. 1. Students age at the beginning of the degree

5 Results

This section shows the results of the descriptive analysis of the data obtained at each stage, that is carried out with the aim of identifying explanatory variables of the students academic performance.

5.1 First Stage

Once the data of this stage is analyzed, it is verified that they are insufficient data to start the work that is intended because they are partial or omitted. The information regarding the qualification access notes and the numerical mark of each subject is incomplete. The time invested in finishing the degree of each student, and the optional subjects, are unknown. It is only fully known the student access method. The results of this stage are not shown because it is decided to perform the analysis with the dataset collected in the second stage.

5.2 Second Stage

Once the information obtained at this stage has been analyzed, of the 2736 students, 880 academic data is discarded because all the information necessary for the research is not available. Therefore, the study begins with the data of the 1856 students who present the complete information related to: gender, geographical origin, access method, age at the ending the degree and number of courses used to finish. The available information is described below.

Gender. The majority of students enrolled are men, as shown in Fig. 2 an Table 6.

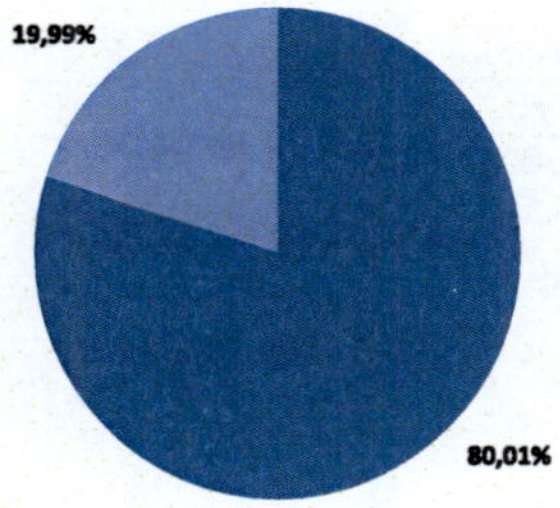

Fig. 2. # students enrolled by gender

Table 6. # students enrolled by gender

Gender	# students
Men	1485
Women	371

Geographical Origin. It is verified that almost all the students are from Autonomous Community of Galicia, as shown in Fig. 3 and Table 7.

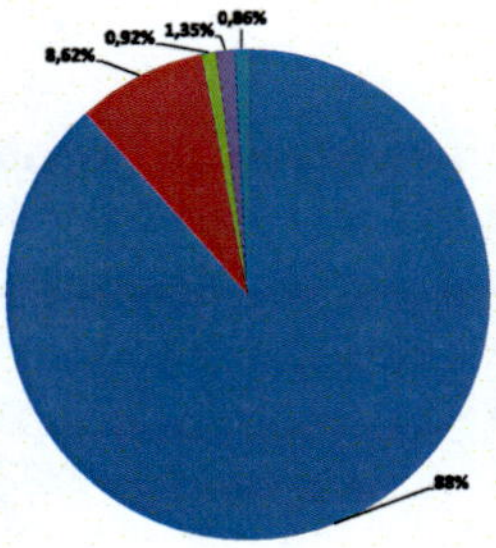

Fig. 3. Geographical origin of the students

Table 7. Number of students according to place of origin

Geographical origin	# students
A Coruña	1638
Lugo	160
Orense	17
Pontevedra	25
Rest of Spain	16

Access Method. Of all the possible access method contemplated in the legislation, the students enrolled mainly come from high school and, lesser extent, from vocational training. Other access method used, a minority, is other university degrees and over 25 years old. The complete data is presented in Fig. 4 and Table 8.

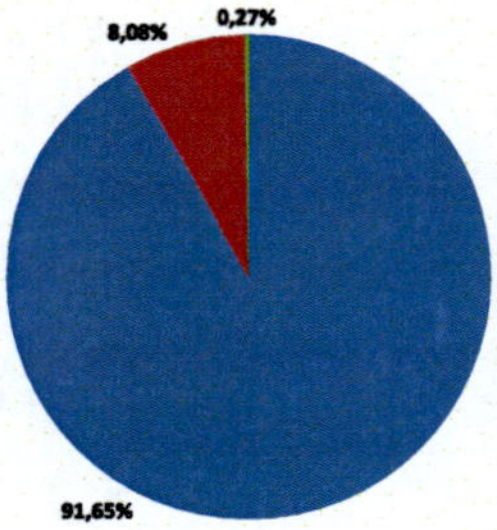

Fig. 4. Degree access method

Table 8. Degree access method

Access method	# students
High school	1701
Vocational training	150
Other university degrees	5
and over 25 years old	

Age at the Beginning of University Studies. The average age of the first registration is 20.86 years. As it can be seen in Table 9, 74.62% of students start the degree with 21 years old or less, and the remaining 25.38% enroll with more than 22 years old. A histogram with this data is shown on Fig. 1.

Table 9. Students age at the beginning of the degree

Age X_i	Absolute frequency n_i	Cumulative absolute frequency N	Relative frequency f_i	Cumulative relative frequency F_i
19	738	738	39.76%	39.76%
20	396	1134	21.34%	61.10%
21	251	1385	13.52%	74.62%
22	174	1559	9.38 %	84.00%
23	92	1651	4.96%	88.95%
24	66	1517	3.56%	92.51%
25	30	1747	1.62%	94.13%
26	24	1771	1.29%	95.42%
27	25	1796	1.35%	96.77%
28	15	1811	0.81%	97.58%
29	16	1827	0.86%	98.44%

Age at the End of Studies. It must be taking into account that, when the dataset is collected, only 962 students (51.83 %) was finished the degree.

The average age at the end of the studies is 26.40 years. Figure 5 shows the histogram of the graduates age.

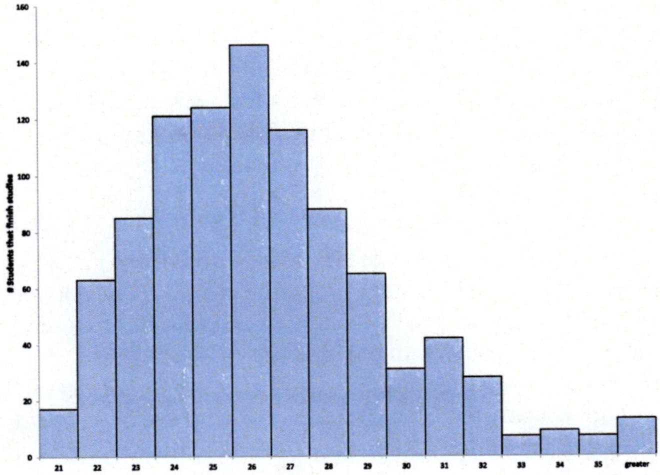

Fig. 5. Graduates age

Number of Courses Used to Complete the Degree. Students use 5.96 years on average to finish the degree. Table 10 shows that only 12.99% of the students finish the degree in the theoretical duration (3 courses), and 52.29% use 6 or more courses to complete their studies.

Table 10. Number of courses used to complete the degree

Courses X_i	Absolute frequency n_i	Cumulative absolute frequency N	Relative frequency f_i	Cumulative relative frequency F_i
3	125	125	12.99%	12.99%
4	157	282	16.32%	29.31%
5	177	459	18.40%	47.71%
6	171	630	17.78%	65.49%
7	128	758	13.31%	78.79%
8	87	845	9.04%	87.84%
9	42	887	4.37%	92.20%
10	37	924	3.85%	96.05%
Greater	38	962	3.95%	100.00%

6 Conclusions and Future Works

The degrees analyzed are studied mostly by men, 80%, so that the need to promote the vocation of engineering in women is evidenced. In order to promote the presence of women in these degrees, it would be convenient to develop specific dissemination and recruitment programs for women. For example, increase the female teachers presence in "STEMbach" type programs, through which it is intended to promote the vocation of students in scientific and technological research.

The analysis of the geographical origin shows that the students enrolled mainly from the provinces of A Coruña (88%) and Lugo (8.62%). Based on this data, several student recruitment programs should be developed: some specific for these areas and others with new recruitment strategies with the aim of expanding the origin of students in the degrees.

Regarding the access method, it is shown that a large majority of the students enrolled from high school (91.68%) compared to who are from vocational training (8.08%). Taking this information as a reference, dissemination and recruitment strategies for students who are studying vocational training cycles could be designed.

The low academic performance resulting in the engineering degrees is verified, since the graduates use an average of 5.96 years to complete the degree. In the

3 academic courses corresponding to the theoretical duration of the degree, only finish 12.99% of the students, and 52.29% use more than 6 courses.

Finally, it should be noted that the data recorded is not sufficient to know and assess the individual academic performance of each student. There is a lack of information in the attempts and marks obtained in each of the subjects and, in many cases, the marks to access the University.

Since academic qualifications are an accurate indicator to assess academic performance [5] and, in practice, most of the research aim to explain the success or failure in university degrees measure academic performance through the marks of the students [2,17]. In order to be able to measure the academic performance of the students, it is necessary to carry out new researches that allow us to expand the method of gathering information, in order to obtain the following data:

- Access mark to the degree.
- Subjects studied by each student.
- Attempts and marks in each subject.

References

1. Bará, J., Córdoba, J., de Luis, R., Hernández, J., Martín, P.: Evaluación transversal del rendimiento académico de las ingenierías técnicas (2002)
2. Cárdenes, N.D., García-Artiles, M.D., Pérez-Sánchez, J.M., Gómez-Déniz, E.: Un modelo de regresión logística asimétrico que puede explicar la probabilidad de éxito en el rendimiento académico. Revista de Investigación Educativa **33**(1), 27–45 (2015)
3. Casteleiro-Roca, J.L., Barragán, A.J., Segura, F., Calvo-Rolle, J.L., Andújar, J.M.: Fuel cell output current prediction with a hybrid intelligent system. Complexity **2019**, 1–10 (2019). https://doi.org/10.1155/2019/6317270
4. Cecilia, A., Costa-Castelló, R.: High gain observer with dynamic deadzone to estimate liquid water saturation in PEM fuel cell. Revista Iberoamericana de Automática e Informática industrial **17**(2), 169–180 (2020). https://doi.org/10.4995/riai.2020.12689
5. Cinalli, D., Martí, L., Sanchez-Pi, N., Cristina Bicharra Garcia, A.: Collective intelligence approaches in interactive evolutionary multi-objective optimization. Logic J. IGPL **28**(1), 95–108 (2020). https://doi.org/10.1093/jigpal/jzz074
6. Hernández, J., Pérez, J.: La universidad española en cifras. Información académica, productiva y financiera de las Universidades Públicas Españolas. Indicadores Universitarios (Curso académico 2002–2003) (2015)
7. Hernández Armenteros, J., Pérez García, J., Hernández Chica, J.: La universidad española en cifras. CRUE, Madrid (2008)
8. Jove, E., Casteleiro-Roca, J., Quintián, H., Méndez-Pérez, J.A., Calvo-Rolle, J.L.: Anomaly detection based on intelligent techniques over a bicomponent production plant used on wind generator blades manufacturing. Revista Iberoamericana de Automática e Informática industrial **17**(1), 84–93 (2020). https://doi.org/10.4995/riai.2019.11055
9. Jove, E., Casteleiro-Roca, J.L., Quintián, H., Méndez-Pérez, J.A., Calvo-Rolle, J.L.: A fault detection system based on unsupervised techniques for industrial control loops. Expert Syst. **36**(4), 1–13 (2019). https://doi.org/10.1111/exsy.12395

10. Jove, E., Casteleiro-Roca, J.L., Quintián, H., Simić, D., Méndez-Pérez, J.A., Luis Calvo-Rolle, J.: Anomaly detection based on one-class intelligent techniques over a control level plant. Logic J. IGPL (2020). https://doi.org/10.1093/jigpal/jzz057
11. Jove, E., Gonzalez-Cava, J.M., Casteleiro-Roca, J.L., Pérez, J.A.M., Calvo-Rolle, J.L., de Cos Juez, F.J.: An intelligent model to predict ani in patients undergoing general anesthesia. In: Proceedings of the International Joint Conference SOCO 2017-CISIS 2017-ICEUTE 2017 León, Spain, 6–8 September 2017, pp. 492–501. Springer (2017)
12. Jove, E., López, J.A.V., Fernández-Ibáñez, I., Casteleiro-Roca, J.L., Calvo-Rolle, J.L.: Hybrid intelligent system topredict the individual academic performance of engineering students. Int. J. Eng. Educ. **34**(3), 895–904 (2018)
13. Marqués, F.: Estadística Descriptiva a través de EXCEL. RC libros (2009)
14. Marrero, A., Méndez, J., Reboso, J., Martín, I., Calvo, J.: Adaptive fuzzy modeling of the hypnotic process in anesthesia. J. Clin. Monit. Comput. **31**(2), 319–330 (2017)
15. Romero Villafranca, R., Zúnica Ramajo, L.R.: Métodos estadísticos para ingenieros. Colección Académica. Editorial UPV (2013)
16. Tejedor, F.J., García-Valcárcel, A.: Causas del bajo rendimiento del estudiante universitario (en opinión de los profesores y alumnos). propuestas de mejora en el marco del eees. Revista de educación **342**(1), 443–473 (2007)
17. Vélez, A., Roa, N.: Factors associated with academic performance in medical students. PSIC. Educación Médica **2**(8), 1–10 (2005)

Special Session: Practical Implementations of Novel Initiatives

Expansion of an Evidence-Based Workshop for Teaching of Artificial Intelligence in Schools

Julian Estevez(✉), Gorka Garate, Jose Manuel Lopez-Guede, and Manuel Graña

University of the Basque Country (UPV-EHU), Donostia, Spain
julian.estevez@ehu.eus

Abstract. This paper presents an expansion of an Artificial Intelligence workshop developed with Scratch for high school students, which already proved its evidence for the demystification and mathematic basis understanding of this discipline. Students will be able to understand and practice with two simple algorithms, following closely the short instructions and mathematical theory provided by teachers.

Keywords: Artificial Intelligence · Scratch · Education

1 Introduction

The widespread knowledge and application of scientific reasoning is a key social aspect in technologically advanced societies [6]. Therefore, it is of great importance that students find its elements in their curricula as early as possible. Nevertheless, high school is a good time because students are mature enough to cope with serious problems. Any effort invested by the education system to foster its mastering by the young students will undoubtedly pay off. It has been acknowledged that there is a quite widespread mistrust of science [4,5], calling for a radical change in the way the younger generations are introduced to science and technology. The education science community widely recognizes that school curricula must move on from traditional expositive classes to more informal, interactive, experimental, and collaborative learning environments. This is specially clear in the case of undergraduate students.

This article presents the expansion of a workshop to promote the understanding and learning of Artificial Intelligence (AI) at schools, previously presented in various articles of this research group [1–3]. The workshop consists of several simple educational exercises, partially based on the usage of Scratch and other simple mathematical formula to enhance the familiarity with this kind of algorithms.

AI is one of the technologies that will transform our society, economy and jobs in a greater extent along the next decades. Some of the most known examples

Á. Herrero et al. (Eds.): ICEUTE 2020, AISC 1266, pp. 331–337, 2021.
https://doi.org/10.1007/978-3-030-57799-5_34

of AI are driverless cars, chatbots, voice assistants, internet search engines, etc. Despite the myth of destructive AI represented in films such as *Terminator* or *I, Robot*, the truth is that nowadays most usual smart algorithms consist of a series of simple rules applied to large series of numbers, and the result of that is called Artificial Intelligence.

British Parliament and other institutions recommend the education from high school of Artificial Intelligence [8], regardless of the pace of development of this technology, in order to cope future technological and social challenges. Main reason is to improve technological understanding, enabling people to navigate an increasingly digital world, and inform the debate around how AI should, and should not, be used.

Our research team obtained evidences that this kind of workshop enhances two capacities: the demystication of AI and the first understanding of what AI implies [1].

The present article is divided in the following sections. First, in Sect. 2, the pedagogical framework and teaching approach used in this activity is described. Next, the new algorithms that will be incorporated to the workshop are described in Sect. 3. In Sect. 4, the Methodology that teachers will use with students is detailed. Finally, in Sect. 5, conclusions of the experiment are analyzed.

2 Methodological Background

There is a wide consensus in computer science literature and community around the difficulty of teaching the basics of AI [9]. This is due, first, to the lack of a unified methodology, and second, to the fact that AI emerges as a mixture of many other disciplines involving skills ranging from very applied thinking to extremely formal reasoning [10].

In order to acquire the skills for the coming of the digital society, different international institutions, such as British Parliament [8], recommend Computational thinking (CT). This way of coping with problems uses techniques typically practiced by computer-scientists, increasingly viewed as an important ingredient of STEM curricula in primary and secondary education [11]. The understanding of the mathematics and structure of simple programming is essential for the first approach to understanding the basics underlying AI.

With these premises, the design of our workshop was made taking into account Design Based Research (DBR) and Experiential Learning Theory pedagogical (ELT) considerations.

Design Based Research (DBR) aims to develop learning pedagogical interventions through a series of refinements of the experiments, focusing the target population, research questions, and intervention implementations, such as software tools, appropriate scaffolding, etc.

The essence of ELT [12] is the iteration over the cycle of concrete experience, observation, abstraction of concepts, and active design of experiences.

3 Content

This section will shortly recall the algorithms developed for the first version of the workshop, described in [2] and designed with Scratch [7] in a workshop with students 16–18 years old. Next, the new algorithms will be introduced.

3.1 Exercises Based in Scratch

Specific algorithms developed with Scratch are:

- K-means
- Neural network

K-means
Starting from a given cloud of points, the aim of this activity is that the student learns how to cluster the points into K groups, by programming an application in Scratch. In order to do the clustering, the points will be classified by colors, and each one will belong to the group that has the closest average or mass center.

Neural Network
A simple neural network with two inputs and an output neuron, is trained with an AND logic gate in an iterative way. In each iteration, students will see the different weights that the neural network gets. Next, and OR logic gate will be used, and as a consequence, students will observe how adjusting parameters change with this new structure.

3.2 Expansion of the Workshop

New exercises are based on the practise of following algorithms:

- Q-learning
- Regression

Q-learning
Reinforcement learning is one of the popular methods of training an AI system. Rather than attempting to fit some sort of model to a dataset, a system trained via reinforcement learning (called an *agent*) will learn the optimal method of making decisions by performing interactions with its environment and receiving feedback. One of the first and most popular algorithms introduced in this area was the Q-learning algorithm. First introduced in 1989, the algorithm involves the agent experimenting with the environment, and receiving a reward for each action. Over time, the agent will learn the optimal actions to take in order to obtain the largest amount of award.

The emphasis is on Q-learning systems since they support off-policy learning, and it represents an easy algorithm to explain AI ideas. For instance, it permits to understand how robot vacuums learn the path to cover efficiently all the surface in a flat.

Q-learning is based on *states* and *actions*. An agent must go from one state to another through different actions, till it maximizes the reward and optimizes the actions to do autonomously. States and actions will easily be understood in Sect. 4.1 with an example.

- For each state-action pair (s, a), initialize the table entry $Q(s, a)$ to zero
- Observe the current state s
- Do iteratively:
  - Select an action a and execute it
  - Receive immediate reward r
  - Observe the new state s'
  - Update the table entry for $Q(s, a)$ as follows:

$$Q(s,a) = Q(s',a') + \alpha \cdot [r + \gamma \cdot (Q_{Max} - Q(s',a'))] \tag{1}$$

Linear Regression. This algorithm is a linear approach to modeling the relationship between a scalar response (or dependent variable) and one or more explanatory variables.

Linear regression is a powerful tool to understand the fitting of mathematical models to the real data in a problem. Thus, concepts like minimum square errors and correlation can be introduced. Moreover, it's a very useful first step to see the data distribution.

In the case of AI, apart from being able to watch the relationship, it is used to show the basics of prediction.

4 Methodology

The methodology of the workshop related to new exercises described in this article, will follow up the same structure detailed in [1]. After a short introduction of what AI is and its implications, these are the steps for Q-learning and regression activities:

- At the beginning, students will receive a file for the fulfilment of the Q-learning exercise. The teachers will give the students a 10 min explanation of the basis of the algorithm, using that file.
- After each explanation, students will have 20–25 min to play and practice in pairs with the formulas already existing in the file and the available data. Teachers will be available to answer questions related to any technical doubt and to give additional explanations.
- The same structure will be repeated for the regression activity.

Next, the details of the exercises and content of the spreadsheets are presented.

4.1 Q-Learning

Q-learning is practised through a maze, inspired by [13], which is a classic approach for the beginners of this algorithm. All the possible steps in the labyrinth have a score, and evolute along the iterations of the algorithm, till the agent finds the optimum path to escape from the scenario. Each room represents the *state*, while the next room to move represents the *action* (Table 1).

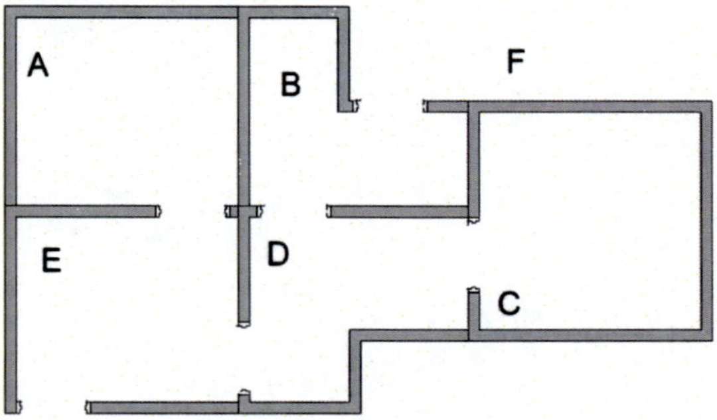

Fig. 1. Scenario for q-learning

Table 1. Q matrix representation of the maze

	Action to go					
Current state	A	B	C	D	E	F
A	-	-	-	-	0	-
B	-	-	-	0	-	100
C	-	-	-	0	-	-
D	-	0	0	-	0	-
E	0	-	-	0	-	100
F	-	0	-	-	0	100

Considering that *F* room is the goal state for the maze and that its score is 100, we can model a Q matrix in order to apply q-learning algorithm.

Students will receive a spreadsheet (i.e., in MS Excel) where the formulas of Q-learning algorithm for the maze shown in Fig. 1 will be represented. Considering that we start the maze in room C, the students will be able to copy, paste and play with the Q-learning formula in Excel, to update the scores of each room till the solution is reached. This exercise is designed for some few iterations.

4.2 Regression

As before, the students will receive a MS Excel sheet with numeric data in 5 columns, and they will be able to practise the linear regression (Fig. 2).

Participants will be able to visualize every possible combination of two columns of data, erase or change the values and observe how the regression line varies depending on that.

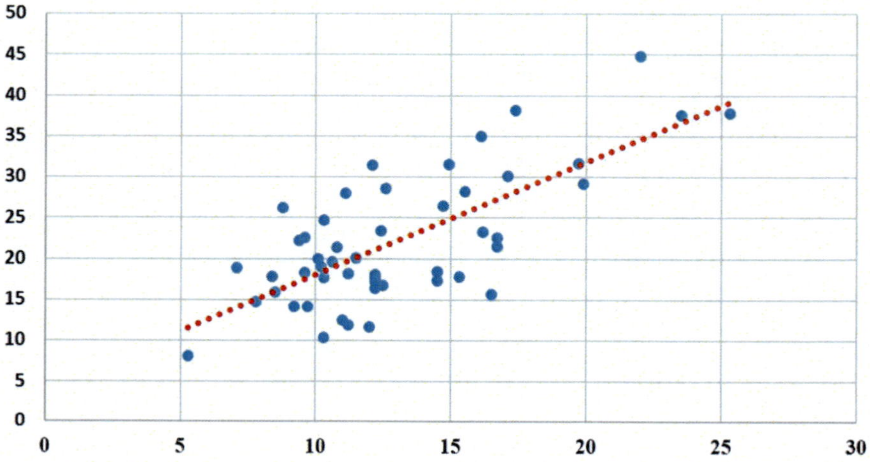

Fig. 2. Regression exercise

5 Conclusions

The paper presents an expansion of an evidence based workshop for teaching Artificial Intelligence foundations to high school students. It consists of easy to implant activities that can be performed at schools with Scratch and spreadsheets like MS Excel. The mathematics have been adapted for 16–18 year-old students mathematical background.

The tasks presented are interactive, so that students can play and practise with the algorithms, following the ELT and DBR learning theories.

Taking into account the work presented in [1], students in this workshop are aimed to acquire the basic knowledge of how to artificially learn a pattern, find the optimum solution, predict a numerical series or cluster a dataset.

The work can be extensible to students of different ages and more AI algorithms can be added to the system.

As a future work, first, the authors should organize the AI workshop several times and measure the grade of the objective achievement in collaboration of pedagogical researchers. Secondly, more programming languages should be explored (GeoGebra, HTML, ShynnyApps). Finally, the development of a team challenge game will be considered for students to engage their learning and practise the acquired skills. This challenge would consist in the resolution of a big problem or puzzle, in which different intuitions, reasonings and algorithms need to be applied to be solved.

References

1. Estevez, J., Garate, G., Graña, M.: Gentle introduction to artificial intelligence for high-school students using scratch. IEEE Access **7**, 179027–179036 (2019)
2. Estevez, J., Garate, G., López-Guede, J.M., Graña, M.: Using scratch to improve undergraduate students' kills on artificial intelligence. In: International Joint Conference: 12th International Conference on Computational Intelligence in Security for Information Systems (CISIS: and 10th International Conference on EUropean Transnational Education (ICEUTE 2019), pp. 312–320. Springer, Cham, May 2019
3. Estevez, J., Elizetxea, X., Garate, G.: Development of a Workshop About Artificial Intelligence for Schools Advances in Intelligent Systems and Computing, vol. 771, 530–535 (2019)
4. Arimoto, T., Sato, Y.: Rebuilding public trust in science for policy-making. Science **337**(6099), 1176–1177 (2012)
5. Haerlin, B., Parr, D.: How to restore public trust in science. Nature **400**(6744), 499 (1999)
6. OECD. Scientific advice for policy making: The role and responsibility of expert bodies and individual scientists. Technical report 21, OECD Science, Technology and Industry Policy Papers (2015)
7. Resnick, M., et al.: Scratch: programming for all. Commun. ACM **52**(11), 60–67 (2009)
8. AI in the UK: ready, willing and able? - Technical report. Select Committee on Artificial Intelligence. House of Lords (2018)
9. Hearst, M.: Improving Instruction of Introductory Artificial Intelligence: Papers from the 1994 AAAI Fall Symposium, November 4–6, New Orleans. AAAI Press, Louisiana (1995)
10. Kumar, D., Meeden, L.: A robot laboratory for teaching artificial intelligence. ACM SIGCSE Bull. **30**(1), 341–344 (1998)
11. Eguchi, A.: Robotics as a learning tool for educational transformation. In: Proceeding of 4th International Workshop Teaching Robotics, Teaching with Robotics, 5th International Conference Robotics in Education Padova (Italy), July 2014
12. Kolb, A., Kolb, D.: Experiential learning theory: a dynamic, holistic approach to management learning, education and development. In: Armstrong, S.J., Fukami, C. (eds.) Handbook of Management Learning, Education and Development (2011). https://doi.org/10.4135/9780857021038.n3
13. Teknomo, K.: Kardi Teknomo's Guidelines on Research and Publication Cooperation (2005). http://people.revoledu.com/kardi/publication/ResearchCollaborationGuide.html

Design of a PBL Experience in the Field of Sustainability for Industrial Informatics

Isidro Calvo(✉), Jose Miguel Gil-García, Estibaliz Apiñaniz, Cesar Escudero, Angel J. García-Adeva, Amaia Mesanza, and María Gastón

Faculty of Engineering of Vitoria-Gasteiz, University of the Basque Country (UPV/EHU), Vitoria-Gasteiz, Spain
{isidro.calvo, jm.gil-garcia, estibaliz.apinaniz, cesar.escudero, angel.garcia-adeva, amaia.mesanza, maria.gaston}@ehu.eus

Abstract. This work describes the design of a PBL experience in the field of sustainability that has been carried out in the subject of Industrial Informatics. Industrial Informatics is a mandatory course taught in the third year of the B.Sc. degree in industrial electronics and automation engineering at the Faculty of Engineering of Vitoria-Gasteiz (UPV/EHU). The proposed PBL experience addresses the design of a real application, based on open-source hardware platforms such as Arduino and Raspberry Pi, aimed at monitoring comfort variables such as temperature and humidity in a distributed way, all over the Faculty. This activity was proposed in the framework of a broader project aimed at improving the sustainability of our University, namely aimed at monitoring the comfort conditions of the Faculty building as well as the performance of the heating system. Therefore, our students may gain experience by analyzing a real project. Also, it could help our students to realize that developing sustainability aimed projects may be a professional venue for their future professional development.

Keywords: Course design and curricula · Project Based Learning (PBL) · Sustainability · IoT · Raspberry Pi · Arduino

1 Introduction

Sustainability is becoming an increasingly relevant topic. Actually, the problem was already identified in the 70s. For example, in a report dating back to 1972 entitled *The limits of Growth* [1], the authors put forward the daunting idea that the natural resources of the Earth were being wasted and, as a consequence, the development of the world would be limited. Later, new concepts were coined. For example, sustainable development was defined as *the development which meets the needs of the present without compromising the ability of the future to meet our needs, ensuring that mankind's use of natural resources and cycles will not lead to reducing the quality of life* [2].

Á. Herrero et al. (Eds.): ICEUTE 2020, AISC 1266, pp. 338–347, 2021.
https://doi.org/10.1007/978-3-030-57799-5_35

The increasing relevance of sustainable development is causing that an increasing number of universities and higher education institutions are starting to include sustainability related topics in their curricula. Sustainable development is a multidisciplinary topic. Actually, some authors consider it as a meta-discipline, since it combines approaches from several disciplines which involve social, economical, scientific and engineering issues [3].

In this scenario, universities should lead the way for introducing sustainable development in the society. They should act as leaders and change drivers ensuring that the needs of present and future generations can be better understood and satisfied. Therefore, students would become the future professionals that would make the transition to more sustainable societal patterns. In order to do so, university leaders and staff must be able to design new paradigms, introduce sustainable development into all courses and curricula and other elements of university activities. In this regard, several declarations have been proposed aimed at introducing sustainability in higher education [4]. As a matter of example, the Declaration of Barcelona, launched in October 2004, calls for multi-disciplinary, systems oriented, critical thinking and participative and holistic education for engineers [5]. Achieving this objective requires that some elements should be reviewed, such as the content of the courses, the teaching strategies, the teaching and learning techniques, evaluation and assessment techniques, etc.

Therefore, in order to promote the objectives of these declarations, new actions must be introduced in the curricula of the universities in order to include sustainability and make students aware of the problem. Some recommended actions are: collaborating with other universities, fostering trans-disciplinarity, making sustainable development an integral part of the institutional framework, creating on-campus life experiences and "Educating the Educators" [4]. In the University of the Basque Country (UPV/EHU), the Campus Bizia Lab is an internal program derived from the Erasmus University Educators for Sustainable Development that promotes sustainability practices in the UPV/EHU according to these principles.

Some authors have identified the key competences that must be supported in the academic programs in order to promote sustainability. These key competences provide a framework for developing student profiles capable of tackling these issues. Since there is some discussion about the meaning of competence, these were defined as a functionally linked complex of knowledge skills and attitudes that enable successful task performance and problem solving. The most relevant sustainability related competences were identified in [6]:

- **System-thinking competence:** It is the ability to analyze and understand complex systems by dividing the problems in separate blocks and analyzing the different interactions among them, such as cascading effects, inertia, feedback loops.
- **Anticipatory competence:** It is the ability to imagine future scenarios, by analyzing, evaluating and visualizing different systems quantitatively and qualitatively.
- **Normative competence**: It is the ability to specify, apply and reconcile with the rules that sustainability approaches should satisfy.
- **Strategic competence:** It is the ability to design and implement interventions, transitions and to define strategies towards sustainability.
- **Interpersonal competence:** It is the ability to motivate, enable and facilitate sustainability. This competence includes advanced skills in communication.

Similar competences to the previous ones were identified as key competences in the field of engineering education, but they need to be adapted in order to be introduced in engineering degrees with sustainability related contents.

Since the sustainability crisis is one of the challenges that engineering education faces for the near future, engineering education institutions have the responsibility to educate engineers that will be able to remediate the existing sustainable problems and prevent future ones [7]. Engineering curricula should create opportunities for students to learn and practice these competences in real-life situations, while breaking with traditional lecturing teaching and compartmentalization of knowledge. Therefore, engineering education needs to change towards an action-oriented curriculum and implement student-centered, experiential, constructive and transformative learning pedagogies such as the Problem Based Learning (PBL) methodology [8]. PBL is an innovative, student-centered learning approach capable of enhancing students' interdisciplinary knowledge, collaboration, problem solving skills, communication, critical thinking and self-directed learning [9]. Some authors have analyzed the synergy between PBL and education for sustainability [10].

This work presents the design of one PBL experience for introducing sustainable development in the B.Sc. degree in industrial electronics and automation engineering (IEAE) at the Faculty of Engineering of Vitoria-Gasteiz (UPV/EHU). The rationale behind it is that these students identify the development of sustainability related Internet-of-Things (IoT) applications as a possible field for their future professional activity. Actually, applications of the kind are at the core of the Industrial Informatics subject, where the PBL activity was designed for.

The proposed PBL activity involves the design of a distributed IoT system aimed at acquiring and monitoring some comfort variables, i.e. temperature and humidity, in one building, namely some classrooms of the Faculty of Engineering of Vitoria-Gasteiz. The PBL activity was designed to involve relatively big groups of students, around 10 students. Therefore, students have to learn how to work in higher size groups, since they are used to work in small groups of around two or three people. However, they are expected to be able to work in higher size, multidisciplinary groups. The PBL activity goes through several stages in which they have to manufacture a new system out of several components.

The layout of the article is as follows. Section 2 is aimed at contextualizing the subject: Industrial Informatics. Section 3 is devoted to introduce the context in which the current PBL activity is embedded. Section 4 describes the proposed PBL activity. Finally, some conclusions are drawn.

2 Description of the "Industrial Informatics" Subject

2.1 Learning Objectives

"Industrial Informatics" is aimed at deepening into the computer concepts used in the modern industrial informatics applications, including communication issues, which with the adoption of the Industry 4.0 and Industrial Internet of Things paradigms are gaining higher relevance. This learning objective is described under TEEO0I10

(*Applied knowledge about industrial informatics and communications*) of the B.Sc. in Electronic Informatics and Automation Engineering. However, other learning objectives for the degree [14] must be also achieved in this subject (See Table 1).

Table 1. Learning objectives of the Industrial Informatics subject according to [14]

Learning Objective	Description
C3	Knowledge about basic and technological subjects. Learning new methods and theories aimed at achieving versatile students adaptable to new situations
C4	Ability to solve problems with initiative, decision making, creativity, critical reasoning and knowledge transmission skills
C5	Knowledge to carry out measurements, complex calculations, valuations, expert's reports, task plans and similar works
C10 (*Trans*)	Ability to work in multilingual and multidisciplinary environments
C12 (*Trans*)	Responsible and organized attitude at work, favoring lifelong learning
C13 (*Trans*)	Applying the scientific methodology strategies: Qualitative and quantitative analysis of the situation and problems, to consider hypothesis and solutions according to the industrial engineering methodology
C14 (*Trans*)	Working efficiently in groups, integrating skills and knowledge to make decisions in the scope of industrial engineering, esp. industrial electronics

2.2 Syllabus

This subsection describes the syllabus of the subject. Some topics are only taught in the classroom, whereas others are only developed by means of the PBL methodology, mainly in the laboratory. The PBL project involves mainly *Programming in C* (Topic 3), and *Design of Embedded Systems* (Topic 4). Due to time restrictions some topics were mainly taught in the classroom, but the acquired knowledge was reinforced throughout the PBL experience. This was the case of the *Introduction* (Topic 1), which introduced concepts related to data acquisition and computer control and to *Introduction to Computer Networks* and *Industrial Communications* (Topics 5 and 6). In the following, there is a brief description of the topics of the "Industrial Informatics" subject:

- **Presentation** ***(Classroom):*** Introduction to the subject. The syllabus, bibliography, objectives and the assessment system are described and explained.
- **Introduction** ***(Classroom):*** Description of different control based applications. The role of the computer in different kinds of industrial systems. Differences between centralized control and distributed control.
- **Operating Systems** ***(Classroom):*** Major objectives of the operating systems. Types of operating systems. Components of an operating system. Role of the kernel. Task scheduler.

- **Programming in C** ***(PBL):*** C programming: Variables. Control flow statements. Modularity and functions. Data structures. Library functions. Operating system API. Decomposition into modules (*top-down* design). Unitary tests. Building complex software out of modules (*bottom-up* design).
- **Programming embedded systems** ***(PBL):*** Introduction to embedded systems. (1) data acquisition; (2) execution of the control algorithm; (3) actuation
- **Introduction to computer networks** ***(Classroom):*** Communication layers. Description of the OSI reference model. Description of the TCP/IP stack.
- **Industrial communications** ***(Classroom):*** Data communications in industrial environments. Automation pyramid. Fieldbuses. Industrial networks.

2.3 Assessment

Table 2 summarizes the assessment followed in this subject. The PBL activity accounts for 45% of the total qualification. However some collaborative activities developed in the classroom (10%) are also closely related to the PBL activity. Another 45% of the total qualification is obtained by means of individual exams. The C programming exam is carried out at the middle of the term, after the finalization of the first subproject, while the final exam covers the rest of the topics of the subject, operating systems, communications, etc. The PBL activity was divided into two subprojects that sought different objectives.

3 Description of the Campus Bizia Lab Project

Campus Bizia Lab Programme (CBL) is an initiative arising from the Erasmus University Educators for Sustainable Development Project, in which the University of the Basque Country/Euskal Herriko Unibertsitatea (UPV/EHU) was involved between 2013–2016. It seeks to trigger a collaborative process between academic staff, service and administrative staff and students (multidisciplinary approach) in order to respond to sustainability challenges within the university.

In particular, the current project is aimed at monitoring the energy consumption of the heating system of the Faculty and identifying discomfort situations, such as Faculty places in which the temperature is not adequate. In the past, some thermal discomfort situations have been detected, which affect the comfort of the entire university community and, of course, the environment. Therefore, in spite of the heating system being in operation, there are cold areas that require the switching on of auxiliary electric radiators. At the same time, there are other areas which are too warm and the members of the university community are forced to open windows even when the heating system is on. These measures are clearly harmful to the environment. This is due to the fact that the current air-conditioning system of the Faculty of Engineering Vitoria-Gasteiz is unbalanced and has a deficient control system. During the last years, modifications have been made to the original design, which have generated malfunctions that involve a waste of primary energy and inefficient distribution of the generated heat. This discomfort and waste of energy requires a study of the current state and a readjustment of the system to return to perform optimally, reducing emissions and air conditioning costs.

Table 2. Summary of the assessment

Activity type	Percentage
Collaborative activities in the classroom	10
Individual C programming exam	20
PBL First subproject	10
PBL Second subproject	25
PBL Project presentation	10
Final exam during the exam period	25

Since the major objective of this project is to identify the causes that give rise to this situation and to propose measures in order to solve it. the following lines of action are proposed to achieve this purpose:

- Evaluation of the current situation of thermal installations.
- Evaluation of thermal comfort.
- Evaluation of the building's thermal insulation.
- Design of an improvement project (sensorial).
- Determination of the viability of the proposal.

4 Design of the PBL Experience on Sustainability

The PBL proposed is aimed at raising awareness about the energy usage, since this is one of the major concerns of Humanity today. Also, it must enforce the learning objectives of the subject as discussed above. More precisely, students were faced with the problem of monitoring the energy consumption of the building of the Faculty of Engineering of Vitoria-Gasteiz, taking into account (1) the behavior of the boiler, (2) the building losses and (3) the behavior of the people. Therefore, bad use of energy might be identified so that the energy consumption could be optimized while keeping the comfort of the university community. As recommended when designing PBL activities, this is a real-world project. Similar systems have been implemented in [11,12] for research purposes.

For that purpose, after analyzing the syllabus of Industrial Informatics, the students identify the topics that can be useful for that purpose, which are basically the whole subject since it focuses mainly on programming, embedded systems and communications. Therefore, they identify that this objective may be attained by designing a distributed IoT system, in which diverse components are involved, such as sensors, smart devices, communications and monitoring devices for keeping and analyzing all captured data. Also, the students must take into account the price of the final system, which must be kept inside certain limits.

Therefore, the final system should be created by integrating diverse components that must be carefully analyzed. These components are: (1) Sensors; (2) Development boards; (3) Communication technologies and (4) Logging and analysis system. Since Industrial Informatics is a 6 ECTS subject students were guided by the instructors in the technology selection. Thus, TMP37 sensors were selected. These are analog temperature sensors that work in the range from 5 ºC to 100 ºC. After evaluating several boards [13], the Arduino Yun was selected since this is a very interesting board that embeds both one ATmega32u4 and one Atheros AR9331 System-on-a Chip. Therefore, students may learn the differences between these. Also, being based on Arduino technology, this board is relatively easy to program with the Arduino IDE environment. In addition, the Arduino Yun integrates WiFi wireless communications, which renders it very suitable to create IoT applications, and an SD slot for keeping the logged files. WiFi technology was selected as communication technology, since this is a technology that may be found everywhere nowadays and, consequently, students should be familiar with it. Finally, students should create a logging system that collects the data from several boards. The proposed platform for the logging system was based on a Raspberry Pi 3 board. Fig. 1 shows the overall IoT system proposed as PBL activity.

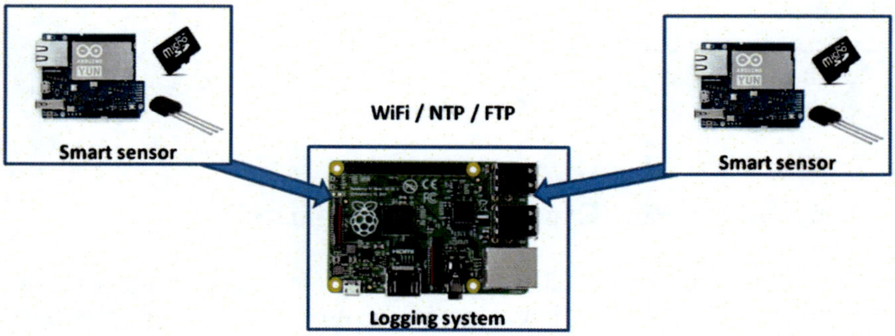

Fig. 1. Overall IoT system proposed as PBL activity

It should be noted that building such as IoT system for monitoring temperature raises several issues involving configuring the devices and protocols, such as NTP (Network Time Protocol) for synchronizing all smart sensors and computers in the distributed system, FTP (File Transfer Protocol) for downloading the data from the smart sensors to the Raspberry Pi based logging system, such as connecting the sensors to the Arduino Yun board, logging the data in the Arduino.

In order to reach the desired goal, the PBL activity was divided into two subprojects. The first subproject, in which groups of two or three people participate, is aimed at creating a simple analysis program that extracts basic information from logged files. This program should be written in C, since this was the programming language that students must learn in this subject. One of the key issues that were involved in this program was the format of the files in which data was logged. In this subproject the data was not collected with the Arduino, but it was a fixed format. The same file format was used as the interface between the Arduino Yuns and the Raspberry Pi.

The second subproject was aimed at implementing the system shown in Fig. 1. This system involves integrating different technologies that execute in different platforms, namely Arduino Yun based smart sensors and Raspberry Pi based logging system.

In addition, several tasks were needed involving cooperation among the students. These tasks are described as follows:

1. Creating a private TCP/IP network based on WiFi
2. Configuration of the Raspberry Pi
   a. Communication configuration for the Raspberry Pi, (IP address, network, mask, etc.)
   b. Configuring a NTP server used to synchronize all Arduino Yuns
   c. Configure an FTP client to downlad the data files kept on the Arduino Yun
3. Configuration of the Arduino Yuns
   a. Communications configuration for the Arduino Yuns, (IP address, network, mask, etc.)
   b. Configure the NTP client to synchronize with the Raspberry Pi
   c. Install and configure an FTP server to make available the files in the SD
4. Programming the Arduino Yun
   a. Reading the temperature from the TMP37 analog sensor
   b. Getting the NTP time from the Arduino Yun
   c. Data logging the temperature data into SD card files, with the NTP and temperature value. These data should be written in the SD card in the same format as defined in the first subproject.
5. Programming the Raspberry Pi
   a. The Raspberry Pi must upload the files obtained from the distributed smart sensor nodes
   b. Analyze the files obtained from several smart sensor nodes and get information

One of the features of the proposed PBL activity is that there are several subtasks that students should perform in order to accomplish the whole project. One of the steps of the project was to identify these tasks. In order to carry out this task, a relatively big size was chosen for the groups. Actually, there were three groups of around 10 people each. This was challenging for the students and the instructors since, at least in our Faculty, group size is typically smaller. For that reason, the instructors had to help students to identify all tasks, so that they could focus on specific tasks of the whole project.

Once the tasks were identified, it was left to the students to organize internally how to solve these tasks. It was recommended by the instructors that for every group of 10 people they create two subgroups depending on the platform, i.e. one for Arduino and another for the Raspberry Pi, so they could concentrate on some tasks. Fig. 2 shows the interactions among all tasks identified at the whole project, according to one of the students group involved in the project. In this picture, green tasks were developed in the Arduino Yuns and orange tasks were developed in the Raspberry Pi. Obviously, the outcome of these tasks had to be properly integrated in order to get the final system. This was quite challenging for the students because they were not used to do so. This way, students followed the bottom-up design approach in order to build the final IoT monitoring system.

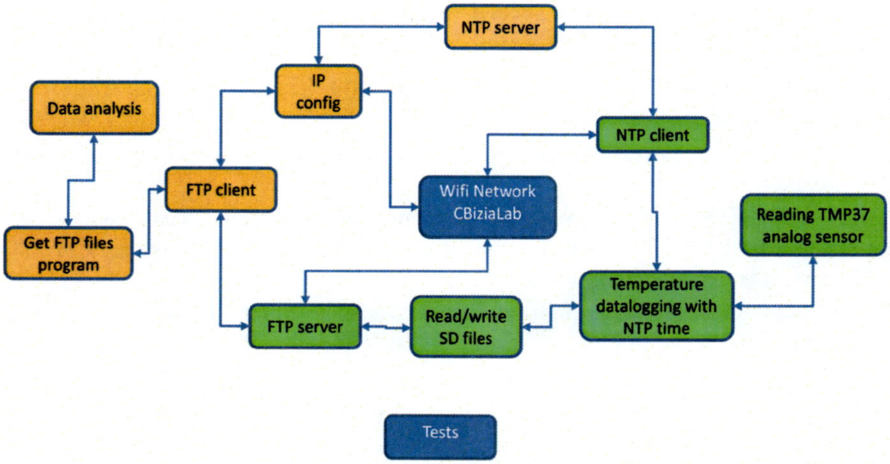

Fig. 2. Interactions among the PBL tasks as described by the students. (Raspberry tasks in Orange; Arduino Yun tasks in Green; Common tasks to all devices in Blue.)

5 Conclusions

This work describes the design of a PBL experience in the field of sustainability that has been carried out in the subject of Industrial Informatics. Industrial Informatics is a mandatory course taught in the third year of the B.Sc. degree in industrial electronics and automation engineering at the Faculty of Engineering of Vitoria-Gasteiz (UPV/EHU). The proposed PBL experience addresses the design of a real application aimed at monitoring comfort variables such as temperature and humidity in a distributed way, over the Faculty. This activity was proposed in the framework of a broader project aimed at improving the sustainability of our University, namely aimed at monitoring the comfort conditions of the Faculty as well as the performance of the heating system. Therefore, our students may gain experience by analyzing a real project. Also, our students may identify that developing sustainability aimed projects as a possible venue for their future professional activities.

The proposed PBL activity allowed students to understand the concepts of the Industrial Informatics. In particular, they learnt concepts about communication networks, programming and creating IoT applications. In addition, the proposed PBL activity was designed for teams of students of around ten students each. This fact enforced the collaboration among the students, since they had to identify all subtasks and integrate partial results in order to get the final system. Students were very motivated because they had to implement a real system.

Acknowledgements. This work was carried out in the framework of the Programme Campus Bizia Lab EHU, supported by the University of the Basque Country (UPV/EHU) and the Basque Government.

References

1. Meadows, D.H., Meadows, D.L., Randers, J., Behrens III, W.H.: The Limits to Growth. Universe Books, New York (1972)
2. World Commission on Environment and Development: Our Common Future. Oxford University Press, New York (1987)
3. Mihelcic, J.R., Crittenden, J.C., Small, M.J., Shonnard, D.R., Hokanson, D.R., Zhang, Q., Chen, H., Sorby, S.A., James, V.U., Sutherland, J.W., Schnoor, J.L.: Sustainability science and engineering: the emergence of a new metadiscipline. Environ. Sci. Technol. **37**(23), 5314–5324 (2003)
4. Lozano, R., Lukman, R., Lozano, F.J., Huisingh, D., Lambrechts, W.: Declarations for sustainability in higher education: becoming better leaders, through addressing the university system. J. Clean. Prod. **48**, 10–19 (2013)
5. EESD04: Declaration of Barcelona, Final version, October 2004. https://www.iau-hesd.net/sites/default/files/documents/declaration_of_barcelona_english.pdf. Accessed 22 Jan 2020
6. Wiek, A., Withycombe, L., Redman, C.L.: Key competencies in sustainability: a reference framework for academic program development. Sustain. Sci. **6**(2), 203–218 (2011)
7. UNESCO: Engineering: Issues, Challenges and Opportunities for Development, p. 392. UNESCO, Paris (2010)
8. Guerra, A.: Integration of sustainability in engineering education: why is PBL an answer? Int. J. Sustain. High. Educ. **18**(3), 436–454 (2017)
9. Graaff, E.D., Kolmos, A.: Characteristics of problem-based learning. Int. J. Eng. Educ. **19** (5), 657–662 (2003)
10. Bocko, P.: Synergy between problem-based learning and educating for sustainability: a review of the literature. In: Papa, R., Saiti, A. (eds.) Building for a Sustainable Future in Our Schools: Brick by Brick, pp. 107–131. Springer, Cham (2016)
11. Ursuţiu, D., Neagu, A., Samoilă, C., Jinga, V.: MODULARITY applied to SMART HOME: from research to education. Lecture Notes in Networks and Systems, vol. 22, pp. 56–67 (2018)
12. Silva, H.E., Coelho, G.B.A., Henriques, F.M.A.: Climate monitoring in World Heritage List buildings with low-cost data loggers: the case of the Jerónimos Monastery in Lisbon (Portugal). J. Build. Eng. **28** (2020). Article No. 101029
13. Plaza, P., Sancristobal, E., Carro, G., Castro, M., Ruiz, E.: Wireless development boards to connect the world. Lecture Notes in Networks and Systems, vol. 22, pp. 19–27 (2018)
14. UPV/EHU, ANECA, B.Sc. in Industrial Electronics and Automation Engineering Degree, Verified compliance ANECA report. https://www.ehu.eus/es/grado-ingenieria-electronica-industrial-y-automatica-alava/verificacion-seguimiento-y-acreditacion. [Only in Spanish]. Accessed Jan 2020

Simple Conception of Photoresistor Cell Current-Voltage Characteristic Measurement by Arduino-Based Platform for Educational Purposes

Abdelaziz El Moussaouy[1,2](✉), Mohammed El Hadi[1], Abdelaziz Ouariach[1], Rachid Essaadaoui[1], Driss Bria[1], and Khalid Laabidi[1]

[1] OAPM Group, Laboratory of Materials, Waves, Energy and Environment, Department of Physics, Faculty of Sciences, Mohammed I University, 60000 Oujda, Morocco
azizel0@yahoo.fr

[2] Regional Center for the Professions of Education and Training, 60000 Oujda, Morocco

Abstract. Recent progress of integration of new technologies in physics education has become very interesting for students, educators and researchers. It allows students to understand the physics phenomena, collect data during experiments, develop their imagination and motivate them in classroom practices. Indeed, the Arduino microcontroller board becomes nowadays one of the principal tools used in learning and teaching physics. It is an open source electronic platform based on easy-to-use hardware and software which allows collecting experimental data and visualizing them on a computer or a Smartphone. In this paper, we have developed a low-cost experiment by using an Arduino-based platform to facilitate the study of Photoresistor Cell Current-Voltage Characteristic. We have found an excellent agreement between the data measurement using our platform and those obtained by the classical experiment. The results obtained show the experimental importance of our Arduino based platform as a tool in educational practices.

Keywords: Physics education · Photoresistor cell · Arduino-based platform

1 Introduction

Physics is an important foundation in science, based on experience and observation of the nature around us [1, 2]. Experimental activities in teaching/learning physical and technological sciences are very interesting, especially when these activities integrate computer tools in education [3, 4]. We can't do a good didactic transposition if we do not allow our learners to extend their creativity in a technologically favorable environment [5]. The Arduino board is a good example used in various fields (mechanics, electronics, robotics, etc.) with an inexpensive, flexible and easily achievable character. Researchers and users of this type of microcontroller have understood the advantages

Á. Herrero et al. (Eds.): ICEUTE 2020, AISC 1266, pp. 348–356, 2021.
https://doi.org/10.1007/978-3-030-57799-5_36

of integrating this card for low-cost data acquisition for laboratory experiments [5–12]. Arduino-based physics experiments encourage the learners to understand physics phenomena. This encouragement play a dominant role for the success of learning [1, 5].

Light Dependent Resistor (LDR) is a type of component that has a change in conductance whose value depends on the light received by it [4, 13]. Photoresistors are made up of high resistance semiconductors which are not protected from light. If we hit the cell with light that has a sufficiently high frequency, the photons absorbed by the semiconductor will cause the electrons to have enough energy to jump into the conduction band. The free electrons produced (and their paired holes) will conduct electricity, thereby reducing its resistance [14, 15].

We have developed a simple Arduino experiment to study the variation of the photoelectric current depending on the voltage applied to the photoresistor for constant radiation intensity, and as a function of the radiation intensity for a constant voltage (Malus law) [13]. The experimental data associated with this experiment is obtained using a DC-DC step converter (2–24 V, 2 A) and an INA219 DC Current sensor connected to an Arduino Nano board.

2 Basic Concepts of the Photoconductivity Phenomena

Photoconduction is the increase of the electrical conductivity in a solid by absorption of light. This phenomenon is observed only for the semiconductors materials. The energy absorbed allows the passage of electron into the conduction band and the generation of holes in the valence band. The conductivity is written by:

$$\Delta\sigma = \Delta n.e.\mu_n + \Delta p.e.\mu_p \tag{1}$$

where Δn and Δp are, respectively, the densities of electrons and holes. μ_n and μ_p are, respectively, their mobility and e represent the elementary charge.

In this work, we used a photoresistor based on Cadmium sulfide CdS which was lit by an incandescent bulb light. The radiation intensity Φ is modified at the photoresistor using two polarizing filters located at the one behind the other. If the two polarizing filters are rotated relative to each other by the angel α, we have:

$$\phi = \phi_0 DCos^2(\alpha) \tag{2}$$

where Φ_0 is the radiation intensity without polarizing filters, and D represents the permeability for positioning in parallel.

In this paper, we have focused on the determination of the current-voltage and current-intensity characteristics of a photoresistor based on cadmium sulfide CdS by the application of a voltage and photoelectric current.

3 Experimental Study Using Arduino-Platform

3.1 Construction of the Photoresistor Cell

Photoresistors are electronic components made up of semiconductors. The cell used in this study is based on cadmium sulfide CdS, whose resistance decreases when the intensity of the light striking them increases. Photoresistors can also refer to LDR or photoconductors (see Fig. 1).

Fig. 1. Picture of photoresistor cell.

3.2 Experimental Setup

Our study is devoted to the determination of current-voltage and current-light radiation characteristics through a photoresistor cell based on cadmium sulphide CdS. The basic experimental setup used consists of a light source 6 V/30 W, a diaphragm, two polarizers, a converging lens (f = +150 mm), a photoresistor cell LDR 05 STE, an adjustable generator DC 0–20 V, and two multimeters as we see in Fig. 2.

By using the polarizers, the α-angle (angle between the two polarizers) is fixed. The light after passing through the two polarizers was converged on photoresistor by converging lens. The current-voltage characteristic is obtained for each α. To plot the light current-radiation characteristic, we fix the voltage U on the generator.

Fig. 2. Image of the experimental setup without Arduino.

3.3 Arduino Measurement

In this part, we have made the same measurements but with a novel method which allows collecting the data using Arduino based platform, in order to compare with the data obtained by classical method. We keep the optical elements used previously and we replace the electrical elements by our platform which contains an INA219 DC Current sensor, a DC-DC step converter (2–24 V, 2 A), and Arduino Nano board as we see Fig. 3.

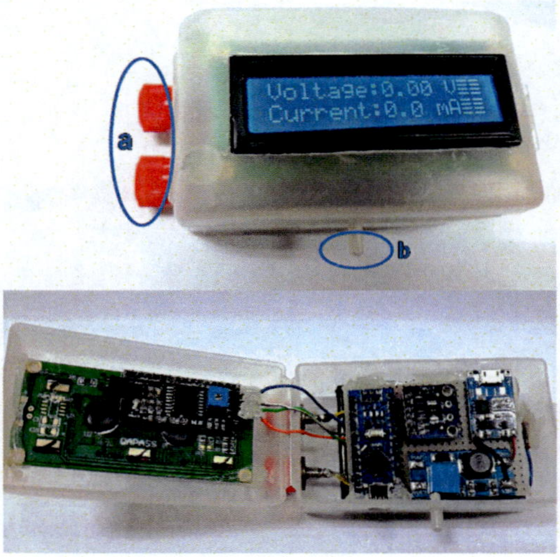

Fig. 3. Image of Arduino-Based Platform: (a) The terminals of the photoresistor, (b) Potentiometer to change the voltage.

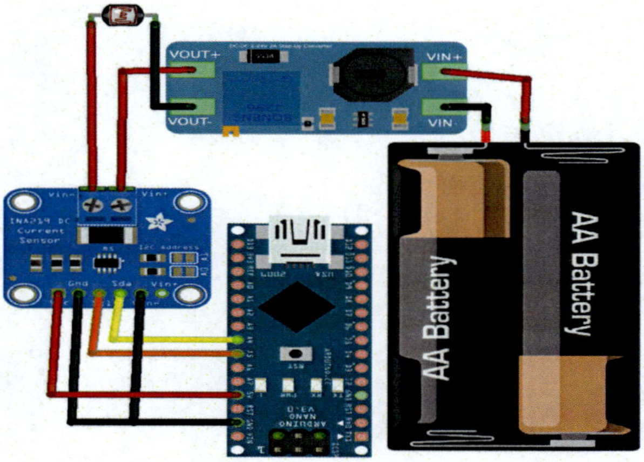

Fig. 4. Sketch of the circuit with Arduino.

To facilitate the assembly, we have presented in Fig. 4 the sketch of the circuit with Arduino. The DC current sensor INA219 is connected to the terminals of the photoresistor cell and to the branches of an Arduino Nano card with male to female connection cables, the circuit is supplied by DC-DC step converter (2–24 V, 2A) (see Fig. 5).

To retrieve the experimental data, we use an Arduino code which is shown in Fig. 6. The Arduino serial monitor displays the experimental measurements after running the code on the motherboard Arduino. The results found are displayed on the computer using an Arduino serial monitor, or on an LCD display as we see in Fig. 3.

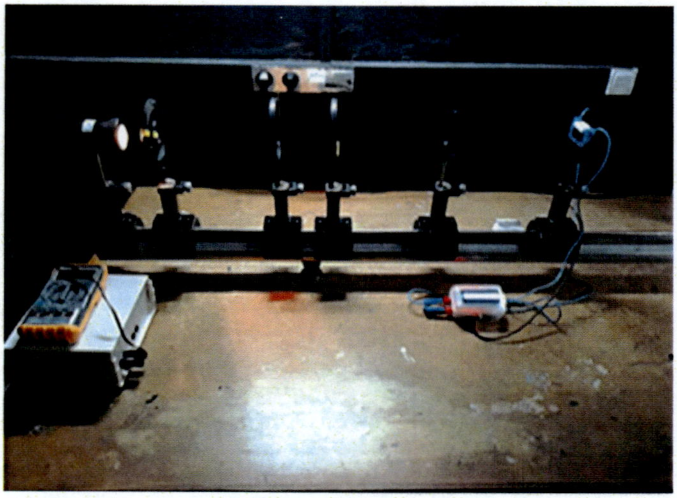

Fig. 5. Image of the experimental setup without Arduino.

```
#include <Wire.h>
#include <Adafruit_INA219.h>
#include <LiquidCrystal_I2C.h>

Adafruit_INA219 sensor219; // Declare and instance of INA219
LiquidCrystal_I2C lcd(0x27,16,2);

void setup(void)
{
  Serial.begin(9600);
  lcd.begin();//Defining 16 columns and 2 rows of lcd display
  lcd.backlight();//To Power ON the back light
  sensor219.begin();
}

void loop(void)
{
  float busVoltage = 0; // Measure in milli volts
  float current = 0; // Measure in milli amps

  busVoltage = sensor219.getBusVoltage_V();
  current = sensor219.getCurrent_mA();

  lcd.setCursor(0,0);
  lcd.print("Voltage:");
  lcd.print(busVoltage);
  lcd.println(" V");
  lcd.setCursor(0,1);
  lcd.print("Current:");
  lcd.print(current,1);
  lcd.println(" mA");
  delay(500);
}
```

Fig. 6. The Arduino code.

3.4 Measurement of Photoelectric Current Iph as Function of Voltage for Constant Radiation Intensity

The values obtained from the experimental method and our Arduino-based platform, are represented on the same graph as we see in Fig. 7.

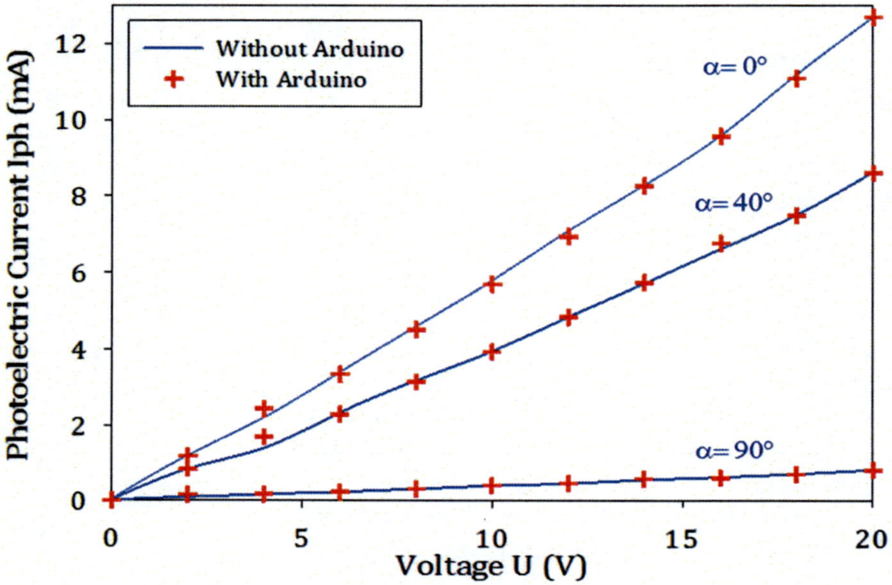

Fig. 7. Variation of photoelectric current Iph as function of voltage for constant radiation intensity.

From the curves representing the variation of the current photoelectric as a function of voltage in the photoresistor cell for a fixed light intensity (for each α-angle), we have found a linear profile for the different values of α-angle. The slopes of these curves represent the value of the cell conductance. Each slope depending on the amount of light received, so the LDR conductance changes with the change in the intensity of the light surround it or those surrounding it. In the dark ($\alpha = 90°$), the LDR conductance reaches a minimum value, and this value increases with the intensity of the light. When a photoresistor cell state receives the maximum of light coming from the light source ($\alpha = 0°$), the conductance LDR lets pass the maximum of the electric current i.e. a very large conductance value. When light energy falls on LDR based semiconductors such as cadmium sulfide, it causes a greater charge release or an increase in electric current. This explains the decrease in strength of the material.

3.5 Measurement of Photoelectric Current I_{ph} as Function of Radiation Intensity for a Constant Voltage U

In Fig. 8, we have plotted the variation of the intensity of the photoelectric current I_{ph} versus $\cos^2 \alpha$ for each voltage at the terminals of LDR with and without Arduino-based platform.

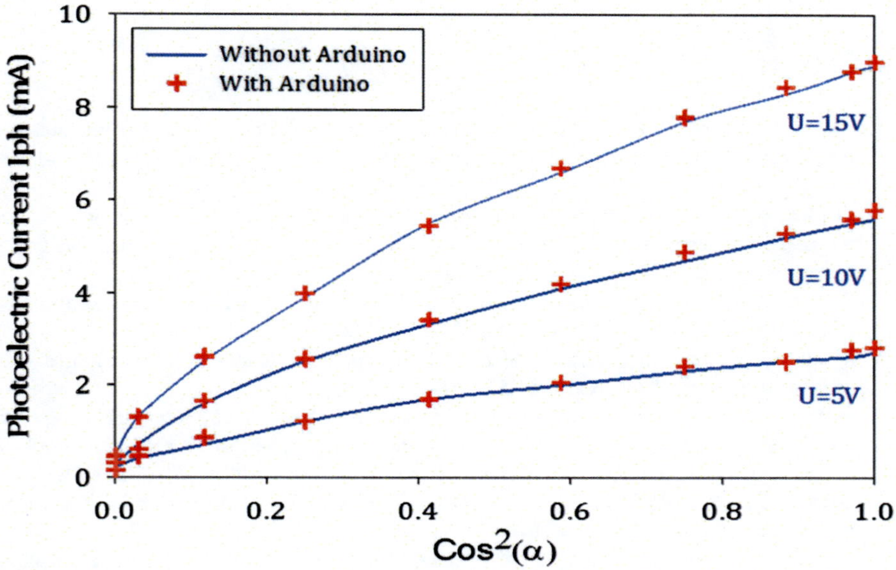

Fig. 8. Variation of photoelectric current Iph as function of radiation intensity for a constant voltage U.

We can note here that the experimental values are in excellent agreement with those found by our Arduino based-platform. In the case of darkness ($\alpha = 90°$), we have observed that the electric current circulating in the photoresistor cell tends to 0 whatever the electric energy given by the generator. Also, the behavior of the graph Iph = f (cos2 α) is a quasi-linear function for the different values of the voltage, since the radiation intensity received by the LDR is proportional to the intensity of the photoelectric current. This intensity of radiation is proportional to the square of cosine the α-angle (the law of Malus).

4 Conclusion

In this work, we have investigated the phenomenon of photoconduction in semiconductor based photoconductance cells using an Arduino Nano Board connected to INA219 DC Current sensor and DC-DC step converter (2–24 V, 2 A). The values obtained by our platform based on Arduino have a very good agreement with the data measurements obtained by the classical experiment. We also conclude that it is possible to check the Malus law in an innovative way, with a low-cost experiment and very accessible to users.

Acknowledgments. This work was supported by the Ibn Khaldoun program of the National Center of Scientific and Technical Research (CNRST) in the field of humanities and Social Sciences (Project No. IK/2018/24).

References

1. Cheon, J., Lee, S., Crooks, S.M., Song, J.: An investigation of mobile learning readiness in higher education based on the theory of planned behavior. Comput. Educ. **59**(3), 1054–1064 (2012)
2. El Moussaouy, A.: Environmental nanotechnology and education for sustainability: recent progress and perspective. In: Hussain, C. (ed.) Handbook of Environmental Materials Management, pp. 1–27. Springer, Cham (2018)
3. Sundara, S.K.V., Suryanarayana, S.: Automatic dual axis sun tracking system using LDR sensor. Int. J. Curr. Eng. Technol. **4**(5), 3214–3217 (2014)
4. Monteiro, M., Stari, C., Cabeza, C., Marti, A.C.: The polarization of light and the Malus' law using smartphones. Phys. Teach. **55**(5), 264–269 (2017)
5. Petry, C.A., Pacheco, F.S., Lohmann, D., Correa, G.A., Moura, P.: Project teaching beyond Physics: integrating Arduino to the laboratory. In: Proceedings of the 2016 Technologies Applied to Electronics Teaching, TAEE 2016 (2016). https://doi.org/10.1109/taee.2016.7528376
6. Galeriu, C., Edwards, S., Esper, G.: An Arduino investigation of simple harmonic motion. Phys. Teach. **52**(3), 157–159 (2014)
7. Schwaab, Q., Belleau, C., Morency, F., Weissx, J.: Low-cost Arduino-based data acquisition system with android mobile interface for undergraduate aerodynamics laboratories. In: 32nd AIAA Aerodynamic Measurement Technology and Ground Testing Conference, pp. 1–15 (2016). https://doi.org/10.2514/6.2016-4323
8. Fernandes, J.C., Sebastião, P.J., Gonçalves, L.N., Ferraz, A.: Study of large-angle anharmonic oscillations of a physical pendulum using an acceleration sensor. Eur. J. Phys. **38**(4), 18 (2017)
9. Uyanik, I., Catalbas, B.: A low-cost feedback control systems laboratory setup via Arduino-Simulink interface. Comput. Appl. Eng. Educ. **26**(3), 718–726 (2018)
10. Tong-On, A., Saphet, P., Thepnurat, M.: Simple harmonics motion experiment based on LabVIEW interface for Arduino. J. Phys: Conf. Ser. **901**(1), 012114 (2017)
11. Galeriu, C.: An Arduino-controlled photogate. Phys. Teach. **51**(3), 156–158 (2013)
12. Ouariach, A., El Hadi, M., El Moussaouy, A., Hachmi, A., Magrez, H., Bria, D.: Development of an educational low-cost teslameter by using Arduino and Smartphone application. Phys. Educ. **55**(3), 033008 (2020)
13. Freitas, W.P.S., Cena, C.R., Alves, D.C.B., Goncalves, A.M.B.: Arduino-based experiment demonstrating Malus's law. Phys. Educ. **55**(3), 4 (2018). 035034
14. Dias Tavares A. Jr., Sosmana, L.P., da Fonsecaa, R.J.M., da Motaa, L.A.C.P., Muramatsub, M.: Using a photo-resistor to verify irradiance inverse square and Malus' laws. In: AIP Conference Proceedings, vol. 992, no. 1, pp. 193–198 (2014)
15. Xiao-Yuan, W., Fitch, A.L., Iu, H.H., Sreeram, V., Wei-Gui, Q.: Implementation of an analogue model of a memristor based on a light-dependent resistor. Chin. Phys. B **21**(10), 108501 (2012)

Dual University Training at the Faculty of Engineering Vitoria-Gasteiz (UPV-EHU)

Jose Antonio Ramos-Hernanz(✉), Inmaculada Tazo-Herran, Ekaitz Zulueta, Amaia Mesanza-Moraza, Ruperta Delgado-Tercero, Javier Sancho-Saiz, Estibaliz Apiñaniz-Fernandez de Larrinoa, and Jose Manuel Lopez-Guede

Faculty of Engineering Vitoria-Gasteiz, University of the Basque Country (UPV/EHU), C/Nieves Cano, 12, 01006 Vitoria-Gasteiz, Spain
josean.ramos@ehu.es

Abstract. One of the implementations of novel initiatives in the field of education that requires more efforts is the entire implementation of a new Bachelor's degree based on dual training. These efforts involve on one hand the design of an extremely experimental degree and on the other hand, the engagement of local companies that must collaborate in a rigorous way with the teaching of the students, since they must give a number of lessons of specific subjects to the hosted students. In this paper, authors describe the implantation process of the Bachelor's Degree in Automotive Engineering at the Faculty of Engineering Vitoria-Gasteiz of the University of the Basque Country (Spain). In the paper the accreditation process carried out in the Basque University System Quality Agency (UNIBASQ) is described, giving some ideas about the information provided for all the points that that were fulfilled to gain the "dual training" accreditation of the agency. This information could be used by other university faculties to be aware and to plan a dual version of any Bachelor's degree in Engineering.

Keywords: Accreditation · Company · Dual training · Quality · University

1 Introduction

Since companies have undergone a series of transformations in recent years and the globalization of the market, the collaboration between the universities and companies is indispensable [1]. For this reason, the relationship between the students and the companies must begin as soon as possible, which will allow the students to achieve, gradually and before finishing their university studies, certain professional skills. Dual university education means that the time needed for students to adapt to working life is shorter. As it is an alternating training between the University and the company, it minimizes the duration of the transition from the educational system to the working environment. This early immersion in the world of companies forces students to adapt their mentality since they will have academic and work responsibilities, which allows them to acquire theoretical knowledge at the same time as a solid company experience [2].

Á. Herrero et al. (Eds.): ICEUTE 2020, AISC 1266, pp. 357–366, 2021.
https://doi.org/10.1007/978-3-030-57799-5_37

Since the education of students is shared between the University and companies, the latter are also responsible for the student's university education. This implies a greater engagement of the companies in the training of their future human resources, ceasing to be secondary actors and becoming main actors in their training.

To be able to offer a dual university education, the university centers have to certification their university degrees, either complete degrees or itineraries of these that are taught in dual format. They will have to meet the requirements set by the accreditation agencies. In the particular case of the Autonomous Community of the Basque Country, on 6 June 2017 it was presented the 1st Basque Strategy for University-Companies Relations 2022 at the Basque Council of Universities. It defines the framework for dual university training by creating the seals of approval for Basque university degrees and masters.

In order to obtain this accreditation, the university centers must adapt their studies to the quality requirements demanded by the quality agencies of the university systems. In the case of the Basque University System Quality Agency (UNIBASQ), a protocol is offered to obtain the recognition of a dual training based on the following regulations [4]:

- Royal Decree 1393/2007, of 29 October, which establishes the possibility of introducing external internships in order to strengthen the commitment to the employability of future graduates. These external practices will have a maximum extension of 25% of the total credits of the degree and should be offered preferably in the second half part of the syllabus.
- The University Student Statute, approved by Royal Decree 1791/2010, of 30th December, indicates in its article 8 the right of undergraduate students to carry out practical, curricular or extracurricular activities, which may be carried out in external entities and in the centers, structures or services of the University.
- The Royal Decree 592/2014, of 11th July, regulates the external academic practices of university students.
- Decree 11/2009, of 20th January, which sets forth the implementation and suppression of official university education leading to the awarding of Bachelor's, Master's and Doctoral degrees.

The remaining of the paper is structured as follows. Section 2 analyses what dual university education is in engineering degrees, analyzing its advantages for the three main agents. Section 3 sets out the criteria that the Faculty of Engineering Vitoria-Gasteiz (FEVG) proposes to the Basque accreditation agency (UNIBASQ) in order to achieve accreditation of the Degree in Automotive Engineering as a dual training degree. Section 4 analyses the Training Project, which explains the distribution of the training and productive activities of the students during their stay in the company. Section 5 reflects the evaluation and monitoring to be carried out of the stays in the companies. Finally, the main conclusions are summarized.

2 Dual University Education

Dual University Education is a set of educational tasks and initiatives that seek to improve the professional qualifications of university students, integrating teaching-learning methods in the University and in the companies to simplify the achievement of professional skills in the workplace [4]. It has the following general objectives:

- The competence development of the students in coherence with the required professional profile and its adjustment to the needs of the companies, that is to say, improving the professionalization and the insertion in a socioeconomic environment.
- The integration of the acquisition of skills in the company in the educational project of the University, innovating the training system.
- The improvement of the teaching-learning process of the students, with a follow-up and evaluation both in the company and in the University.

In turn, it has the following specific objectives:

- The acquisition of techniques and skills, which due to their nature and characteristics, require means, organizations and productive structures that are only found in real scenarios.
- To contribute to complement the knowledge, skills and abilities acquired in the educational institution through stays in the workplace.
- To apply the knowledge acquired in the educational centers in real production or service provision situations.
- To promote the sense of autonomy, creativity and responsibility, enabling the student to learn to seek solutions and solve professional problems that arise in the workplace.
- To know the organization of the companies and the labor relations that take place in them.
- To facilitate the relationship and the exchange of information between the education and the productive systems.

Dual university training is a relatively new teaching modality in Spain, although in other European Union countries such as in Germany or France it is already consolidated [5].

The three main actors in this teaching modality [6,7] are:

- The University, as the link between companies and students, is responsible for interacting with companies, and for the student achieving the necessary set of skills in Engineering from the academic point of view.
- The company or entity is in charge of making the working reality visible and that the students obtain the necessary transverse competences from the professional point of view.
- The students, who will obtain during their studies the academic and professional knowledge.

The advantages of dual training for each one of the agents are [ibidem]:

Advantages for the University:

- Continuous updating of knowledge.
- New forms of collaboration: conferences, visits, research projects…

Advantages for the company:

- Participation in the selection of candidates.
- Reduction of integration times.
- Improvement of the strategic planning for the renewal of engineers.

Advantages for the students:

- Obtaining of some financial resources.
- It improves the labor insertion.
- Proactive role in the learning process, motivated and competent

3 Evaluation Criteria for the Accreditation of the Degree in Automotive Engineering

Centers or universities may request accreditation of their studies for a complete degree or a dual itinerary (only part of the student body enrolled in the degree will follow the dual format). The evaluating agency will take into account the information provided by the universities for the following sections.

3.1 Title Description

This section describes the studies that are presented for accreditation, listing the number of students that could take the dual training, the number of credits that will be taken in collaboration with the companies and the form of collaboration.

In the case of the Degree in Automotive Engineering, 40 places will be offered in the classroom, with a study plan of 240 credits of which 22.5% will be taught within the collaborating entities, 54 credits. The rest up to the 25% required for accreditation will be 60 h of classroom training given by the companies within the subjects of the degree through technical talks that complement these subjects.

3.2 Justification of the Curriculum Leading to an Official Degree

The reason for the use of the dual training methodology in the degree proposed for accreditation must be justified, indicating the benefits that will be obtained by the different agents that participate in its realization. The competencies that the student will acquire at the university and in the company will be indicated, analyzing how the coordination between both is carried out.

In order to accredit the achievement of these competences, a Joint Commission will be established, composed of the Vice Director of Relations with Companies, the Coordinator of the Degree in Automotive Engineering and a professor who teaches the subject of Projects Management at the University. On the company side, there will be two tutors from the companies involved in the training of the students.

3.3 Skills to Be Acquired by Graduates

The student who completes a dual degree will acquire some skills at the university and others during his or her stay in the company. For accreditation, it is necessary to justify how these skills will be acquired. These competencies are explained in the Verified Report of the Degree in Automotive Engineering.

During the time that the student is at the university faculty, the competencies will be acquired through active learning methodologies, such as Project Based Learning, Reverse Classroom, Cooperative Work, etc. The degree is divided into the following modules:

- Basic Training Module (60 ECTS)
- Engineering Fundamentals Module (42 ECTS)
- Automotive Technology Module (42 ECTS)
- Vehicle Manufacturing Module (24 ECTS)
- Optional module (18 ECTS)
- External practices (3rd and 4th course) (42 ECTS)
- End of Degree Project (12 ECTS)

Once the Basic Training Module has been acquired, the students will have an individualized training adapted to the specific needs of the companies in the sector, complementing their training with the rest of the subjects of the degree.

3.4 Access and Admission of Students to the Entities

Degrees with mention of dual training must develop a procedure that guarantees that students can take the credits determined by the curriculum in collaboration with the companies.

The process of assigning students to companies will be carried out by the students, the companies and the FEVG, and consists of the following steps:

1. An information meeting will be held with students at the beginning of the second term of the second course to explain the characteristics of the external practices.
2. A list of the companies with which FEVG will sign an agreement will be made public at the end of April or beginning of May, with the number of places requested by each company and with the area or areas of the company in which the student will carry out his/her training.
3. A list of the students will be made, in order of choice of places, taking into account specific academic criteria.
4. From this list, the students will choose a maximum of three options.
5. The FEVG will offer companies 3 students for each of the positions on offer.
6. Companies should select the candidate, considering specific academic criteria and one interview.

The preferential link between the student and the entity will be the part-time contract with formative link based on:

- Law 11/2013 of 26th July (actions to support entrepreneurs and stimulate growth and job creation)
- A legal bond that implies remuneration to the student and an insurance that can be assimilated to the rest of the employees

A commitment agreement will be signed between the student, the FEVG and the Entity. It contains the competencies and activities that the student must develop in the entity, as well as the rights and obligations that all the signatories have.

Taking as a reference Decree 83/2015, of 2nd June, which establishes the Dual Vocational Training in the Alternating Regime in the Autonomous Community of the Basque Country, and not having at the moment a Decree that regulates the University Dual Training, the minimum remuneration that the students will receive during the time they stay in the company cannot be, preferably, less than the monthly value established in the year that corresponds to the minimum interprofessional wage, in proportion to the time they stay in the company. In all cases, the monthly amount received by the students shall not be less than 60% of that value established for the minimum interprofessional salary. Thus, for example, by 2018 the minimum amount to be received by students was 515.13 euros per month (pro rata for extra payments).

The company, as long as it complies with the requirements, may apply the incentives of the part-time contract with formative linkage that implies a reduction of up to 100% of the company Social Security contribution for common contingencies.

3.5 Lesson Planning

During the first two years, the training will be carried out completely at the University, however, from the first year the students will have training in collaboration with the companies within the context of specific subjects.

The degree is defined in such a way that from the third year, the activities in the classroom and in the entity coexist.

3.6 Academic Staff

The academic tutors responsible for the students will have a profile focused on engineering, having a recognized professional experience, their research is related to automotive technologies, have extensive experience in tutoring voluntary practices in companies or collaborate on industrial doctorates that are related to the automotive sector. As training for tutoring, part of the academic staff of the FEVG has carried out different conferences such as the "Dual Training in Higher Education" or "Dual University Training, Study of Five Cases".

3.7 Material Resources and Services

For the accreditation of a degree as dual, it must be ensured that all students enrolled can perform external practices, having a sufficient number of practices offered by entities that shows a recognized prestige and solvency.

Both for the tutoring and for the evaluation by the entity tutor, as well as by the academic tutor, the necessary resources will be available.

The University of the Basque Country (UPV/EHU) Virtual Campus has an eGela Portfolio platform that allows the development of an electronic portfolio, interconnected through the virtual classroom management service (https://egelapi.ehu.eus). With this tool, students have a personal area with files in which they can store, among other things, their Curriculum Vitae, create blogs, views or web pages to save their ideas, tests, reflections, evaluations and feedback that both the academic tutor and the company tutor provide.

3.8 Quality Assurance System

The Faculty of Engineering Vitoria-Gasteiz, within its Internal Quality Assurance System, has the procedure: 2.2. PES-D07 Dual Training, included within the subprocess 2.2 of development of Higher Education, in the process 2.

4 Training Project

The company must analyze its needs (positions, generational change, etc.) and a company-university meeting will be held to assess the internship proposal requested and the suitability of the skills. As a result, the training project will be defined, and the distribution of training and production activities will be agreed upon.

The training project will be a document containing the following points:

- Company data.
- Job description, project and position where the learning will take place.
- For the academic year 2019/2020 (3rd year) and 2020/2021 (4th year): the general description of the Professional Development Plan to be carried out by the student will be from November 2019 to June 2021 (minimum of 1,350 total hours, equivalent to 54 ECTS). (Stays + Final Degree Project).
- Time distribution and calendar.
- Competences to be acquired by the student in each of the periods, foreseen in the Report of the Degree in Automotive Engineering.
- Student profile preferences.
- Language(s) to be used in the workplace.
- Other information: Location where the activity will take place, Duration and hourly dedication of the Program and Remuneration foreseen.
- Description of general activities to be developed for the acquisition of competences.
- Attention plan for students by the Faculty and company tutor.

5 Evaluation and Monitoring

5.1 Evaluation

The evaluation requires the monitoring of the students throughout their stay in the company by the academic tutor and the company instructor, and it will be based on the different reports issued by them and the report made by the students. A summary is shown in Table 1.

Table 1. Student assessment.

Stays	Documents	Rating
Phase I	Report that summarizes the technical and organizational fundamentals of the company	Academic tutor: 100%
Phase II	Report with a concrete and detailed description of the tasks and activities carried out	Academic tutor: 50% Tutor/company: 50%
Phase III	Technical report describing the characteristics of the intervention carried out in the company	Academic tutor: 50% Tutor/company: 50%
Phase IV	Report of the activity developed in the company	Academic tutor: 50% Tutor/company: 50%
	End of Degree Work	FDP Advisor and FDP Court: 100%

5.2 Monitoring

The figures of the academic tutor and the company tutor are key to the monitoring of the student training. It is important that the company instructors have a training in pedagogical aspects, as well as the tutors of the University have training/experience in company aspects. Monitoring and evaluation meetings of the dual training will be established with each one of the participating companies.

1. Preliminary meeting: During October, before starting the stay in the company, it is important that the academic tutor and the person responsible of the Company Relations at the FEVG, make a presentation visit to clarify doubts, establish expectations and ratify the professional training project.
   Thus, the team responsible for the project, the tutor and the instructor will have to agree on a training project that takes into account different levels of participation in the company (observation, application, taking responsibility), the professional profile and the activities or tasks to be carried out.
2. Coordination/evaluation meetings: For the correct follow-up of the student's training, a joint meeting will be held between the entity, the University and the student at the beginning of the period of stay in the entity in order to establish the Professional Development Plan (PDP). Another one will take place at the end of each stay to analyze and evaluate the level of achievement of the objectives set out in the initial Plan.

3. Reporting: At the end of each stay, a presentation/defense of the report will be made, which will serve to evaluate the results obtained, analyzing the difficulties and establishing new objectives for the following period. It should be a reflexive practice for which there is a "monitoring portfolio" in which the objectives and evaluations have been collected.

The evaluation of the dual modules will involve both the person tutoring the teaching in the classroom and the person tutoring the teaching in the entity, allowing the acquisition of the established learning results. For the correct evaluation of the students' training, they will be able to collect all the activities they carry out in a portfolio that will be hosted in the University's eGelaPortfolio application.

6 Conclusions

In the coming years, university institutions will have to be much more sensitive to social and company requirements due to the demands and needs of society. The precariousness of work, as well as the existing youth unemployment, added to the transformation in the companies due to the Industry 4.0 paradigm will imply changes in the way of preparation of the university students.

In October 2017, the evaluation agency UNIBASQ published the "Accreditation Protocol for Dual Training for University Degrees", [3,8]. This paper presents a summary of the arguments presented by the FEVG to obtain the accreditation of the Degree in Automotive Engineering according to this protocol.

The key factors for the success of this teaching modality are the following:

- Importance of the professional profile reference
- Importance of visualizing the training path in the company
- The importance of flexible training at the university
- Importance of the company tutor
- Importance of the academic tutor
- Importance of the student's professional project
- Importance of student monitoring
- Importance of the relationship between the student, the company tutor and the academic tutor

The majority opinion of the collaborating companies is that this type of study is the most suitable for engineering companies and should be applied to all of them. It is not understood that there may be students who have not seen or worked in a company during their training period in the teaching institution, which unfortunately is still common in non-dual engineering degrees. The companies consider that the voluntary extra-curricular internships that are usually done in these non-dual grades are short and do not give time for the students to adapt to the companies.

For dual training to be successful, it is necessary for companies to be actively involved in supporting this form of teaching with the acceptance of students for their compulsory work experience.

Acknowledgment. The Regional Development Agency of the Basque Country (SPRI) is gratefully acknowledged for economic support through the research project "Refrigeración de dispositivos de alto flujo térmico mediante impacto de chorro" (AIRJET), KK-2018/00109, Programa ELKARTEK.

References

1. Vilalta, J.M.: ¿Es posible (y deseable) la formación universitaria dual? Blog de Studia XXI (2017)
2. ACUP: Promoción y desarrollo de la formación dual en el sistema universitario catalán. Asociación Catalana de Universidades Públicas Universitat de Lleida (2015)
3. Unibasq: Protocolo para la obtención del reconocimiento de formación dual para títulos universitarios oficiales de grado y máster (2017)
4. Fundación Bertelsmann: Guía básica sobre la formación profesional dual (2014)
5. Rego Agraso, L., Barreira Cerqueiras, E.M., Rial Sánchez, A.F.: Formación profesional dual: comparativa entre el sistema alemán y el incipiente modelo español. Revista Española de Educación Comparada **25**, 149–166 (2015)
6. Dual Training: Guía metodológica alumnos formación profesional dual (2015)
7. Durán López, P., Santos Primo, J.R., Gil Pérez, R.: Guía de la formación dual. Cámaras de Comercio, Ministerio de Educación, Cultura y Deporte y Fondo Social Europeo (2017)
8. Nuevo: Protocolo para la obtención del reconocimiento de formación dual para títulos universitarios oficiales de Grado y Máster. https://www.unibasq.eus/wp-content/uploads/2019/11/Nuevo_Protocolo_Unibasq_DUAL.pdf

Electromobility Laboratory: A Contribution for Student Participation in Higher Education

Felipe A. Nuñez-Donoso[1] and Jose Manuel Lopez-Guede[2(✉)]

[1] Department of Innovation, Technological University of Chile INACAP, Campus Osorno, Alcalde Rene Soriano Borquez 2382, Región de Los Lagos, Osorno, Chile
felipe.nunez@ayn.cl

[2] Faculty of Engineering Vitoria-Gasteiz, Department of Systems Engineering and Automatic Control, University of the Basque Country (UPV/EHU), Nieves Cano 12, 01006 Vitoria-Gasteiz, Spain
jm.lopez@ehu.es

Abstract. Academic programs are trying to implement edge technologies on their Bachelor's and Master's degrees offerings. In this paper authors present the experience of a dedicated electromobility research project and its alternative use for academic purposes. A case study is selected in a university in Chile, where the design of a new laboratory that serves for e-mobility research and for careers activities is described as an example of a long-term academic strategy. The numerous activities derived from a governmental funded academic research represents a parallel opportunity for continuous student education on high tech fields, in such a way that active student participation is highlighted as a successful academic knowledge transfer. In a period of nine months, over nineteen students from Bachelor's Degree in Electrical Engineering and Bachelor's Degree in Automotive Engineering have direct participation, through student internships and four final degree projects are being developed.

Keywords: Electromobility · Higher education · Career implementation

1 Introduction

Electromobility (e-mobility) is a term associated with a new transport paradigm, based on electrical energy sources. It is estimated that it will have a worldwide disruptive impact on the automotive market, with a growth like the internal combustion automobile industry at the beginning of the 20th century. On the last decade, the development of e-mobility has been exponential as is shown in Fig. 1, starting at developed countries with the recent appearance of electric car models with similar or better performances than conventional ones [1, 2].

For emerging countries, such as Chile in Latin America, an electromobility related industry represents an historic opportunity for developing a growing market based on local technology, human capital training and urban life quality improvement [3]. Nevertheless, the absence of a Chilean automotive car-building related industry diminishes the possibility of locally generated knowledge and represents a high barrier

Á. Herrero et al. (Eds.): ICEUTE 2020, AISC 1266, pp. 367–374, 2021.
https://doi.org/10.1007/978-3-030-57799-5_38

for major investments on newer technology like e-mobility. This issue represents a strong restriction for any university studies planning that pretends to integrate e-mobility related contents on their academic courses and Bachelor's degrees for future local labor and professional opportunities.

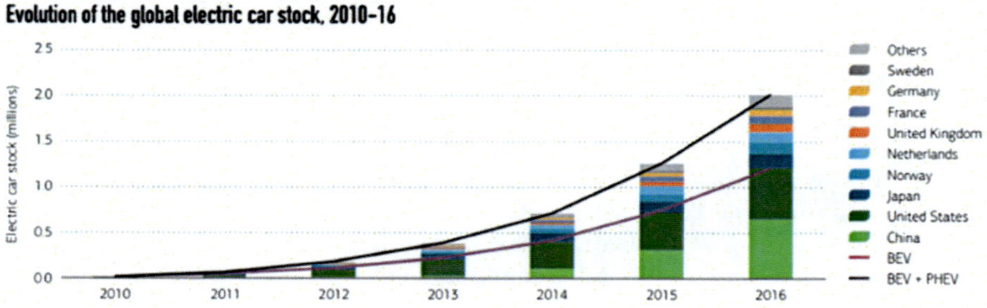

Fig. 1. Global electric car stock evolution on the last decade [4].

The Chilean government has recently granted research funding for e-mobility projects [5]. At first, the main scopes for this type of projects are to achieve the technical objectives for their hypotheses, on behalf of the possible technology transfer for economic activities that are not necessarily educationally oriented.

The Chilean higher education system is divided into technical and professional estates represented by colleges and universities separately or may be present in the same institution [6]. The latter is the case with Technological University of Chile (INACAP), that has college and university degrees as an integrated educational system, with 2-, 4-, 5- or 6-year degree programs. INACAP has more than 50 years of experience in Chile, with 26 campuses throughout all the country, from Arica in the extreme North to Punta Arenas in the extreme South [7]. The campus located in the city of Osorno has several careers related to technology, like Bachelor's Degree in Electrical Engineering, Bachelor's Degree in Computer Engineering and Bachelor's Degree in Automotive Mechanics, among others.

INACAP is accredited by the National Accreditation Commission and has a strong commitment to its students and the development of the country [8]. Its Learning by Doing methodology with a focus on technology, allows to put knowledge into practice from an early stage in laboratories, workshops and field trips [9]. Academic programs are always being updated and new areas like electromobility are part of a future offering for the next decade, according to the institutional strategic planning [10]. Automotive Mechanics both as technical and engineering careers are present in INACAP and have major facilities for conventional car maintenance studies. Integration of e-mobility equipment is part of the intention for all associated laboratories.

The objective of this paper is to present the different contributions to high education teaching that are possible due to the implementation of research project activities and facilities. In Sect. 2 the educational model is presented and its correlation with practical experiences for students and faculty members, through the novelty implementation process, from the necessity to the physical laboratory implementation. Section 3 shows the academic contributions through the student participation as an opportunity case for knowledge-skill transfer from research to teaching fields on academic programs. Section 4 summarizes the main conclusions of the paper and expected future work.

2 Novelty Implementation

2.1 Educational Model

Higher education models usually suggest strong theorical components in their academic programs, especially in Science and Engineering fields, where additional practical skills are also necessary for future professional and productive economic activities. INACAP's educational model on Training by Competencies focuses on the student's comprehensive education through the Learning by Doing pedagogical approach. The aim is to achieve the highest levels in the teaching-learning process, having in mind the harmonious development of the student's personality as well as the increase of their knowledge [11]. A methodology based on Learning by Doing is part of the Active and Cooperative Learning paradigms, that group several methods, all based on the responsibility and participation of students in their learning [12, 13]. Cooperative Learning is a philosophy in which learning activities are planned to look for positive interdependence of students [14, 15]. In this sense, a high-quality higher education model needs to integrate practical experiences that reflect real life situations where the students will be exposed later in their professional lives. Since technology advancements are exponential, it is difficult to design an academic model that will keep constantly updated to the latest technology.

A possible alternative is to complement graduate student formation (Bachelor's degrees) with postgraduate research projects that usually are investigating cutting edge technologies and paradigms that may become future economy niches. This involvement may create an optimum environment for the development of future professionals and provide them with the necessary skills and knowledge for their success. It forms graduates who are entrepreneurs, creative, innovative, and with a vocation for the public service. In this sense, the graduate should have a global vision and, at the same time, should be able to commit with the national development and assume a positive leadership. The educational model of INACAP seeks to train for the development of job skills. To do this, it is strongly based on the needs of the productive sector, delivering response to the demands of today's society.

Therefore, the necessity to count with the proper infrastructure, materials and equipment for practical activities for graduate students to replicate future working scenarios becomes a constant challenge for the higher education model, and have boosted the actions that have been taken under the educational project described in this paper.

2.2 Material Obtaining

In order to achieve the expected learning objectives through a practical approach, it is necessary for the students to count with the proper equipment and materials, especially in technology fields of study. Education institutions usually struggle with budget restrictions and financial delays that slow down laboratory implementations. Keeping up to date with technology becomes a constant challenge for long term academic programs. A possible alternative, may be to use facilities that already have dedicated funding for the latest technology.

In 2018, a government research fund was granted to the INACAP, for the development of a computer algorithm for battery optimization on electric vehicles [16]. The research required a specialized car repair shop and laboratory for electric car restoration and battery management respectively. The project team was in Osorno, where the university campus did not have any e-mobility equipment. It only had an automotive maintenance facility that may serve has a starting point.

A joint venture with other universities and local companies integrates this project association, that include activities in Santiago and Osorno. An elaborate coordination effort is necessary to achieve the goals proposed and a strong academic collaboration is required. Therefore, the future benefits of the project results must be in accordance to a probable retribution to the academy.

New laboratory facilities, equipment, tools, and especially human capital development are among secondary objectives to look forward. An important student collaboration is there for suggested as a complementary project strategy.

2.3 Laboratory Implementation

The new facility is part of the automotive area and represents a major improvement for the overall academic experience. It consists of a closed space for new electrical instrumentation for battery managing, insulated car services tools, car elevator crane and proper safety delimited areas as shown in Fig. 2.

Safety delimited areas, for electrical and mechanical procedures according to local legal rules, are implemented with working procedures. Electrical battery charging stations and elevated cars are among the major safety concerns when working with outside personnel, specially students.

3 Academic Contributions

3.1 Student Participation

As any funded action, there is permanent academic staff considered from the beginning. Nevertheless, sometimes there is room for younger members and new collaborations. That is the case here, where a minimum of nineteen students have direct relationship with the laboratory on complementary activities.

As an important byproduct of this electromobility facility, there are four final degree projects being developed. Two of them are directly connected to the e-mobility and a third one on a related electrical subject, serving all of them for Bachelor's degree in Electrical Engineering with career ending requirements on topics such as electrical motor alternatives, battery management systems and e-chargers. A fourth final degree project is conducted for a Bachelor's degree in Risk Prevention on safety procedures for e-mobility labs, based on the implementation experience of this laboratory.

Fig. 2. Exclusive laboratory for electromobility teaching.

Alternative studies conducted by students may also be useful for the main project itself. Although, the project's results do not rely on the student's final degree outcome, any technical contribution may be welcomed. Students usually have more time and less constraints that may open new perspectives to an issue that is probably biased by more experienced academic personnel.

Additionally, these studies conducted by students, may be of interest for the university faculty members that have direct participation as student advisors and look forward for the possible results or procedures that may later be implemented on class activities.

Student internships in the recently implemented laboratory are also an interesting opportunity for alumni development, where a dozen students have already applied for internships as a voluntary choice. Safety laboratory procedures, tasks reports, schedule checking and research assistantships are among the main student contributions to the project, that also serve for their personal skill set and career enhancement.

Bachelor's degree in Automotive Engineering students are first inducted into the safety and hazard elements present in laboratory conditions. These may have been presented as general rules on their formal classes, but in this case, there are hazards on specific and real situations. Safety standards are a major concern while working with students on a non-curriculum academic project. Any accident may derive on legal issues that have to be consider on the student insurance policy.

Student participation is conducted as a permanent strategy. Pedagogical procedures are being filmed by the academic staff to be more consistent for future student incorporations as is shown in Fig. 3. Laboratory procedures and equipment manuals, online and written material, technical lectures and experts' presentations are among the activities and necessary materials to keep improving.

So far, the student's incorporation to the project has been a successful strategy for the research, the project team, the university and the students participating. Many comments regarding new student participation requests for the remaining project period, due to positive comments from all the involved, give a promising vision of the future of academic integration with high end research for Chilean technical university institutions.

Fig. 3. Student participation in the e-mobility laboratory.

3.2 Career Program Impact

Many higher education institutions in Chile are planning e-mobility subjects on their Bachelor's and Master's degree programs. However, there are few experts in the country and new knowledge is required.

As future benefits, this research project experience has set up a solid starting point for e-mobility formal studies. A new career specialized on electromobility is considered by INACAP starting on the year 2021. The presented project contributes to give a head start on this field on a campus located in southern Chile, that otherwise may not had been a first choice for new university studies implementation.

Among the major technical topics on electromobility to be considered in this new Bachelor's degree field are battery storage systems, range anxiety issues, electric charging, autonomous driving, artificial intelligence programing, remote sensing and control, driving safety, car manufacturing, and many social and economic relationships with urban development and smart city ecological strategies.

4 Conclusions and Future Work

New academic programs are trying to implement edge technologies on their Bachelor's and Master's degree programs. In this paper the experience of an electromobility research laboratory and its alternative implementation for educational purposes is presented. A case study is selected in a university that has the largest geographical distribution and the highest number of students in Chile. Being a centralized country where most of the Universities are in Santiago, this study avoids this issue and takes place in a campus located in the southern part of the country. Accordingly, the potential benefits of this implementation may be relevant to the local community.

In the paper, we have presented the experiences and accomplishments in student participation through the Learning by Doing approach, in the context of the resources offered by a project implementation. The facilities built and the associated activities, offer the opportunity for students to elaborate their final degree projects with Active and Collaborative Learning techniques in trending technology topics and equipment. It is shown that higher education institutions may benefit from complementary activities derived from a governmental funded project. This represent an opportunity for continuous education on high technology fields for students and trained faculty members who can access funding for edge technology research. It is demonstrated that pedagogical tools, methodologies and the appropriate technologies integration, help get students involved in their learning process to achieve the best academic results. Significant value may be obtained through the integration of pedagogical methods and practical scenarios on early student participation in high responsibility projects.

As future work, it is expected to use the new laboratory for research and for student programs as an example of a long-term academic strategy. Involve more students and increase the number of activities is upcoming. Final degree projects with official subject titles may be presented. Local enterprises interested on e-mobility issues are possible future participants. Continued research development and new funding are necessary for active student participation on high technology fields in higher education institutions.

Acknowledgments. The work on this paper has been partially supported by the FONDEF ID 18/10379 grant of the call "Concurso Idea de Investigación y Desarrollo 2018" from the Chilean National Commission of Research in Science and Technology CONICYT, support that is highly appreciated.

References

1. Rezvani, Z., Jansson, J., Bodin, J.: Advances in consumer electric vehicle adoption research: a review and research agenda. Transp. Res. Part D Transp. Environ. **34**, 122–136 (2015)
2. Jensen, A.F., Mabit, S.L.: The use of electric vehicles: a case study on adding an electric car to a household. Transp. Res. Part A Policy Pract. **106**, 89–99 (2017)
3. Ministerio de Energía, Gobierno de Chile: Chile ya se adapta a la Electromovilidad (2017). http://energia.gob.cl/electromovilidad. Accessed 3 Feb 2020
4. International Energy Agency: IEA analysis based on EVI country submissions, completed by EAFO (2017a), IHS Polk (2016), Marklines (2017), ACEA (2017a, 2017b) and EEA (2017)
5. Revista Electricidad: Las cuatro tendencias en I + D para la electromovilidad en Chile (2019). http://www.revistaei.cl/2019/07/04/las-cuatro-tendencias-en-id-para-la-electromovilidad-en-chile/. Accessed 3 Feb 2020
6. Ministerio de Educación de Chile MINEDUC: Objetivos y organización de la educación superior. https://www.ayudamineduc.cl/ficha/objetivos-y-organizacion-de-la-educacion-superior-5. Accessed 3 Feb 2020
7. Universidad Tecnológica de Chile INACAP: Why study at INACAP, Dirección de Asuntos Internacionales (2012). http://www.inacap.cl/tportalvp/pruebav2/extranjeros/international-exchange-students/why-study-at-inacap-2. Accessed 3 Feb 2020
8. Universidad Tecnológica de Chile INACAP: Acreditación Institucional, Dirección de Gestión de la Calidad (2012). http://portales.inacap.cl/acreditacion/acreditacion-inacap. Accessed 3 Feb 2020
9. Universidad Tecnológica de Chile INACAP: Expertos estadounidenses analizan el enfoque pedagógico de Aprender Haciendo en seminario de INACAP, Vicerrectoría de Innovación y Postgrado (2012). http://www.inacap.cl/tportalvp/portal/noticias/noticias-portal/expertos-estadounidenses-analizan-el-enfoque-pedagogico-de-aprender-haciendo-en-seminario-de-inacap. Accessed 3 Feb 2020
10. Universidad Tecnológica de Chile INACAP: INACAP firma compromiso público-privado para impulsar la electromovilidad, Escuela de Mecánica y Electromovilidad (2019). https://portales.inacap.cl/noticias/destacadas/2018/INACAP-firma-compromiso-publicoprivado-para-impulsar-la-electromovilidad. Accessed 3 Feb 2020
11. Universidad Tecnológica de Chile INACAP: Master in Pedagogy in Higher Education Degree, Centro de Innovación en Educación (2012). https://www.inacap.cl/tportalvp/?t=153&i=2&cc=10097&tm=2. Accessed 3 Feb 2020
12. Bonwell, C., Eison, J.: Active learning: Creating excitement in the classroom, ashe-eric higher education report no. 1 (1991)
13. Felder, R.M., Brent, R.: Active learning: an introduction. ASQ High. Educ. Brief **2**(4), 1–5 (2009)
14. Felder, B.R., Richard, M.: Cooperative learning in technical courses: Procedures, pitfalls, and payo s (1994)
15. Felder, R.M., Brent, R.: Effective strategies for cooperative learning. J. Coop. Collab. Coll. Teach. **10**(2), 69–75 (2001)
16. Comisión Nacional de Investigación Científica y Tecnológica CONICYT: Inauguran Primer Laboratorio de Electromovilidad en la Región de Los Lagos, Fondo de Fomento al Desarrollo Científico y Tecnológico FONDEF. https://www.conicyt.cl/fondef/2019/06/27/inacap-osorno-inaugura-primer-laboratorio-de-electromovilidad-en-la-region-de-los-lagos/. Accessed 3 Feb 2020

Special Session: Innovation in Computer Science Higher Education

Classroom Improvement Cycle in Architecture by Means of Problem-Based Learning

Maria-Victoria Requena-Garcia-Cruz(✉) and Antonio Morales-Esteban(✉)

Department of Building Structures and Geotechnical Engineering, University of Seville, Seville, Spain
{mrequenal, ame}@us.es

Abstract. This paper presents a teaching innovation proposal applied to the subject of Soil Mechanics and Foundations, which is taught on the third course in the Degree in Fundamentals of Architecture. This is a very technical subject, with a high level of difficulty and density of concepts. Therefore, in order to improve the students' learning, experimentation is necessary. This experimentation consisted in a Classroom Improvement Cycle (CIC) applied to the "soil surveying" topic during two sessions of four and two hours of duration, respectively. The cycle is based on the alternative didactic model, which is focused on problem-based learning with brief contributions by the professor. A sequence of activities has been developed, with the last activity being the most important one, as it is based on the resolution of a real problem. The results of this cycle have been satisfactory. The students have assimilated the sequences and improved their learning. However, it demands more effort from both the students and the teachers.

Keywords: Soil Mechanics and Foundations · Degree in Fundamentals of Architecture · University teaching · University teaching experimentation · Problem-based learning

1 Brief Description of the Context

The subject of Soil Mechanics and Foundations (SMF) is taught in the second term of the third year of the degree in Fundamentals of Architecture (Plan 2012) at the School of Architecture of Seville. It is a very technical subject, with a high level of difficulty and density of concepts. Furthermore, it is the first time that students are confronted with the complexity of studying soil behavior. Despite its close relationship with the Structures subjects taught during the degree, these subjects do not delve into the study of foundations and soil behavior. Therefore, the students have a relatively basic level of knowledge before taking the course. In general terms, the course focuses on the following issues:

Á. Herrero et al. (Eds.): ICEUTE 2020, AISC 1266, pp. 377–387, 2021.
https://doi.org/10.1007/978-3-030-57799-5_39

- Identification of soil types, foundation typologies, retaining structures, improvement and underpinning systems, and excavations.
- Design, calculation, analysis and definition of soil structures, foundations, retaining structures, pathology and underpinning.
- Foundation safety according to regulations.
- Study and analysis of soil behavior.

Given the complexity of the subject, teaching experimentation is necessary to improve student learning. Therefore, this paper presents a proposal for teaching innovation made during the seventh and eighth sessions of the academic year 2018–2019. The duration of the cycle was four and two hours during the first and second session of the cycle, respectively. This experiment consisted of a Classroom Improvement Cycle (CIC) applied to the topic "Soil surveying". This topic was chosen because it encompasses most of the conceptual contents of the subject.

The cycle is divided into three parts: a practical part carried out in the laboratory; a brief theoretical introduction combined with an activity; and, finally, a main and longer practical activity. The latter has been designed according to the problem-based learning method.

The number of students in the class is small, 24 in total. To carry out the activities, the students are divided into groups of three or four members. The classroom have non-fixed chairs and tables to allow and encourage group work.

2 Pre-design of the Classroom Improvement Cycle

This CIC aims to bring students closer to the real problems they will have to face during their professional life: the preparation of a soil survey campaign and the analysis of the information in a geotechnical investigation. In order to do this, the student must be able to interpret and apply the relevant regulations.

2.1 Map of Contents

The contents of the design of the improvement cycle are of a conceptual, procedural and attitudinal nature. All of them have been considered for the forecast of the activities carried out in the two sessions of the cycle. Contents are organized in a map (Fig. 1) and are as follows:

Conceptual:

- Description and analysis of a geotechnical investigation.
- Identification of the necessary information to be included in a geotechnical investigation.
- Description of a soil survey campaign.
- Distinction between deep and superficial soil surveying.
- Distinction between on-site, laboratory, borehole and sampling tests.

Procedural:

- Preparation of the index of the documentation in a geotechnical investigation.
- Handling, analysis and interpretation of regulations.
- Application of the regulations for determining the survey campaign.
- Simulation and preparation of a survey campaign for a real case.
- Observation of the performance of on-site and laboratory tests.

Attitudinal:

- Teamwork. The student learns to respect his or her classmates, respects their turn to speak, listens to their opinions and values them.
- Decision making effort.
- Awareness of the architect's responsibility in the preparation of the document and in determining the solution for the foundation of the work.
- Understanding the architect's need to collaborate with the company in charge of the investigation.

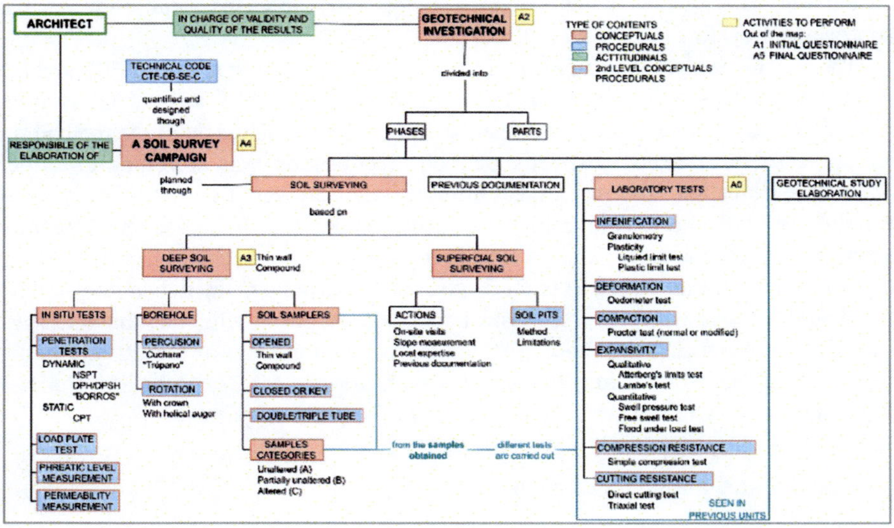

Fig. 1. Map of contents, including the activities to be carried out.

The map highlights the contents seen in other topics and the conceptual topics of first and second level. The latter includes procedural contents, since they deal, to a greater extent, with the process of carrying out the tests.

2.2 Possible Methodological Model

The methodological model proposed for this CIC focuses on students. Specifically, it is based on the Problem-Based Learning (PBL) method [4] and the one proposed by D. Finkel [3]. The first method entails active and cooperative learning. Through problem solving, students must be able to seek the theoretical tools needed to find the solution. This model favours students' autonomous work and teamwork. In addition, it strengthens the students' commitment and responsibility towards the group, which are two of the most necessary qualities to carry out a career as an architect. Likewise, the model proposed by Finkel presents the so-called "conceptual workshops". During these workshops, students also work on the resolution of a problem based on a real case, with brief contributions and guidance from their professor.

To organize the CIC, a sequence of activities based on the contents established in the course has been carried out. For this purpose, the indications of De Alba and Porlán [1] have been considered. The sequence of activities is shown in more detail in the section corresponding to the application of the cycle. This is divided into six activities, four of which are main activities and two secondary ones. The four main ones have been planned according to the students' active learning. The other two activities are carried out to evaluate student learning before and after the application of the improvement cycle

The general methodological model is based—as previously stated—on PBL and the "conceptual workshops". However, the main activity of the cycle is to develop an experience that is more specific and features the practical application of the method. In the first activity, students must carry out soil identification tests according to a brief technical explanation. The second and third activities are based on a brief theoretical introduction by the professor and on the students' work with a basic document for their future professional activity. The fourth activity focuses on a problem based on a real case. In order to solve it, students must look for the necessary theoretical tools while working in groups, where they discuss and share ideas about what they have learned individually. As in Finkel's model [3], the professor limits him or herself to guiding them by participating occasionally in the groups while they debate. At the end, students must hand in a document with the resolution of the problem posed.

2.3 Initial and Final Questionnaire

In order to assess the initial and final state of the students' knowledge, questionnaires are designed and carried out. Both include the same questions in order to be able to really assess the students' learning in relation to the issues explored. These questions cover the five first-level conceptual contents of the subject: the definitions of geotechnical investigation, survey campaigns and types of soil surveying (deep, surface or through laboratory tests). At the end of the session, students should be able to adequately answer the five questions posed.

The questionnaires are evaluated by means of a learning scale, following the guidelines established in Rivero and Porlán [6]. To this end, four tiers have been developed; they represent a ranking from lower (stages 0) to higher (stages 3) level of knowledge. Level of knowledge 0 groups erroneous, blank or meaningless answers.

Level 1 groups answers that show a very basic level of knowledge without any reasoning. Tier 2 encompasses correct answers that include very basic reasoning. Tier 3 includes answers that show high and reasoned knowledge on the issue. The questions included in both questionnaires, as well as the types of answers, are shown in Table 1.

Table 1. Questions, tiers and answer groups included in the questionnaires.

Q1. What is a geotechnical investigation?	
Tier 0	Student does not answer the question or answers wrongly at random
Tier 1	Student vaguely defines the concept of geotechnical investigation
Tier 2	Student knows and describes what a geotechnical investigation entails in a very basic way
Tier 3	Student knows and describes what a geotechnical investigation entails in a clear and concise way
Q2. Would you know how to carry out a soil survey campaign?	
Tier 0	Student does not answer the question or answers wrongly at random
Tier 1	Student is vaguely familiar with the process to be followed to carry out a soil survey campaign
Tier 2	Student has some notions regarding how to carry out a soil survey campaign and defines the process in a simple way
Tier 3	Student knows how to carry out a soil survey campaign and defines the process properly
Q3. What is the architect's responsibility in a soil survey campaign and a geotechnical investigation?	
Tier 0	Student does not answer the question or answers wrongly at random
Tier 1	Vaguely recognizes the role or job of the architect
Tier 2	Student indicates that the architect has a responsibility for carrying out a survey campaign and a geotechnical investigation, but does not offer any reasoning
Tier 3	Student indicates that the architect has a responsibility for carrying out a survey campaign and a geotechnical investigation and, furthermore, offers his or her reasoning
Q4. What do deep and superficial soil surveys entail?	
Tier 0	Student does not answer the question or answers wrongly at random
Tier 1	Student vaguely identifies the questions asked
Tier 2	Student acknowledges some simple differences in the questions but does not describe them
Tier 3	Student recognises all the differences and can describe them
Q5. What are the differences between on-site, borehole, sampling and laboratory tests?	
Tier 0	Student does not answer the question or answers wrongly at random
Tier 1	Vaguely identifies some of the questions asked
Tier 2	Differentiates some differences and defines some of them vaguely
Tier 3	Clearly defines and differentiates the questions asked

3 Application of the Classroom Improvement Cycle

First, this section shows the sequence of programmed activities, including a summarized account of the actual development of the sessions and, second, the process considered for the evaluation of student learning.

3.1 Sequence of Programmed Activities

The sequence programmed for the application of the cycle is divided into six activities: four are main activities and two are secondary. The general methodological model is based on the students' active learning, which has been considered for the design of activities A0, A2, A3 and A4. This last activity (A4) is also a more specific way to experience and apply Finkel's "conceptual workshops". Activities A1 and A5 serve to assess the level of learning before and after the application of the improvement cycle.

A0. Visit to the Laboratory. For this activity, students visit the School of Architecture's Geotechnical Engineering laboratory.

- Scheduled time: 120 min during the first session.
- Description: the purpose of this activity is to give students a first-hand experience of the reality of the lab. To do so, they carry out the tests together with the technicians. First, the professor gives a brief description of the different properties of the soil types, for which there are 22 samples. At the same time, students go through the samples to observe them, touch them and even smell them. They can add water to check the sample's permeability or cohesion. The purpose is for them to be able to identify different types of soil at a glance. Next, different tests are carried out: a regular Proctor test, a simple compression test, and the liquid and plastic limit tests. Then, students perform a granulometry test and are shown the oedometric, triaxial and permeameter test equipment. Like the direct cutting test, these three tests are not carried out due to time constraints.
- Contents: conceptual, those discussed from the beginning of the course; procedural, handling of regulations and carrying out tests; and attitudinal, understanding the architect's need to collaborate with the investigation's company.
- Actual development: the atmosphere was very positive, since this is a very different activity from the usual ones and is also carried out outside the classroom. The students' attitude reflected a much higher level of interest judging by the amount of questions they asked.

A1. Initial Questionnaire. From this activity onwards, all activities are carried out in the classroom where teaching normally takes place.

- Scheduled time: 30 min during the first session.
- Description: this activity is carried out to determine the students' level of knowledge prior to the CIC. Although it is an initial questionnaire, the activity is carried out after the laboratory visit.
- Contents: all the conceptual concepts included in the map.

- Actual development: during the activity, the atmosphere was quite relaxed. One of the difficulties analyzed was that they did not take it seriously and did not answer sincerely. However, the students' performance was more than satisfactory since they understood it was an activity to improve the teaching of the subject.

A2. On-site, Borehole and Sampling Tests. Deep vs. Superficial Soil Surveying

- Scheduled time: 45 min during the first session.
- Description: the purpose of this activity is for students to learn to differentiate between superficial and deep surveying. To this end, it has been approached as a short theoretical class focusing on the different types of on-site, borehole and sampling tests. Slides and videos on how each test is performed are used. This short activity is necessary to carry out the following activity.
- Contents: conceptual, distinction between deep and superficial soil surveying and between on-site, laboratory, borehole and sample taking tests; procedural, observation of carrying out the tests; and attitudinal, decision-making and awareness of the architect's responsibility in determining the building's foundation.
- Actual development: by reducing the theoretical load and using new tools, students seemed more interested. It was a real step forward, since the brief theoretical input focused on the importance of the subject.

A3. Geotechnical Investigation

- Scheduled time: 30 min during the first session.
- Description: the aim is for students to work with a document necessary for creating an architectural project. They are given (in pdf) a geotechnical investigation, which they must analyze by simulating a job they must perform during their professional activity. From the analysis, they must determine the parts into which it is divided, how it is done and the tests it includes. The teacher, based on the students' contribution, creates the index of the document on the blackboard and concludes by highlighting the importance of the document.
- Contents: conceptual, description and analysis of the geotechnical investigation and identification of the required information; procedural, preparation of the index of the geotechnical investigation; and attitudinal, teamwork, respect for classmates and speaking time, listening to the opinions of classmates, and assessing and becoming aware of the architect's responsibility in the preparation of the document.
- Actual development: during the activity, students were encouraged to participate actively, even those who do not normally participate. No time was lost due to computer problems, which were one of the expected risks of the activity.

A4. Carrying Out a Soil Survey Campaign

- Scheduled time: 15 and 90 min for the first and second sessions, respectively.
- Description: this activity has been conceived as a real problem to work on from a small problem-based learning experience. The students in groups will have to face a problem that architects find in any project. The activity consists of conducting an on-site soil survey campaign for a real building (already built). During the first

15 min of the first session, they are given the statement and divided into working groups. The groups have three or four members and are assigned an identification number corresponding to the building on which they shall have to do the work. During this first session, the professor briefly comments on the statement. The sections they will have to answer are:

- Planimetry and photographs of the building. The classification of the building and the type of land must be included.
- Determination, distribution and justification of the survey points to be carried out.
- Determination, distribution and justification of the types of penetration tests and surveys to be carried out.
- Determination and justification of the geotechnical profile of the land planned according to the "*ITE Sevilla*" document.
- Determination, explanation of the process and justification of the types of laboratory tests to be carried out (for each tier).
- Determination and justification of the test results obtained according to the document "*ITE Sevilla*".

Previously, as out-of-class work, each member is responsible for resolving a section to become an "expert". During the second in-class session, students work in groups to solve the questions, telling each other what they have studied and analyzed on their own. In order to carry out this work, students must use the applicable regulation, CTE-DB-SE-C [5]. As the tests are not going to be carried out, the students will obtain this information from Annex I.2 of the document "*ITE Sevilla*" [2]. This document includes maps and test results on the properties of the different strata normally found in Seville. This work should be delivered one week after this session.

- Contents: conceptual, description of a soil survey campaign; distinction between deep and superficial soil surveying and between on-site, laboratory, borehole and sampling tests; procedural, handling, analysis, interpretation and application of the regulations for determining the survey campaign, and simulation and preparation of a real-case campaign; and attitudinal, all of those listed in the map of contents.
- Actual development: the work climate in the classroom was intense. The majority of the students had worked at home and had some doubts. However, many questions arose after sharing ideas with their classmates. The professor's involvement was focused on helping the students if they did not make progress in solving the problem. The main difficulty was that one of the groups had not worked at home and its members could not keep up with the class. The time scheduled for the second session was 90 min; however, it had to be increased to 120 min. Students were interested and liked that it was a real case.

A5. Final Questionnaire.

- Scheduled time: 30 min during the second session.
- Description: the purpose of this activity is to analyze the development of the students' learning by comparing the results with those of the initial questionnaire by

means of learning tiers. In order for the comparison to be real and the results not to be affected, students are asked the same questions included in the initial one.

- Contents: all the contents included in the map.
- Actual development: students answered the questionnaire more easily than the previous one. However, they stressed that this activity should be done after the final delivery of A4.

3.2 Evaluation of Student Learning

The evaluation of the students' learning has been carried out through observation, comparing their initial and final knowledge and assessing the work delivered corresponding to A4. In total, 24 students were surveyed. Figure 2 shows the students' progress for each of the questions analyzed. In addition, the average score obtained for each of the questions according to the initial and final questionnaire is included.

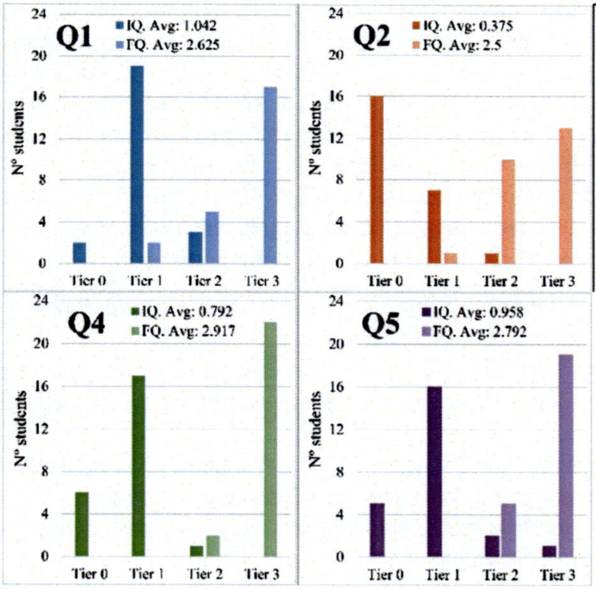

Fig. 2. Initial and final learning tiers in the different questions of the questionnaire.

As it can be observed in the graphs, students have generally significantly improved their level of knowledge regarding the questions asked. In the first questionnaire, most answers have been placed in tiers 0 and 1. Only some of them could be grouped in tier 2. It can be noted that students have a higher initial level of knowledge in Questions 1 and 4. These are simpler questions that, through logic and general knowledge, can be answered. However, less positive data have been obtained in the more technical questions. Particularly, Question 2 obtained the worst results. Regarding Question 3, it seems that students have some knowledge about the architect's role and responsibility in this subject.

As for the results of the final questionnaire, these are more than satisfactory. However, students do not define well the answer to Question 2. The most substantial improvement took place in Question 3. Students largely understood the responsibility of an architect in a soil survey campaign and a geotechnical investigation. After all, this is one of the main concerns of this cycle.

4 Evaluation of the Classroom Improvement Cycle

Overall, the result of implementing the improvement cycle has been quite satisfactory. Students have positively assessed the activity. However, they are aware that much more time is needed to carry it out. Furthermore, it requires greater effort from both professor and student. The professor must prepare the class and the students must find, by themselves, the theoretical tools necessary to resolve the problem.

There were only a few isolated cases in which the CIC did not help to improve the students' level of knowledge and learning. Even so, it can be said that the students assimilated the sequence of activities perfectly and that they improved their learning.

For future improvement cycles, the alternative didactic model applied in the subject of Soil Mechanics and Foundations is shown to be an improvement for student learning. The implementation of problem-based learning fosters students' reading, analysis and interpretation skills, which are necessary for the development of any professional work. Furthermore, it helps them to face a real problem on their own, which they should be able to solve in the not-so-distant future. The didactic principles that have guided the experience and that should be kept in the future are:

- Why teach: to ensure the development of meaningful learning; to contribute to the development of the 'learning to learn' competence.
- What to teach: general hypothesis of progressive knowledge development; learning from maps of contents by subject including conceptual, attitudinal and procedural contents.
- How to teach: promoting active participation; starting from the level of student development; problem-based learning and coherent sequence of activities.
- Evaluation: focused on the development of student knowledge; carried out through the use of various monitoring instruments.

References

1. de Alba, N., Porlán, R.: La metodología de enseñanza. In: Porlán, R. (ed.) Enseñanza Universitarla. Cómo mejorarla, Morata, Madrid, pp. 37–54 (2017)
2. Colegio de Arquitectos de Sevilla: Anejo I. 2: Mapa geotécnico básico de la ciudad de Sevilla. In: Protocolo de Inspección Técnica de Edificaciones (2010)
3. Finkel, D.: Experiencias que enseñan: crear esquemas para el aprendizaje. In: Dar clase con la boca cerrada, pp. 153–186. Publicaciones de la Universidad de Valencia, Valencia (2008)

4. Justo Moscardó, E., Delgado Trujillo, A., Vázquez Boza, M., Branda, L.: Implementation of problem-based learning in structural engineering: a case study. Int. J. Eng. Educ. **32**(6), 2556–2568 (2016)
5. Ministerio de Fomento de España: Documento Básico de Seguridad Estructural Cimientos (DB-SE-C) (2007). https://www.codigotecnico.org/images/stories/pdf/seguridadEstructural/DBSE-C.pdf. Accessed 13 May 2019
6. Rivero, A., Porlán, R.: La evaluación de la enseñanza universitaria. In: Porlán, R. (ed.) Enseñanza Universitarla. Cómo mejorarla, Morata, Madrid, pp. 73–92 (2017)

Check for updates

Problem Generalization for Designing Recursive Algorithms

Diana Borrego(✉), Irene Barba, Miguel Toro, and Carmelo Del Valle

Departamento de Lenguajes y Sistemas Informáticos, Universidad de Sevilla, Sevilla, Spain
{dianabn,irenebr,migueltoro,carmelo}@us.es

Abstract. This paper focuses on the difficulty for university students to acquire, within computational thinking, the skills to solve certain problems through recursion. The acquisition of this type of reasoning is essential to understand the different problem solving techniques that are based on recursive algorithms, such as divide and conquer or dynamic programming. Therefore, first, the generalization of problems is proposed as a strategy for designing recursive algorithms. As a second step, that generalization is formalized through a specification sheet that contains different fields that correspond to the characteristics that are relevant to solve a problem recursively.

Keywords: Recursive algorithms · Computational thinking · Problem generalization

1 Introduction

Computer science university students have the need to develop computational thinking, which is the mental process used to formulate problems whose solutions can be carried out by a computer. Computational thinking includes skills such as modeling and breaking down problems, processing data, creating algorithms and generalizing them.

There are several definitions related to computational thinking in the literature. Reviewing publications about this topic, Wing et al. [6], who made the term computational thinking famous, present the following definition: "Computational thinking involves solving problems, designing systems, and human behavior understanding, by drawing on the concepts fundamental to computer science. Computational thinking includes a range of mental tools that reflect the breadth of the field of computer science."

Therefore, students must acquire computational thinking to be able to carry out problem solving through modularization, top-down analysis, bottom-up analysis, recursion, and so on. Particularly, the understanding and application of this last concept of recursion [3,4] requires the students to make an extra mental effort. They usually understand and learn more easily the iterative formulations of problems. However, to understand the recursive formulation, a special type of thinking is needed, and therefore to be able to use recursive thinking [7]. This presents a difficulty that makes a mental predisposition necessary that is not always frequent or easy to reach depending on each person.

In detail, the recursive resolution of a problem is raised when the specification of its solution is made based on the solution of other problems of the same nature but a

Á. Herrero et al. (Eds.): ICEUTE 2020, AISC 1266, pp. 388–398, 2021.
https://doi.org/10.1007/978-3-030-57799-5_40

smaller size. Therefore, every recursive definition has (i) base case(s), encompassing and resolving cases where the resolution of the problem is direct and does not require the use of the function being defined; (ii) recursive case(s), which are those whose resolution does require the use of the function being defined; and (iii) size of the problem, which is a cardinal indicating the dimension of the problem to be solved, and that decreases in each subproblem. Thus, a recursive algorithm is one that expresses the solution of a problem in terms of invoking itself.

As an example, the definition and recursive algorithm for the resolution of the factorial of a number n are shown below.

$$n! = \begin{cases} 1 & n = 0 \\ n * (n-1)! & \mathrm{n} > 0 \end{cases}$$

```
N f(Integer n){
    N s;
    if{n == 0}{
        s = 1;
    }else {
        s = n * f(n-1);
    }
    return s;
}
```

Algorithm 1: Recursive algorithm for calculating the Factorial number (in Java)

Not all recursive definitions are suitable for building an algorithm. To be adequate, it must have: (1) at least one base case; (2) in each recursive step the size should be reduced (which implies that the execution of the algorithm will end).

Taking all this into account, and to facilitate the acquisition and application of recursive thinking, this paper proposes to carry out two steps prior to the design of a recursive algorithm to solve a given problem, as shown in Fig. 1. Specifically, these steps are:

1. Generalization of the problem through a set of problems [1]. More precisely, generalizing consists of adding properties to the original problem to consider it a particular case of a wider set of problems.

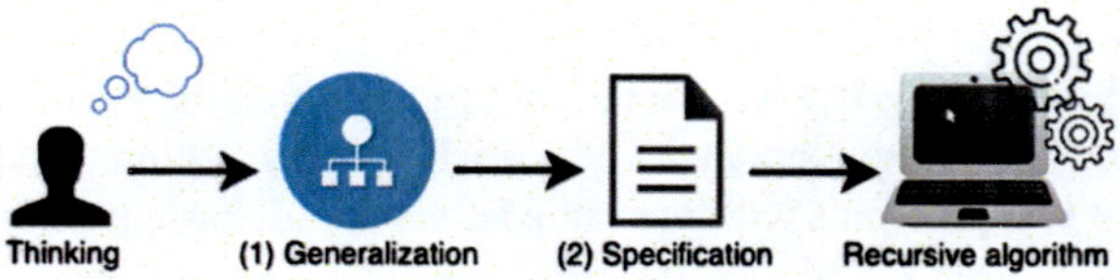

Fig. 1. Proposed approach for designing recursive algorithms

2. Specification of the solution to the problem through a predefined specification sheet based on the previous generalization made. This sheet contains different fields that correspond to the characteristics that are relevant to solve this problem recursively.

Note that the goal of the proposed approach is not to provide a formal method but to facilitate the acquisition of skills related to the design of recursive algorithms. For this, we use our experience in the design of types in the context of object-oriented programming to conceptualize the population of problems that are generated when designing recursive algorithms, i.e., when the generalization is carried out. In such context, the correct definition of the size of the problems (that is a derived property of such problems) is key when analyzing the correctness of the generalization performed. To be more precise, when generalizing a problem for designing a recursive algorithm, it is necessary to verify that there is an initial problem, and that all the problems are ordered by size.

The remainder of the paper is structured as follows: Sect. 2 presents the generalization of problems through sets of problems, Sect. 3 includes the specification sheet of the problem solution by recursion along with some illustrative examples, and Sect. 4 presents a critical discussion of the proposed approach. Finally conclusions are drawn and future work is proposed in Sect. 5.

2 Problem Generalization

This section presents the proposed approach: Sect. 2.1 explains the concept of set of problems in the context of problem generalization, Sect. 2.2 details how such set of problems can be used for problem generalization, and, lastly, Sect. 2.3 includes the general schema of recursion that is related to the proposed problem generalization.

2.1 Set of Problems

In a recursive definition it is always necessary to start from a set of problems $P = \{p, p_0, p_1, \ldots\}$. In such set each problem is characterized by a set of properties $x = \{x_0, x_1, \ldots\}$. Each property of a set of problems can be categorized as:

- Shared: shared properties are defined as properties of the set of problems that take the same value for all problems of such set.
- Individual: individual properties are defined as properties of the set of problems that take different values for each problem of such set.

Both shared and individual properties can be categorized as *basic* or *derived properties*. The inclusion of the latter, unlike the former, is optional since their values can be calculated from the values of the basic properties. Derived properties are typically included for the sake of clarity. It is important to clarify that the basic individual properties are the ones that unequivocally identify each problem within the set of problems, i.e., there are not two problems with the same values for all basic individual properties.

Moreover, each problem of a set of problems has a specific size, that is a derived individual property. To be more precise, the size of a problem is defined as an integer greater than or equal to zero that is related to the complexity of the problem (i.e., how far is the problem from a case base). There are different ways of defining the size of a problem. However, it is necessary to check the correctness of such definition by considering that the size of a problem should (1) be a function over some of the properties of the problem, (2) be reduced as a base case is approached.

The problems of a set of problems can be categorized as:

- Base case of a set of problems *P*: a problem of a set of problems is a base case when it has a direct solution (i.e., without performing a recursive call). A problem that is a base case has a small size. Moreover, there can exist several base cases within the same set of problems.
- Recursive case of a set of problems *P*: a problem of a set of problems is a recursive case when it does not have a direct solution (i.e., it is necessary to perform a recursive call to solve it). The solution of a recursive problem is defined by combining the solutions related to problems of smaller sizes (i.e., subproblems).
- All problems, in the set of problems of interest, verify a predicate $d(x)$. This predicate establishes the domain of valid problems.

2.2 Problem Generalization

The key idea to solve a problem in a recursive way is based on expressing its solution in terms of operations over solutions to a set of problems of smaller size. For this, in many cases it becomes necessary to imagine that set of problems from the original problem (i.e., *problem generalization*).

To be more precise, the problem generalization consists of the following steps:

1. Adding, if necessary, a set of additional properties to those that already had the problem and classify them into basic, derived, individual, shared, etc.
2. Defining the domain of problems of interest through a predicate.
3. Defining the size of a problem.
4. Stating the base cases and the recursive cases.
5. Specifying how each recursive case is reduced to subproblems with a smaller size.
6. Specifying how each recursive case's solution is related to that of its subproblems.
7. Specifying the solution of base case.
8. Indicating which of the problems in the problem set corresponds to the original problem.

2.3 General Structure of a Recursive Algorithm

In order to design a recursive algorithm for solving a problem with parameters x (of type X) it is typically advised to perform 2 functions. The first one (i.e., function f, cf. Algorithm 2) receives as input parameters of the initial problem to solve x, and performs an initial call to a second function g (cf. Algorithm 3). The latter is a recursive function that receive as input parameter both the parameters of problem to solve x and the additional properties that are considered after problem generalization (i.e., y of type Y). From now on the properties of the generalized problem will be both the parameters of the original problem and the new properties added. Function f is optional. There are cases where the function g is sufficient and f does not appear. This situation is found when it is possible to consider a broad set of problems by varying within a domain the values of the parameters. One function calls the other as depicted in Algorithm 2.

```
S f(X x) {
    return g(x,y_0);
}
S g(X x, Y y){
    ...
}
```

Algorithm 2: Functions in a recursive algorithm (in Java)

The general structure of a recursive algorithm is shown in Algorithm 3.

```
S f(X x) {
    return g(x,y_0);
}
S g(X x, Y y){
   S s;
   if(b(x,y){
        s = sb(x,y);
   } else {
        Y y_0 = sp_0(x,y);
        S s_0 = g(x,y_0);
        Y y_1 = sp_1(x,y);
        S s_1 = g(x,y_1);
        ...;
        Y y_{m-1} = sp_{m-1}(x,y);
        S s_{m-1} = g(x,y_{m-1});
        s = c(s_0,s_1,...,s_{m-1},x,y);
   }
   return s;
}
```

Algorithm 3: General structure of a recursive algorithm (in Java)

As follows, concepts related to Algorithm 3 are detailed:

- x: parameters of problem of type X.
- S: type of the solution.
- y: additional properties (of type Y) for the set of problems that are included after problem generalization.
- $size(x,y)$: function that returns the size of the problem (x,y).
- $d(x,y)$: predicate establishing the domain of (x,y).
- $b(x,y)$: predicate that shows if a problem is a base case

- $sb(x,y)$: function that returns the solution to a base case.
- $np(x,y)$: function that returns the number of subproblems related to (x,y).
- $m = np(x,y)$: the number of subproblems
- $y_0, y_1, ..., y_{m-1}$: variables of type Y that store the subproblems related to (x,y). To be more precise, y_i refers to the i-th problem related to (x,y).
- $sp_0(x,y), sp_1(x,y), ..., sp_{m-1}(x,y)$: functions that return the subproblems related to (x,y). To be more precise, $sp_i(x,y)$ returns the the i-th problem related to (x,y).
- $s_0, s_1, ..., s_{m-1}$: variables of type S storing the solutions to the subproblems related to (x,y). Thus, s_i refers to the solution to the i-th problem related to (x,y).
- $c(s_0, s_1, \ldots, x, y)$: function that returns the solution to x by combining the solutions to its subproblems.

3 Specification Sheet for the Problem Generalization

This section details the sheet that is proposed for specifying the problem generalization previously explained (cf. Sect. 2.2). The parts that are included in such specification sheet are explained as follows:

- Types: information related to the types that are relevant to solve the problem, e.g., the type of the solution.
- Shared properties: shared properties of the set of problems that is obtained after the problem generalization.
- Individual properties: individual properties of the set of problems that is obtained after the problem generalization.
- Size: size of the problem, which is a function over the properties of such problem.
- Initial instantiation: initial value that is given to the individual properties.
- Base cases: it gives information about whether a problem is a base case. In case there is more than one condition that causes a problem to be considered a base case, one per line is specified.
- Solution for base cases: solutions for the problems that are base cases. If there is more than one base case condition, the solution is provided to each of them on different lines, using the operator $\rightarrow$.
- Number of subproblems: number of subproblems that need to be solved to obtain the solution to the original problem.
- Subproblems: subproblems that need to be solved to obtain the solution to the original problem. When determining them, it is only necessary to specify the changes that occur in the individual properties with respect to the original problem, since the shared properties remain the same for all problems. Different subproblems are also specified in different lines, with the operator $\rightarrow$. In case that only one subproblem needs to be solved, its is determined through the condition to select which one, with the format $c_i \rightarrow s_i$. On the other hand, when there are more than one subproblem, a superscript is added to distinguish the referred one ($\xrightarrow{0}, \xrightarrow{1}, \ldots$).
- Combine: function that obtains the solution to the original problem by combining the solutions that are obtained for the subproblems.

As can be observed, there are many similarities between the elements of the specification sheet and the elements of the general structure of a recursive algorithm (for details cf. Table 1).

Table 1. Relation between the specification sheet and the recursive algorithm.

Specification sheet	General structure of a recursive algorithm
Types	S, X, Y
Shared properties	
Individual properties	
Domain	$d(x,y)$ predicate
Size	$size(x,y)$ function
Initial instantiation	y_0
Base cases	$b(x)$ predicate
Solution for base cases	$sb(x,y)$ function
Number of subproblems	$np(x,y)$ function
Subproblems	$sp_0, sp_1, \ldots$. functions
Combine	$c(s_0, s_1, \ldots, x, y)$ function

3.1 Running Examples

This section includes how the proposed approach is applied to two illustrative examples. For each example, (1) the specification sheet related to the problem generalization and (2) the related recursive algorithm are detailed.

Example 1. Calculating the Fibonacci numbers [5]. The Fibonacci number is recursively defined as follows:

$$f(n) = \begin{cases} n & n \leq 1 \\ f(n-1) + f(n-2) & \text{otherwise} \end{cases}$$

Regarding such definition, a recursive algorithm can be designed (cf. Algorithm 4) according to the problem generalization that is specified in Table 2.

```
N fib(Integer n){
    N s;
    if(n<=1){
        r = n;
    } else {
        Integer y_0 = n-1;
        s_0 = fib(y_0);
        Integer y_1 = n-2;
        s_1 = fib(y_1);
        s = s_0+s_1;
    }
    return s;
}
```

Algorithm 4: fib - Recursive algorithm for calculating Fibonacci numbers (in Java)

Table 2. Specification sheet for solving the Fibonacci problem.

Fibonacci	
Types	Solution: Integer
Shared properties	
Individual properties	n: Integer
Domain	$n \geq 0$
Size	n
Initial instantiation	
Base cases	$n \leq 1$
Solution for base cases	n
Number of subproblems	2
Subproblems	$(n) \xrightarrow{0} (n-1)$ $(n) \xrightarrow{1} (n-2)$
Combine	$s_0 + s_1$

Example 2. Binary search in a sorted list [2]. The binary search algorithm is an efficient search method, which finds the position of a target value within a sorted list. To do this, the algorithm divides the list by its middle element into two smaller sublists, and compares the target value with the middle element. If they are equal, the search ends. If the target value is lower, it must be (if present) in the first half of the list, in the second half otherwise. The process continues until the target is found, or until it reaches an empty sublist due to the target value is not in the list (returning -1), which are two base cases conditions.

For this problem, the following shared properties are considered: (1) list of elements *ls*, (2) element that is searched *m*, and (3) size of the list of elements, that is a derived property *n*. To generalize, the following individual properties are established: (1) index *i* that indicates the starting position of the sublist considered for the current subproblem, (2) index *j* that indicates the ending position of the sublist considered for the current subproblem, and (3) the middle position *k*, a derived property, between *i*, *j*. As detailed, the problem only invokes one of the two possible subproblems.

More details are given in Table 3, and the related algorithms are depicted (cf. Algorithm 5). As can be observed, the main algorithm (cf. *bs* in Algorithm 5) performs the initial call to the recursive algorithm (cf. *bs_g* in Algorithm 5) according to the initial instantiation of the individual properties.

4 Discussion

The proposal presented in this paper presents several advantages. To be more precise, on the one hand, the fact of analysing the solution to a problem before carrying out the implementation itself facilitates error isolation as well as error detection. On the other hand, the generalization of problems and the related specification sheet allow for establishing a systematic way of designing a recursive algorithm for solving a specific problem.

Based on our experience as teachers of programming subjects in the first university-level courses, students are reluctant at first to specify the generalization of problems through the specification sheet. On the one hand, this is something new for them, and therefore, a period of learning and training is required. In this period, students acquire the necessary knowledge regarding the meaning and usefulness of each of the properties that are included in the specification sheet. On the other hand, specifying the problem generalization is more time-consuming than directly implementing the solution. Moreover, the application of the proposed approach makes sense for problems that presents certain characteristics. To be more precise, the proposed approach makes more sense in the case that designing an efficient recursive algorithm for solving a problem is not trivial for the students, i.e., previous analysis is required (e.g., cf. Example 2). However, there are some problems, i.e., those problems whose definition itself is recursive, that

Table 3. Specification sheet for solving the Binary search problem.

Binary search	
Types	Solution: Integer
Shared properties	ls: *List* $< E >$ e: E n: Integer (derived - size of ls)
Individual properties	i: Integer j: Integer k: Integer (derived), $k = (j+i)/2$
Domain	$i \in [0,n) \&\& j \in [i,n]$
Size	$j-i$
Initial instantiation	$(0,n)$
Base cases	$e = ls[k]$ $j-i < 1$
Solution for base cases	$e == ls[k] \rightarrow k$ $j-i < 1 \rightarrow -1$
Number of subproblems	1
Subproblems	$e > ls[k] \rightarrow (k,j)$ $e \leq ls[k] \rightarrow (i,k)$
Combine	s_0

presents a recursive algorithm that can be directly implemented in an easy way without requiring a previous analysis (e.g., cf. Example 1).

Therefore, in summary, through the use of specification sheets, students are able to address complex problems in less time, since they are guided and assisted in the reasoning necessary to reach the solution.

```
Integer bs(ls,e){
   Integer n = ls.size();
   return bs_g(ls,e,n,i,j,(j+i)/2);
}
Integer bs_g(ls,e,i,j,k){
   Integer s;
   if(j-i < 1) {
       s = -1;
   } else if(e == ls[k]) {
       s = k;
   } else {
       Integer n1, nj, nk;
       if(e < ls[k]) {
           ni = i;
           nj = k;
           nk = (ni+nj)/2;
       } else if(e > ls[k]){
           ni = k;
           nj = j;
           nk = (ni+nj)/2;
       }
       s = bs_g(ls,e,ni,nj,nk);
   }
   return s;
}
```

Algorithm 5: Recursive algorithm for the binary search (in Java)

5 Conclusions and Future Work

In this paper, a methodology for solving recursive problems is proposed, so that the acquisition of this skill within computational thinking is easier for computer science university students. For this, the generalization of problems is proposed as a strategy for the design of recursive algorithms. After this generalization, and based on it, a formalization is made by filling in a defined specification sheet that contains different fields encompassing the characteristics that are relevant to solve a problem recursively. Through this step-by-step methodology, students can understand and learn recursive reasoning more quickly and easily.

As future work, we intend to make proposals regarding the specification sheets of the solution using techniques such as dynamic programming, or backtracking. It is also proposed to conduct a study about the acceptance of this type of techniques by the students, and the impact that their use has on the grades obtained.

References

1. Algoritmos iterativos y recursivos. https://drive.google.com/file/d/1qF7eDq0a7MvrFTbfWVJF7qKb9LZTrezb/view. Accessed 21 Jan 2020
2. Knuth, D.E.: The Art of Computer Programming, Volume 3, Searching and Sorting. Addison-Wisley, Reading, MA (1973)
3. Ljung, L., Söderström, T.: Theory and Practice of Recursive Identification. MIT press, Cambridge (1983)
4. Shih-hua, H.: Theory of recursive algorithms I. Sci. Sinica **9**, 843–875 (1960)
5. Vorobiev, N.N.: Fibonacci numbers. Birkhäuser (2012)
6. Wing, J.M.: Computational thinking. Commun. ACM **49**(3), 33–35 (2006). https://doi.org/10.1145/1118178.1118215
7. Zapata-Ros, M.: Pensamiento computacional: Una nueva alfabetización digital. Revista de Educación a Distancia (RED) (September 2015). https://doi.org/10.6018/red/46/4

Analysis of Student Achievement Scores via Cluster Analysis

Viviana E. Jiménez Chaves[3], Miguel García-Torres[1(✉)], David Becerra Alonso[2], Francisco Gómez-Vela[1], Federico Divina[1], and José L. Vázquez-Noguera[3]

[1] Division of Computer Science, Universidad Pablo de Olavide, 41013 Seville, Spain
{mgarciat,fgomez,fdivina}@upo.es

[2] Universidad Loyola Andalucía, Seville, Spain
dbecerra@etea.com

[3] Universidad Americana, Asunción, Paraguay
{viviana.jimenez,jose.vazquez}@americana.edu.py

Abstract. In education, the overall performance of every student is an important issue when assessing the quality of teaching. However, in the traditional educational system not all students have the same opportunity to develop their academic skills in an efficient way. Different teaching techniques have been proposed to adapt the learning process to the student profile. In this work, we analyze the profile of students according to their performance on academic activities and taking into account two different evaluation systems: work-based assessment and knowledge-based assessment. To this aim, data was collected during the fall semester of 2019 from a physics course at Universidad Loyola Andalucía, in Seville, Spain. In order to study the student profiles, a clustering approach combined with supervised feature selection was applied. Results suggest that two student profiles are clearly distinguished according to their performance in the course in both evaluation approaches. These two profiles correspond to students that pass and fail the course. The output of the analysis also indicates that there are redundant and/or irrelevant features. Therefore, machine learning techniques may be helpful for the design of effective activities to enhance the student learning process in this physics course.

Keywords: Educational Data Mining · Clustering · Feature selection.

1 Introduction

The last decades have seen an increasing interest on the application of Information Technologies to educational systems, given the large amount of data in this field [2]. The analysis of this type of data has provided new opportunities to improve the educational system from different points of view, such as: student performance, subject evaluation processes or identification of key competencies

Á. Herrero et al. (Eds.): ICEUTE 2020, AISC 1266, pp. 399–408, 2021.
https://doi.org/10.1007/978-3-030-57799-5_41

in the learning process among others [5]. In this context, the analysis of students' academic records could identify the causes of academic failure and, where appropriate, indicate the steps needed to mitigate them [3].

In addition, these data cannot be handled manually due to the large number of students enrolled in a given educational system every year (e.g. 1,575,579 students in the 2018–2019 academic year in the Spanish University System [11]). As a result, data mining techniques emerge as a powerful tool for in-depth data analysis in the field of educational systems. In particular, the application and development of data mining approaches in the field of education is called Educational Data Mining (EDM) [20]. EDM is capable of performing pattern identification, clustering, and predictive tasks for characterisation of student's student assessments, behaviour and merit, educational functions among others [4,15,19].

In the field of EDM, researchers have widely applied clustering methods such as K-means or Hierarchical clustering [8]. Clustering methods are able to split the input data into different groups based on common features. These approaches provide an overview of the studied data, thus being able to carry out a reduction in the data size or to identify groups of interest [6]. For example, significant variables that affect and influence the performance of weak students can be identified using this technique [8]. Clustering techniques allow using multiple attributes to identify similar groups in an unsupervised way (i.e. to find groups of similar learners). In the work by Wook et al. [24] the authors proved that clustering methods can improve and accelerate the creation of a cluster of students with similar learning styles. Another example can be found in the work presented by Perera et al. in [16]. This work concerns with students working in teams, the authors performed the extraction of data collected to characterise the work of stronger and weaker students. A clustering approach was applied to find clusters of similar teams and similar individual members. In addition, sequential pattern mining was applied to extract sequences of frequent events.

Another relevant approach in the field of EDM is the application of Feature Selection (FS) techniques [22]. These techniques are applied in search for the most relevant subset of characteristics in a data set [18,26]. FS algorithms remove irrelevant and/or redundant attributes from educational datasets and thus facilitate the identification of key attributes and reduce the dimensionality of the input dataset [25]. In this sense, the work by Punlumjeak and Rachburee [17], proposed a comparison of four feature selection techniques: genetic algorithms, support vector machine, information gain, minimum and maximum relevance algorithms with supervised Classifiers such as Naive Bayes, decision tree, k-nearest neighbour and neural networks. The results achieved show that minimum and maximum relevance feature selection methods with 10 features give the best result on 91.12% accuracy with k-nearest neighbour Classifier. Another work by Acharya and Sinha [1] defined a student data set with 309 records and 14 features collected by a survey from various graduation level students majoring in Computer Science under University of Calcutta. Different feature selection algorithms were applied on this data and the best results were achieved by correlation based feature selection (CFS) with 8 features. Finally, in the paper by

Velmurugan and Anuradha [22], the authors compared the performance of various feature selection techniques using WEKA [13] in the prediction of students' performance on the final semester exams using different Classification algorithms such as J48, Naive Bayes, Bayes Net, IBk, OneR, and JRip. The results obtained by the authors show that the accuracy of IBK is 99.680% which is found to be higher than other Classifiers over the CFS subset evaluator.

In this work, student achievement scores data is studied applying cluster analysis. We also apply clustering to data after applying a feature selection approach. Results achieved are promising and may be helpful to design effective activities for a course.

The rest of the paper is organized as follows. In Sect. 2, the data and evaluation are introduced. Machine Learning techniques are described in Sect. 3. Experiments are presented in Sect. 4 and, finally, conclusions in Sect. 5.

2 Methodology

This section describes the data and the evaluation systems.

2.1 Data Collection

The data used for this work was collected from the Physics course given during the fall semester of 2019 at Universidad Loyola Andalucía, in Sevilla, Spain. This course, taught in 3 months, covers the following units:

- **Unit 1: Free vectors**. Reviewing the basic algebra of vectors that was already studied during high school. It includes operations with vectors, basic geometry, and how this is applied to physics.
- **Unit 2: Kinematics of a point particle**. Reviews prior knowledge about the basic motion of particles. As new material it includes the use of the Frenet Trihedron.
- **Unit 3: Dynamics of a point particle**. Reviews prior knowledge about Newton's laws of motion. Emphasis is made in the correct application to exercises.
- **Unit 4: Dynamics a system of particles**. Reviews prior knowledge about the interaction of point particles. Conservation laws for momentum and energy are put to practice.
- **Unit 5: Sliding vectors**. Covers new material for the students on the algebra of sliding vectors and the moment vector fields they generate.
- **Unit 6: Rigid body kinematics**. Applies the knowledge on sliding vectors to the three-dimensional motion of solids.
- **Unit 7: Planar motion of rigid bodies**. Studies the specific case of solids that move in two dimensions. This is applied to the motion of machine parts.
- **Unit 8: Cartesian tensor**. Introduces tensors as a means to explain how the moment of inertia of a rotating solid is obtained.

- **Unit 9: Rigid body dynamics**. This units uses everything learned up to it, and explains Newton's laws of motion on solids. It also covers the corresponding conservations of energy and angular momentum.

During the course, a series of activities are proposed. These provide a series of evaluation items that here will be used as features. The features generated are the following:

- **eInstruction** [23]. The features *eInstructionT1*, *eInstructionT2*, *eInstructionT7* and *eInstructionT8*, refer to activities proposed for units 1,2,7 and 8, respectively.
- **Lab reports**. For units 1–5, laboratory experiments are evaluated: *LabP1*, *LabP2*, *LabP3*, *LabP4* and *LabP5*.
- **EduZinc Homework** [7]. The students are expected to complete some physics exercises at home. The variables for this case are *HwT1*, *HwT2*, *HwT3*, *HwT4* and *HwT7* refer to Homework activities proposed during the course for some specific units.
- **EduZinc Questionnaires**. The EduZinc in-Class questionnaire platform was used for units 1, 2, 3 and 6. These features are named *TestT1*, *TestT2*, *TestT3* and *TestT6*.
- **Mid-term questions**. Includes questions from units 2 to 5. The features collected are *ExT2*, *ExT3*, *ExT4* and *ExT5*.
- **Final questions**. It includes questions from units 6 to 9. Unit 9 has two questions due to the relevance of the knowledge given. The features are *ExT6*, *ExT7*, *ExT8*, *ExT9* and *ExT9b*.

All these variables are to be compared against a Class variable that will simply be *yes* or *no* to whether students passed the course.

2.2 Evaluation Systems

Two different evaluation systems are considered in this work. The first one is called work-based assessment (WbA), and emphasis is placed on the daily effort done during the semester. The final exam only amounts to 30% of the total grade. However, students must score beyond 70% (all activities included) in order to pass.

The second one, called knowledge-based assessment (KbA), is aimed either at students who already have a working knowledge of the course material, or for those who didn't apply themselves on a daily basis (as a last resort). The final exam in KbA amounts to 50% of the total grade. Students must score beyond 50% (all activities included) in order to pass. The KbA is also an option for students with limited time available due to some reason such as work, family, etc.

3 Machine Learning Techniques

On this section we define the Machine Learning techniques used in this work. First, the clustering task is presented and, then, the feature selection problem is described.

3.1 Clustering

In clustering, a set $\mathcal{A}$ of n objects described by d features X_j, $j = 1, 2, ..., d$ is given. a_{ij} denotes the value of the feature X_j for the object $\mathbf{a_i}$. The aim in clustering is to group the n objects into p clusters $C_1, C_2, ..., C_p$ such that objects in the same cluster are more similar to each other than to those in other clusters.

The clustering algorithm used in this work is k-means [14], a popular and centroid-based technique. The objective of k-means is to minimize the Euclidean distance between each object and the centroid of the cluster it belongs to. The function to minimize is the squared error function.

In order to estimate the optimal number of clusters, the Average Silhouette (AS) approach [21] was applied. The Silhouette gives information about which objects lie within their cluster and which are between clusters. AS computes the average silhouette over a range of possible values for k. The optimal number of clusters is the one that maximizes AS.

3.2 Feature Selection

Feature selection methods try to find the most useful features to carry out a task. Let $\mathcal{T}$ be a sample of size n described by d features $\mathbf{X}$ and let $\mathcal{Y}$ be the target variable. Given $J(S)$, the objective function to evaluate the goodness of a subset of features S, the purpose of feature selection is to find the subset of features $S \subseteq X$ that optimizes $J(S)$.

In this work we use, as $J(S)$, the Correlation-based Feature Selection [12] (CFS) measure. CFS evaluates subsets in such a way that assigns higher values to subsets that include features highly correlated with the Class but uncorrelated to each other. As search strategy, Scatter Search [10] (SS). SS is a population based metaheuristic that uses intensification and diversification mechanisms to evolve solutions. The parameters of the SS where set according to the best combination found in [9].

4 Experiments

In this section the different experiments conducted are explained. Such experiments can be summarized as follows:

- Sect. 4.1 shows the analysis of the 2019–2020 academic year using clustering techniques is presented.
- Sect. 4.2 explains the predictive power of the features involved.
- Sect. 4.3 includes the discussion of the results in the context of education.

4.1 Cluster Analysis

In order to set parameter k, which refers to the number of clusters, the AS Method was applied. As it can be seen in Fig. 1, the method finds that the value for k that maximizes the average silhouette is $k = 2$. Therefore, the cluster analysis is performed for this value of k.

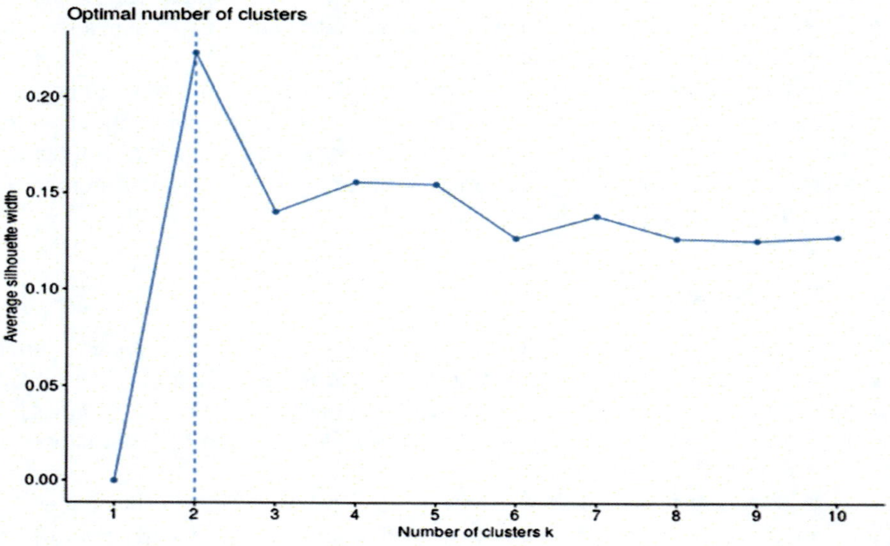

Fig. 1. Average Silhouette width computed for 1–10 clusters.

The clustering results identify two student profiles. One of them with high scores in the activities and the other with low ones. Hence, such profiles are studied according to a passing or failing score in both evaluation types. Table 1 presents the confusion matrix resulting from crossing the clustering label with the passing or failing score. This result suggests that profiles found by k-means match with such score.

Table 1. Confusion matrix associated to the descriptive models for WbA and KbA.

	WbA			KbA	
	No	*Yes*		*No*	*Yes*
1	22	2	1	24	0
2	6	39	2	10	35

The match between clusters and the student's profiles for WbA is shown in Fig. 2. For visualization purposes, the data was projected in two dimensions using Principal Component Analysis. Both profiles are clearly distinguished. KbA results are similar; so the plot is not shown to comply with article size requirements.

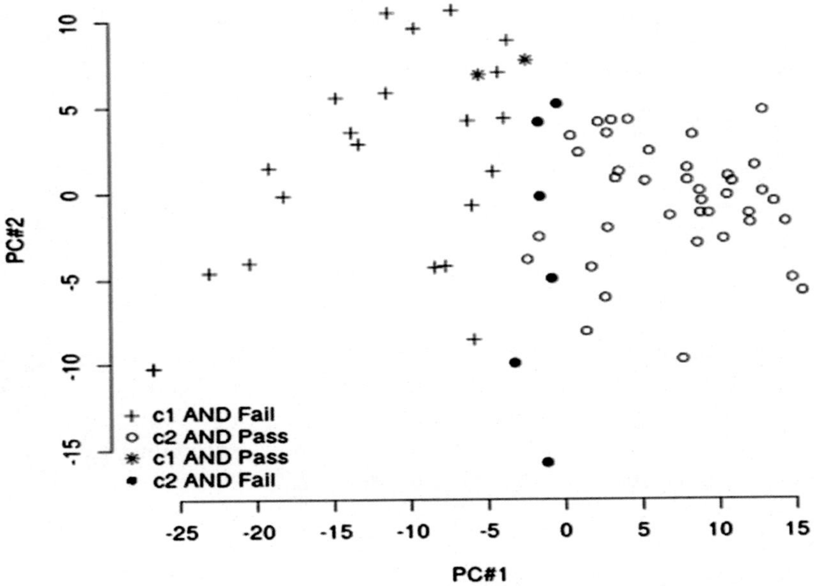

Fig. 2. Match between clusters and student profiles for continuous evaluation.

4.2 Feature Selection

In this section the features involved in the study are analyzed. As seen in the previous section, the clustering descriptive model identifies student's profile according to the score. Hence, a supervised feature selection method is applied. Then, the descriptive models with the feature selected is studied and compared with the previous model. For a more detailed information on the features here described, and both forms of evaluation, refer to Sect. 2.1.

The Features selected by a Scatter Search with WbA and KbA are shown below

$$S_{WbA} = \begin{cases} eInstructionT2 \\ eInstructionT7 \\ LabP5 \\ HwT2 \\ HwT3 \\ HwT4 \\ TestT2 \\ ExT3 \\ ExT4 \\ ExT5 \\ ExT7 \\ ExT9 \\ ExT9b \end{cases} \qquad S_{KbA} = \begin{cases} HwT2 \\ HwT3 \\ HwT4 \\ ExT3 \\ ExT4 \\ ExT5 \\ ExT7 \\ ExT9 \\ ExT9b \end{cases}$$

It is worth stressing that with WbA the subset of features selected is greater than with KbA. Furthermore, all the features selected for the WbA case are also included in the KbA solution.

Performing a similar cluster analysis and matching according to score, the descriptive models induced are similar as seen in Table 2.

Table 2. Confusion matrix associated to the descriptive models for WbA and KbA using features selected by SS.

WbA			KbA		
	No	*Yes*		*No*	*Yes*
1	25	4	1	28	0
2	3	37	2	6	35

4.3 Discussion

The clustering analysis identifies two different student's profiles. These profiles match students with high scores and passing grades with those with low scores and failing. The robustness of the profiles found is noteworthy since it was also found after applying feature selection. Therefore, these results suggest that it is not necessary to know the performance on all proposed activities to identify the profile of a given student. Furthermore, this knowledge may be helpful for the teacher to identify performance issues of some students.

Homework activities as well as the mid-term and final exam questions have shown to contain useful information to profile students, specially for the KbA. The other kinds of activities have been more relevant when WbA was applied.

The course lecturer intended (using WbA and KbA) an evaluation system that would more likely produce high and low grades instead of intermediate. However, he did not expect that this would in turn produce such degree of clustering-based predictability, indicating the most relevant the activities created (represented here as selected features). This has led to many insights regarding the course's material and activities.

5 Conclusions

In this work, the student achievement scores have been analyzed using clustering and feature selection techniques. Results are promising but still require more work to make this analysis useful in an academic environment.

Future work will include competencies in the analysis. This information will allow to improve the assessment of student needs.

References

1. Acharya, A., Sinha, D.: Application of feature selection methods in educational data mining. Int. J. Comput. Appl. **103**(2) (2014)
2. Aldowah, H., Al-Samarraie, H., Fauzy, W.M.: Educational data mining and learning analytics for 21st century higher education: a review and synthesis. Telematics Inform. **37**, 13–49 (2019)
3. Algarni, A.: Data mining in education. Int. J. Adv. Comput. Sci. Appl. **7**(6), 456–461 (2016)
4. Anjewierden, A., Kolloffel, B., Hulshof, C.: Towards educational data mining: using data mining methods for automated chat analysis to understand and support inquiry learning processes. In: International Workshop on Applying Data Mining in e-learning, ADML 2007 (2007)
5. Bakhshinategh, B., Zaiane, O.R., ElAtia, S., Ipperciel, D.: Educational data mining applications and tasks: a survey of the last 10 years. Educ. Inf. Technol. **23**(1), 537–553 (2018)
6. Bapu, G.K., Deshmukh, M.P.R., Ashok, M.B., Shamrao, S.P., Tanaji, S.G.: Clustering moodle data as a tool for profiling students. Int. J. Adv. Res. Comput. Eng. Technol. (IJARCET) **4** (2015)
7. Becerra-Alonso, D., Lopez-Cobo, I., Gómez-Rey, P., Fernández-Navarro, F., Barbera, E.: A new tool to create student material for automated grading. Distance Education (2020, in print)
8. Dutt, A., Aghabozrgi, S., Ismail, M.A.B., Mahroeian, H.: Clustering algorithms applied in educational data mining. Int. J. Inf. Electron. Eng. **5**(2), 112 (2015)
9. García-López, F.C., García-Torres, M., Melián-Batista, B., Pérez, J.A.M., Moreno-Vega, J.M.: Solving the feature selection problem by a parallel scatter search. Eur. J. Oper. Res. **169**(2), 477–489 (2006)

10. García López, F.C., García Torres, M., Moreno Pérez, J.A., Moreno Vega, J.M.: Scatter search for the feature selection problem. In: Conejo, R., Urretavizcaya, M., Pérez-de-la Cruz, J.-L. (eds.) Current Topics in Artificial Intelligence, pp. 517–525. Springer, Heidelberg (2004)
11. Spain Goverment: Datos y cifras Sistema Universitario Español (2019). http://www.educacionyfp.gob.es/dam/jcr:2af709c9-9532-414e-9bad-c390d32998d4/datos-y-cifras-sue-2018-19.pdf. Accessed 20 Feb 2020
12. Hall, M.A.: Correlation-based Feature Subset Selection for Machine Learning. Ph.D. thesis, University of Waikato, Hamilton, New Zealand (1998)
13. Holmes, G., Donkin, A., Witten, I.H.: Weka: a machine learning workbench. In: Proceedings of ANZIIS'94-Australian New Zealnd Intelligent Information Systems Conference, pp. 357–361. IEEE (1994)
14. Lloyd, S.P.: Least squares quantization in pcm. Technical report, Bell Laboratories (1957)
15. Peña-Ayala, A.: Educational data mining: a survey and a data mining-based analysis of recent works. Exp. Syst. Appl. **41**(4), 1432–1462 (2014)
16. Perera, D., Kay, J., Koprinska, I., Yacef, K., Zaïane, O.R.: Clustering and sequential pattern mining of online collaborative learning data. IEEE Trans. Knowl. Data Eng. **21**(6), 759–772 (2009)
17. Punlumjeak, W., Rachburee, N.: A comparative study of feature selection techniques for classify student performance. In: 2015 7th International Conference on Information Technology and Electrical Engineering (ICITEE), pp. 425–429. IEEE (2015)
18. Ramaswami, M., Bhaskaran, R.: A study on feature selection techniques in educational data mining. J. Comput. **1**(1) (2009)
19. Rodrigues, M.W., Isotani, S., Zárate, L.E.: Educational data mining: a review of evaluation process in the e-learning. Telematics Inform. **35**(6), 1701–1717 (2018)
20. Romero, C., Ventura, S.: Educational data mining: a review of the state of the art. IEEE Trans. Syst. Man Cybern Part C (Appl. Rev.) **40**(6), 601–618 (2010)
21. Rousseeuw, P.J.: Silhouettes: a graphical aid to the interpretation and validation of cluster analysis. J. Comput. Appl. Math. **20**, 53–65 (1987)
22. Velmurugan, T., Anuradha, C.: Performance evaluation of feature selection algorithms in educational data mining. Perform. Eval. **5**(02), 131–139 (2016)
23. Ward, D.: einstruction: Classroom performance system [computer software]. EInstruction Corporation, Denton, TX (2007)
24. Wook, M., Yahaya, Y.H., Wahab, N., Isa, M.R.M., Awang, N.F., Seong, H.Y.: Predicting ndum student's academic performance using data mining techniques. In: 2009 Second International Conference on Computer and Electrical Engineering, ICCEE'09, vol. 2, pp. 357–361. IEEE (2009)
25. Zaffar, M., Hashmani, M.A., Savita, K.: Performance analysis of feature selection algorithm for educational data mining. In: 2017 IEEE Conference on Big Data and Analytics (ICBDA), pp. 7–12. IEEE (2017)
26. Zaffar, M., Savita, K., Hashmani, M.A., Rizvi, S.S.H.: A study of feature selection algorithms for predicting students academic performance. Int. J. Adv. Comput. Sci. Appl. **9**(5), 541–549 (2018)

Knowledge Modelling for Ill-Defined Domains Using Learning Analytics: Lineworkers Case

Guillermo Santamaría-Bonfil[1,2](✉), Hector Daniel Díaz-Rodríguez[3], Gustavo Arroyo-Figueroa[2], and Rafael Batres[4]

[1] CONACYT-INEEL, Reforma 113, 62490 Cuernavaca, Morelos, Mexico
[2] Instituto Nacional de Electricidad y Energías Limpias, Reforma 113, 62490 Cuernavaca, Morelos, Mexico
{guillermo.santamaria,garroyo}@ineel.mx
[3] Universidad de Colima, Av. Universidad 333, 28040 Colima, Colima, Mexico
hdiaz6@ucol.mx
[4] Laboratorio de Decisiones Intelligentes, Instituto Tecnológico y de Estudios Superiores de Monterrey, Cuernavaca, Mexico
rafael.batres@tec.mx

Abstract. Representation of knowledge used by E-learning systems to modulate learning processes plays a key role in its effectiveness. In ill-defined domains where training is carried out using an apprenticeship model such as the area of Technical and Vocational Education and Training, building a Knowledge Model is not straightforward. In such areas the knowledge model heavily depends on the journeyman tacit expertise which is spread across text documents such as manuals, books, reports, competency descriptions, among others. Hence, in this work it is proposed to employ Learning Analytics for building a Knowledge Model from text documents used in lineworkers vocational education. The model is organized by declarative, procedural, and competency layers. Each of these contains a semantic networked built from extracted concepts, and the semantic relations between concepts is obtained using the Normalized Web Distance. Initial results shows that building knowledge models for ill-defined domains is promising, although more experimentation is required.

Keywords: Knowledge Model · Ill-defined domain · Learning Analytics · Natural Language Processing · Normalized Web Distance

1 Introduction

E-Learning and most Intelligent Tutoring Systems (ITS) have been devised to provide any given individual, more or less personalized support for learning about a given knowledge domain. Delved ITS have ranged from classical expert

Á. Herrero et al. (Eds.): ICEUTE 2020, AISC 1266, pp. 409–418, 2021.
https://doi.org/10.1007/978-3-030-57799-5_42

systems, to Machine Learning (ML) algorithms and hybrid models, all with different levels of success in real applications.

The main component of any ITS is knowledge, and defining the Knowledge Model (KM) is the most crucial part in its design. To choose the appropriate KM it is necessary to determine the nature of the knowledge in question, how it is constructed, and how is validated and evaluated [12]. To address these, the epistemological analysis and the computational implementation are carried out. The epistemological analysis consists in building Domain Models (DMs) and Problem Models (PMs), which corresponds to facts and declarative knowledge (e.g. concepts, properties, relations) and the inference model (e.g. Process to solve a problem), respectively. The selection of the computational implementation is consequence the epistemological analysis. For instance, the declarative knowledge is well-suited to be represented by general-purpose languages such as ontologies, semantic networks, or conceptual graphs, already equipped with formal proof theories. In practice, the distinction between DM and PM remains debatable since ITS developers only need to focus on specific tasks and the task-relevant DM subset [2].

ITS are effective in well-defined knowledge domains and tasks (e.g. arithmetic or geometry), where general-purpose languages such as First-Order Logic or other systematic method can be used to build facts and validate solutions [12]. Alas, most human knowledge domains and problems are ill-defined [11]. These lack of a strong theory uniquely specifying, on one part, the concepts, procedures, and their relationships, and on the other, how to validate solutions through a formal procedure [2,9,11]. Further, these have large and complex solution spaces that prohibits identifying all possible solutions. Even their verification heavily depends on human experts, which not always attain consensus on all domain solutions [9,11].

There are several forms to circumvent the *ill-definedness* issue for building an ITS: comparing between expert and novice behaviors using cognitive models, mapping facts and relations to production rules within expert systems, partial models based on Machine Learning (ML), and mixtures of these approaches [2]. All of these involve the *re-characterization* of the KM, from ill to well-defined [2,9,11]. The *re-characterization* of a KM assumes that relevant concepts and relations are already available, i.e. provided by educational experts [2]. Yet, defining KM building blocks in this way is expensive, slow, and error-prone [21]. To ameliorate the burden of defining concepts and relations, semi-automatic extraction of these from text documents can be carried out using Learning Analytics (LA) techniques [4,6,13,20,21]. In short, LA is the application of a multitude of areas (e.g. cognitive science, data science, pedagogy, etc.) and techniques (e.g. Machine Learning (ML), Text Mining, Natural Language Processing -NLP-, among others) for the analysis and modelling of educational data [15]. Using LA techniques, educational text documents can be cleaned and processed to extract concepts and properties such as morphology or semantics, to pose concept representations and relations between these, and to infer new knowledge [19].

Among the LA approaches, hierarchical clustering allows to discover *topics* by assigning terms into hierarchical groups using similarity metrics. When the objects are *literal*, concepts can be compared as strings using standard similarity distances (e.g. euclidean distance) in terms of bag of words, n-grams, or any other model for strings representation. Nevertheless, when the objects are *non-literal* such as names or ideas, these do not contain information within themselves. Rather, they acquire meaning based on common sense and human knowledge background [18]. In this cases, we can employ the information contained in the World Wide Web (WWW) as a proxy of knowledge background, which compensates for its unstructured information by its large size. This idea is exploited by the Normalized Web Distance (NWD), which employs WWW search engines to obtain semantic similarity information between non-literal concepts [4,18].

Therefore, in this work it is proposed to build a knowledge model using a hierarchical clustering along with the NWD, which is well suited to capture semantic knowledge for ill-defined domains. Particularly, it is discuss how to model knowledge into a hierarchical concept graph with separated knowledge layers, using as benchmark the Technical and Vocational Education and Training (TVET) required for lineworkers. The rest of the paper is organized as follows: Sect. 2 presents the TVET and lineworkers case; Sect. 3 describes the proposed approach for building a KM for ill-defined domains based on clustering and the NWD; Sect. 4 elaborates the experiments on building the domain model for the study case; finally, conclusions and future work are discussed.

2 Materials

This work is focus on power systems technicians, namely Lineworkers, which are in charge of carrying out maintenance and repair tasks over the power systems infrastructure. Hence, first it is introduced the paradigm where lineworkers training happens (TVET) and the competency-based training. Then, the the Apprenticeship model used by lineworkers is described. Finally, we detail the text documents employed to build the KM which belong to a virtual reality training system called ALEn^{3D} $- MT$.

2.1 TVET, Competency-Based Training, and the Apprenticeship Model

TVET corresponds to the programs developed by countries in accordance to their industrial development requirements, that aims to train and educate individuals for carrying out skilled labor to be employed in industry, agriculture, and service sectors among others [8]. Amidst the TVET programs that exist stand those oriented to energy education which are increasing day by day [8,17]. A key aspect of TVET is that it is moving towards a competency-based approach. The Competency-based Education (CbE) proposal raised around the year 1960 and focuses on program goals and work objectives. Hence, CbE is organized around predefined abilities or skills that are considered as the outcomes of the curriculum. CbE emphasizes abilities, flexibility regarding learning times, and establishing clear goals to learners (to mention a few) [3].

However, vocational education is more complex in comparison with traditional education due to its knowledge base, characterized by a double theory/practice: students must complete formal and technical training in the specialized area, and they must obtain practical experience through an apprenticeship. In this sense, they not only need to acquire theoretical content, but also tacit knowledge. This approach is called the *apprenticeship model* since it combines classroom learning with direct work experience. This approach is used by the construction industry as part of the industry certification used for carpenters, electricians, lineworkers, plumbers, heating/venting/air conditioning (HVAC) technicians, to mention a few [1,7].

Finally, the apprenticeship model for lineworkers, pairs a novice worker with a journeyman for a given lapse of time, such that novices acquire knowledge and skills about tacit knowledge from experts. Apprentices must course up to five years of classroom and work hours to obtain an industry certification [1,7]. Still, training electricians for live-line work is not standard, and differs at the levels of countries, regulatory institutions, and companies [10,14].

2.2 Text Documents from a Lineworker Virtual Reality Training System

Lineworkers are part of the staff required by a public or private utility for maintaining and repairing its transmission and distribution infrastructure. Lineworkers preparation is highly complex given it involves physical preparation to stand very demanding physical activities, knowledge on physics and electromagnetism, theoretical and tacit knowledge regarding maintenance maneuvers execution, as well as hygiene and safety industrial rules [14,15].

The electrical industry has invested in the development of e-learning technologies for training its Lineworkers. Among the most effective technologies stand out Virtual Reality Training Systems (VRTS) since they offer the simulation of realistic environments without the risks of real life. The system $\text{ALEn}^{3D} - MT$ is a 3D non immersive VRTS used by the main Mexican utility to train lineworkers in maintenance maneuvers for the power distribution infrastructure. It contains 43 maintenance maneuvers, from rescuing an electric shock injured lineworker to the replacement of several pole structures. The maneuvers employ a total of 211 different tools and equipment to carry out the maintenance maneuvers. These are contained in a tools catalogue, with names that range from single words such as *hammer* to multi words such as *synthetic resin insulator*.

3 Methods

E-Learning engineers are confronted many times with cognitive and pedagogical knowledge contained into text documents whose important features are unknown to them. Thus, to build a KM from text sources, first we need to define the concept *model*. In this sense, by *model* we refer to a *concept map* built upon concepts organized hierarchically (i.e. hierarchical clustering) using a semantic distance measure inspired in the Kolmogorov complexity concept, namely Normalized Web Distance (NWD) [18]. The proposed KM has a layered top-down organization, and for each layer a model is obtained. In the top layer stands the Competency Model (CM) which defines very broadly competencies and skills in regards to a curricula for a specific job. Then, the PM and DM model are presented, keeping both in separated layers as suggested in [12]. The general process is depicted in Fig. 1: first, a text mining pre-processing step is carried out to standardize texts, remove stop words, and extract concepts; then, a Hierarchical Clustering is used to construct a concept map for each layer using the NWD. It is worth to mention that, depending on the technicality of each knowledge domain, units and special acronyms may not be removed during the text pre-processing due its importance.

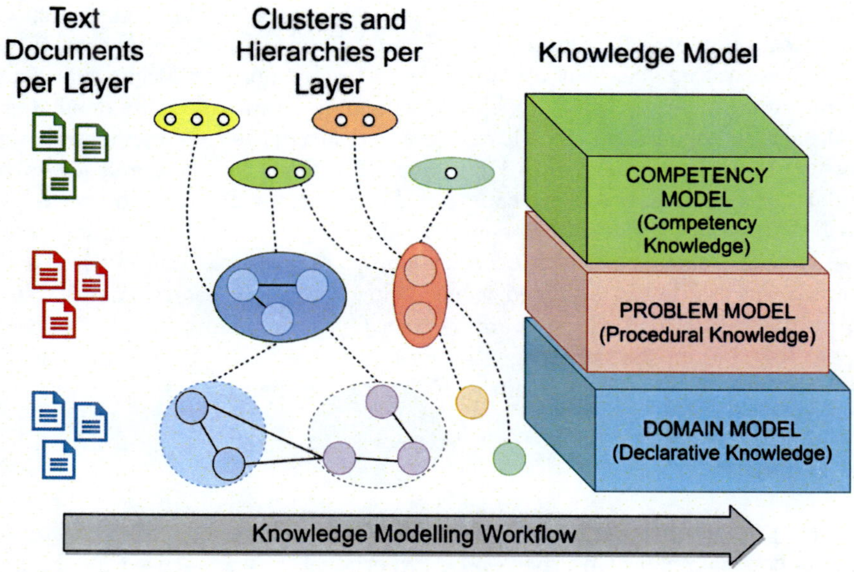

Fig. 1. Overall workflow for knowledge Modelling from text documents. Documents are organized per layer (DM/PM/CM), and semantic relations is given by the proposed clustering approach.

3.1 Normalized Web Distance

The Normalized Information Distance (NID) is a universal similarity measure. It can be applied to compare similarity between two objects based on the Kolgomorov complexity. It satisfies the identity, symmetry, and triangle inequality, but unfortunately it depends on the uncomputable Kolgomorov complexity, hence is uncomputable too. Besides this, there are *non-literal objects* such as names given to ideas and concepts. These are words that do not contain information within themselves, and acquire their meaning by their contexts and human knowledge. Consider for instance the word *honor*, the sequence of letters or word itself does not provide any information regarding what *honor* means.

We may approximate the Kolgomorov complexity of a *non-literal object* using the World Wide Web (WWW) and the Normalized Web Distance (NWD). If W is the set of all web pages in WWW, and $\mathbf{x} \subseteq W$ is the set of pages that contains the search term x, we can approximate its universal probability such as $\frac{log(1)}{\mathbf{m}(\mathbf{x}|\subseteq W)} = K(\mathbf{x} \mid \mathbf{x} \subseteq W) + O(1)$, where K is the Kolgomorov complexity, and m is a universal lower bound estimated by the distribution of x in the W. Then, the NWD between two objects $x, y \in W$ is defined as

$$NWD = \frac{\max\{\log f(x), \log f(y)\} - \log f(x,y)}{\log N - \min\{\log f(x), \log f(y)\}}, \tag{1}$$

where $f(x)$ is the number of pages containing term x, $f(x,y)$ is the number of pages containing both x and y, and N is a large constant corresponding to the total number of indexed pages. The NWD was originally devised to use Google® search engine, but nowadays any search engine or large text corpus can be employed. Yet, the NWD is not a similarity metric due to its dependence on the search engine, and since it does not satisfies neither the symmetry property nor the triangle inequality [18]. Thus, the NWD is considered a relative similarity measure between two concepts. An important difference between NID and NWD is that, the former is constrained to $0 \leq NID \leq 1$, obtaining values close to zero for objects which are very similar and one for very dissimilar. In contrast, the NWD can obtained values below zero if the search engine counts are untrustworthy and $f(x,y) > \max\{f(x), f(y)\}$, or infinite if objects do not appear together on the same page, but do occur separately. Nevertheless, NWD is constrained to the same domain as NID for convenience.

4 Results

Now we present the results for building a KM by the construction of a hierarchical concept map using the NWD. Experimentation is focused on the text documents used by an E-learning system in the training of lineworkers. In particular, in the tools used by the main Mexican utility lineworkers for the 43 maintenance maneuvers carried out over its power distribution infrastructure. Since we are only focusing on tools, we are building the lowest layer of the KM, the Domain Model. Regarding the HC algorithm, we employed an agglomerative approach using the *complete-linkage* criteria to find similar clusters (although the latter is an arbitrary selection). In the same fashion we only employed 5 clusters, but the number of optimal clusters can be determined by any other means. To measure the NWD we employed two WWW search engines, namely *Bing*® from Microsoft® and *Google* from the same name company. It is worth to notice that, on one hand, tool names are in Spanish, and on the other, to improve and constrain the semantic search we included as keyword the word *liniero* (lineworker in Spanish) as suggested in [18]. Similarly, to improve discussion we only show results for the 30 first tools (sorted alphabetically) and capped NWD to three (i.e. $NWD \leq 3$) instead of ∞. Figure 2 presents the results for both search engines, on the left side the HC corresponding Bing, whereas on the right the HC corresponding to Google; clusters are depicted by colored branches.

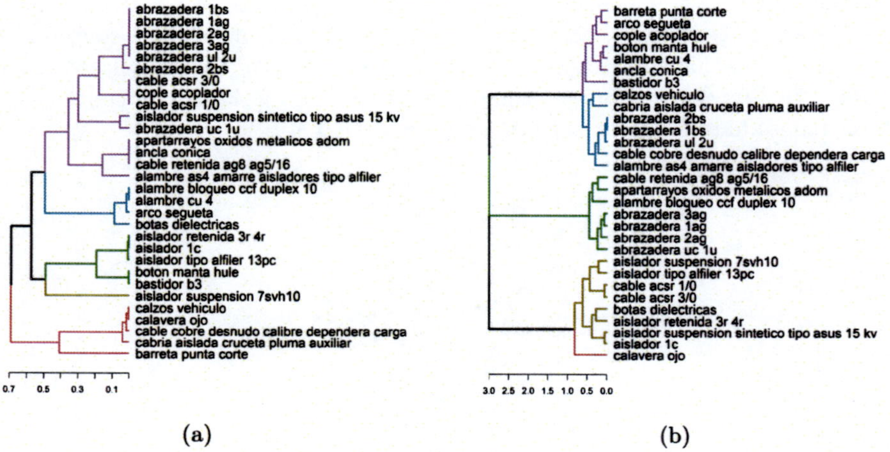

Fig. 2. Hierarchical Concept Maps: on (a) the dendogram built using the Bing search engine, whereas on (b) the dendogram built using Google search engine. The 5 clusters are shown in colors.

Observe in Fig. 2 that while for the Bing search engine, all distances and hence clusters, can be considered *similar* (i.e. $NWD_i \leq 1$), for Google some clusters can be considered totally dissimilar (i.e. $NWD_i \geq 1$) due to the complete-linkage criterion. This means that in Bing search engine there are documents where all tool pairs exists, while in Google this is not the case. An interesting case to compare is *braces* (*abrazaderas* in Spanish). There are seven different types of braces used by lineworkers in accordance to the $ALEn^{3D} - MT$ system, and are shown at the top of the Bing hierarchy. In the case of the Bing search engine, all braces are within the same cluster, and most of them have a $NWD \sim 0$. This means that using Bing braces are considered equal. In contrast, for the hierarchy obtained by Google, braces are separated in two clusters, which notably coincide with some of the tools required together by specific maintenance maneuvers.

4.1 Comparison Between Search Engines

For elaborating further the similarities and differences between DMs built with the two search engines we employ a Tanglegram [5]. This is a plot to visually compare two hierarchical clusterings and is employed in systematic biology for contrasting phylogenetic trees [16]. A tanglegram places two rooted trees horizontally facing each other by the leaves. Items in leaves are matched and connected by *connectors.* Branches that contain a unique subsets of items are highlighted with dashed lines, and common subtrees in both hierarchies are presented in solid lines. Another interesting property of tanglegrams is that is possible to measure the quality of the hierarchies alignment called *entanglement.* Although there are several functions of entanglement [16], its basic function is to measure entanglement by penalizing crossings using a L-norm function. This function is between 1 (full entanglement) and 0 (no entanglement). In our case, a lower entanglement coefficient corresponds to a good alignment between concepts in HCs, although this does not guarantee trees equivalence or even topological similarity. The tanglegram for our study case is shown in Fig. 3, and for convenience tools names have been replaced by numbers alphabetically.

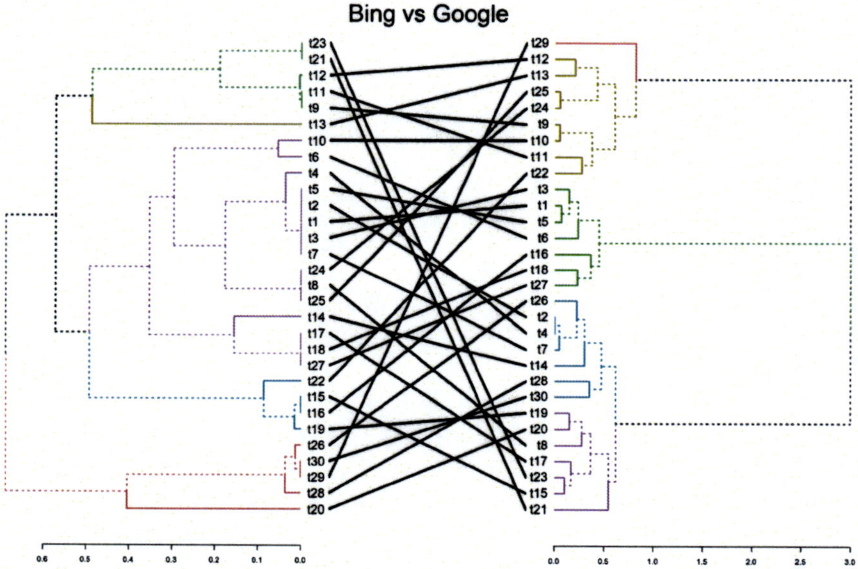

Fig. 3. Tanglegram of the two hierarchies for comparing between Bing and Google search.

Observe that in our case, at leaves level several sub trees repeat (solid branches). In contrast, items combinations at the level of clusters (and above) are depicted by dashed lines, pointing out that most combinations of elements are not present in the other tree. This disagreement between Bing and Google hierarchies can be corroborated by the entanglement function using a $L^1 - Norm$ resulting in 0.43 of entanglement. In this sense, only pair of tools or the group of tools that have a very similar name can be considered equal or highly similar between both hierarchical concept maps. Nevertheless, it is important to stress that this behavior is given by the linkage criterion. It is expected that different criteria will spawn different hierarchical structures and different levels of agreement. For instance, using the same experimental configuration (i.e. only

30 tools, 5 clusters, $L^1 - Norm$, etc.) but replacing the linkage criterion by the *average* obtains an entanglement of 0.35, alas, with no warranty of increasing the topological similarity.

5 Discussion and Future Work

In this work is proposed a method for building a Knowledge Model from text documents based on LA for ill-defined domains. In particular, the CbE of lineworkers which has a high dependence on the apprenticeship model in which, the educative information is only available through text documents (e.g. manuals, books, ad-hoc systems, and so on) or as tacit knowledge in the expert minds. The proposed methodology builds a KM from available text documents and the aid of web search engines to obtain a semantic similarity between objects (e.g. tools) in the domain model, namely the NWD. Employing the NWD a hierarchical clustering can be built and used as a hierarchical concept map, in which cluster elements are part of the same topic. This will allow to retrieve similar objects that may be presented to a student during a training session in regards to its similarity. Moreover, this a promising way of modelling KMs for CbE that is carried out on the apprenticeship model. In this sense, the proposed model applies not only to Mexican lineworkers training, but to any other lineworker training program (regardless of their nation), or even any other educational setting in which hierarchical concept maps have to be obtained for *non-literal objects*. Nevertheless, there are several drawbacks. First, we still need to determine what does *hierarchical relations* between objects stand for. This is, by mapping DM clusters to components of above layers (i.e. PM or CM) to relate it with topics and skills. Second, the semantic web search can be improved (or biased) by using keywords, thus, there is not a unique NWD similarity matrix per search engine. Similarly, using different linkage criteria for the hierarchical model leads to different hierarchies. Thus, an ensemble clustering method can be used to bring down to consensus several clusterings into a single one. Finally, experts intervention is inexorable due to the requirement of KM validation, yet it is expected that such validation will be reduced considerably.

Acknowledgements. GSB thanks the Cátedra CONACYT program for supporting this research.

References

1. Dickerson, M.: A case study of four female electrician technicians in a male-dominated occupation. Ph.D. thesis, University of South Florida (2016)
2. Fournier-Viger, P., Nkambou, R., Nguifo, E.: Building intelligent tutoring systems for ill-defined domains. SCI, vol. 308, pp. 81–101 (2010)
3. Frank, J., Snell, L., Cate, O., Holmboe, E., Carraccio, C., Swing, S., Harris, P., Glasgow, N., Campbell, C., Dath, D., Harden, R., Iobst, W., Long, D., Mungroo, R., Richardson, D., Sherbino, J., Silver, I., Taber, S., Talbot, M., Harris, K.: Competency-based medical education: Theory to practice. Med. Teach. **32**(8), 638–645 (2010)

4. Fujihara, S., Batres, R.: A micro-genetic algorithm for ontology class-hierarchy construction. Int. J. Computat. Linguist. Appl. **7**(1), 51–65 (2016)
5. Galili, T.: dendextend: An R package for visualizing, adjusting and comparing trees of hierarchical clustering. Bioinformatics **31**(22), 3718–3720 (2015)
6. George, G., Lal, A.: Review of ontology-based recommender systems in e-learning. Comput. Educ. **142**, 103642 (2019)
7. Hardcastle, A., Jull, P., Hanson, S.: Workforce challenges of electric power employers in the pacific northwest. Technical report, Pacific Northwest Center of Excellence for Clean Energy, October 2013
8. Kacan, E.: Renewable energy awareness in vocational and technical education. Renew. Energy **76**, 126–134 (2015)
9. Le, N., Loll, F., Pinkwart, N.: Operationalizing the continuum between well-defined and ill-defined problems for educational technology. IEEE Trans. Learn. Technol. **6**(3), 258–270 (2013)
10. Lovrenčić, V., Miletić, Z.: Periodical training of linemen is a guarantee for competent and safe live working. In: 11th International Conference on Live Maintenance, pp. 1–8 (2014)
11. Lynch, C., Ashley, K., Pinkwart, N., Aleven, V.: Concepts, structures, and goals: Redefining Ill-definedness. Int. J. Artif. Intell. Educ. **19**(3), 253–266 (2009)
12. Nkambou, R.: Modeling the domain: An introduction to the expert module. SCI, pp. 15–32. Springer, Berlin, Heidelberg (2010)
13. Peña-Ayala, A.: Learning analytics: A glance of evolution, status, and trends according to a proposed taxonomy. Wiley Interdiscip. Rev. Data Mining Knowl. Discov. **8**(3), e1243 (2018)
14. Santamaría-Bonfil, G., Escobedo-Briones, G., Perez-Ramírez, M., Arroyo-Figueroa, G.: A learning ecosystem for linemen training based on big data components and learning analytics. J. Univ. Comput. Sci. **25**(5), 541–568 (2019)
15. Santamaría-Bonfil, G., Ibáñez, M.B., Pérez-Ramírez, M., Arroyo-Figueroa, G., Martínez-Álvarez, F.: Learning analytics for student modeling in virtual reality training systems: Lineworkers case. Comput. Educ. **151**, 103871 (2020)
16. Scornavacca, C., Zickmann, F., Huson, D.: Tanglegrams for rooted phylogenetic trees and networks. Bioinformatics **27**(13), 248–256 (2011)
17. U.S. Department of Energy: Transforming the Nation's Electricity System the Second. Technical report January, U.S. Department of Energy (2017)
18. Vitányi, P., Balbach, F., Cilibrasi, R., Li, M.: Normalized information distance. In: Information Theory and Statistical Learning, pp. 45–82. Springer US, Boston, MA (2009)
19. Wong, W., Liu, W., Bennamoun, M.: Ontology learning from text: A look back and into the future. ACM Comput. Surv. **44**(4), 1–36 (2012)
20. Zouaq, A., Nkambou, R.: Building domain ontologies from text for educational purposes. IEEE Trans. Learn. Technol. **1**(1), 49–62 (2008)
21. Zouaq, A., Nkambou, R., Frasson, C.: Document semantic annotation for intelligent tutoring systems: A concept mapping approach. In: Florida Artificial Intelligence Research Society Conference, pp. 380–385 (2007)

School Success and School Dropout in Portuguese Polytechnic Higher Education (Case Study)

Elsa Rodrigues(✉)

Polytechnic Institute of Beja, Rua R. Pedro Soares, 7800-295 Beja, Portugal
elsa.rodrigues@ipbeja.pt

Abstract. This study is based on students attending a polytechnic higher education. It was concluded that, regardless of their social and ethnic origin, their behavior at certain times intersects with age, choices, motivation, success and failure, among others. Polytechnic Institutes offer several courses in different areas, such as higher vocational technical courses (2 curricular years), bachelor's programs (3 curricular years). For this study, we select some specific courses from the School of Technology and Management of the Polytechnic Institute of Beja, namely: 1) the course Web Technologies and Mobile Devices, the course Networks and Information Systems, Computer Engineering (Bachelor's Degree). These 3 courses were chosen based on their common point, digital skills; 2) Legal Services (face-to-face), Legal Services (e-learning), these courses are in the area of law, and due to the difference in teaching methodology. The most important aspect of these choices were the scientific areas, as they are literally opposite. This survey intends to verify if school success/failure depends on age and/or choice of scientific area. To what extent are the student's characteristics combined with the course ones linked to failure/achieving success? How to act, what initiatives can we take to tackle failure and enhance success? The study focused on document analysis, analysis of statistical indicators, provided by the quality office of the Polytechnic Institute of Beja and on observation in the classroom. During the investigation, we concluded that age is not a decisive factor for school failure, while the scientific area is debatable, as long as courses in the same area are not analyzed.

Keywords: Higher education · Courses · Success · Failure

1 Higher Polytechnic Education Training Offer

In Portugal, higher education is organized in a binary system composed of the University and Polytechnic subsystems. The university subsystem highlights the offer of solid scientific training, combining efforts and skills in the areas of teaching and research. While the polytechnic subsystem, where 15 public institutes are inserted, aims, in particular, technical training, professionally oriented, and applied research, creation of knowledge of a professional nature [1].

In the past, the Polytechnic Institutes of Portugal were limited to teaching bachelor's degrees. However, due to the Bologna process, it is now possible to teach

Á. Herrero et al. (Eds.): ICEUTE 2020, AISC 1266, pp. 419–427, 2021.
https://doi.org/10.1007/978-3-030-57799-5_43

undergraduate, master's and Higher Professional Technical Courses (CTeSP). Regarding the CTeSP, they were first implemented with the Decree-Law No. 43/2014, of March 18 [2], which suffered alterations by the Decree-Law no. 63/2016, of September 13. This decree states that these courses are a new type of higher education, however without granting an academic degree. In this way, the Polytechnic Institutes offer several courses in different fields in different study cycles, such as the Higher Professional Technical Courses (CTeSP) (2 curricular years), the bachelor's degrees (3 or 4 curricular years) and the master's degrees (1 to 2 curricular years).

Our study focuses on some courses taught at the School of Technologies and Management (ESTIG), namely: Web Technologies and Mobile Devices (TWDM), Networks and Computer Systems (RSI), Computer Engineering, Legal Services (face-to-face), Legal Services (e-learning). The choice of CTeSP courses and the bachelor's degree in Informatics Engineering was based on digital skills; in the case of Legal Services (taught either face-to-face or e-learning), the reasons why we chose this specific course were: first, its scientific field is completely opposite to the Informatics Sciences one; and second the teaching methods is quite different.

This survey aims at ascertain if school success/failure depends on age and/or choice of scientific area. To what extent are the student's characteristics combined with the course's ones linked to failure/achieving success? How to act, what initiatives can we take to tackle failure and enhance success?

2 Study Methodology

This study involved a multiple methodology, involving fundamentally: documentary analyses; analysis of statistical indicators and, essentially, institutional databases. The methodological plan comprehended quantitative [3] and qualitative [4] approaches as well as inductive and deductive procedures. Which implied carrying out a structural and institutional analysis.

The structural analysis was carried out by collecting quantitative indicators, which were given by the Polytechnic Institute of Beja (IPBeja), based on information provided by the Office of Academic Services (IPBeja) combined with data from official sources, such as the General Directorate of Education Higher. The Polytechnic Institute of Beja annually collects the information provided by the Ministry of Science, Technology and Higher Education, which is subsequently treated in alignment with the Institution's own data.

According to [5], higher education is like a subsystem that establishes social relationships between agents, establishes these relationships in a way that is inseparable from the structural features and vectors of transformation of global society, hence the importance of structural analysis for understanding success and failure. However, the author warns us of the need to cross the differentiation between scientific areas, which correspond to different teaching-learning strategies and practices, measuring success and failure at school with different requirements.

In turn, the institutional analysis forced us to look inside the institution as a mirror, allowing us to analyse the role of the institution in relation to students. Each student has his own characteristics, which will allow him to adapt with greater or lesser difficulty to the course and the institution, this degree of integration can promote success/failure. According to [6], and [7] the scientific area of studies, the pedagogical practices, the size of the classes, the evaluation processes, the timing of the evaluation moments and the student support structures, will influence success/failure. In this way, methodological procedures were carried out, involving case studies, particularly, documentary analysis regarding existing documents, at the institute, concerning the courses previously mentioned.

3 Parity versus Dissimilarity

In order to analyse the common trades and differences found we started by investigating what led students to choose certain courses. Afterwards, age, motivation, demotivation, success rate and failure rate were analysed.

3.1 Methodologies for Access to Portuguese Higher Education

Portuguese students, when finishing the cycle of studies concerning basic education, must follow regular education or professional education. If they choose regular education, they must choose one of the 4 courses in the scientific-humanistic areas [8], lasting 3 academic years, namely: Science and Technology; Socioeconomic Science; Languages and humanities; Visual arts. At the end of this three-year cycle, these students take 4 national exams, the Portuguese language exam being common to all students. The remaining exams depend on the scientific areas to which students intend to attend higher education.

After receiving the grades of the national exams, students apply for higher education, with a specific application average. According to this average, they may or may not be accepted in the course they placed in 1st option [9], as you can see through the website of the General Directorate of Higher Education.

In addition to the national competition for access to higher education, there is another form of access, special applications that allow students over the age of 23 to attend higher education. In the last case, the process goes through the higher education institutions themselves, submitting the applicants to written tests and an interview for admission. Pointing out, that according to data from the IPBeja Quality Office, most of these students are classified as working students. Moreover, according to the same source, 90% of students choose courses in the area of their professional functions. For example: computer technicians choose the Computer Engineering course; administrators in notorious offices and 1st instance assistants choose the Legal Services course; administrative staff in accounting offices, choose the Business Management course. The remaining 10% choose their courses based on their preferences (callings).

Students wanting to attend CTeSP courses must complete 12 years of regular mandatory school or Professional Technological Specialization and Professional Certification, or a Bachelor's Degree [3]. According to the data selected in the survey developed by the IPBeja Quality Office, which includes 48 applications to the CTeSP in the Information Technology field. Regarding these applicants, 80% of them came from Professional Technological Specialization and Professional Certification and the remaining 20% from the regular via.

3.2 Age of the Students

To ascertain, if maturity has an influence on success, we selected the variable "age" of the students. Thus, in the period under analysis, the majority of students in the Networks and Information Systems and Web and Mobile Technologies courses are between 20 and 23 years old [10].

Concerning 1st cycle courses, we analyse both the ages of the students of the Computer Engineering course and the Legal Services (face-to-face) course. Moreover, we found out that 43% of students are between 20 and 23 years old. Students under 20 and over 28 have the same contribution, of 17%. While students aged between 24 to 27 years old represent 12% of all students [10].

Due to the fact that the course has two different modalities (face-to-face and e-learning), we also chose to analyse the ages of the students attending Legal Services. We can observe that, regarding the face-to-face modality half of the students is between 20 to 23 years old, 36% is under 20 years old, 10% is over 28 years old and 4% is between 24 to 27 years old. Concerning the other modality (e-learning), we can see that the majority of students is over 28 years old (85%), students aged between [10].

3.3 Success, Failure Rate and Dropout

Success Rate and Failure Rate. It should be mentioned that, concerning the two courses of CTeSP, there is a time difference of 1 academic year in the two courses. The Web Technologies and Mobile Devices course started the academic year in 2015/2016, while the Network and Computer Systems course started the academic year in 2016/2017.

We decided to start analyse the success rate of CEeSP courses, as during observation in the classroom teachers have indicated serious difficulties of students performing well. Putting this aside, let us begin with the Networks and Information Systems course. As it can be seen from the Table 1 the success rate is 21.43%.

Table 1. Success rate of networks and information systems course

Course	Duration of the cycle of studies	Enrolled 1st year 1st time academic year Graduates in 2016/2017	Graduates in the academic year 2017/2018				Total of graduates	Success rate
			N Years *	N * Years + 1	N * Years + 2	>=N * years + 3		
Networks and information systems	2	14	3	0	0	0	3	21.43%

Regarding the CTeSP- Web Technologies and Mobile Devices course, the success rate is 47.37%, see Table 2 [10].

Table 2. Success rate of web technologies and mobile devices

Course	Duration of the cycle of studies	Enrolled 1st year 1st time academic year Graduates in 2015/2016	Graduates in the academic year 2017/2018				Total of graduates	Success rate
			N Years *	N * years + 1	N * years + 2	>=N * years + 3		
Web technologies and mobile devices	2	19	9	0	0	0	9	47.37%

Let us now analyze the data related to the bachelor's degrees mentioned above.

Observing the Legal Services Course (face-to-face) I have obtained the following results regarding to its success rate and completion time, this data is provided by the IPBeja quality office (see Table 3) [10]. We have a success rate of 66.70%. This indicates that during the 3 years of the course, most students successfully completed all course units.

Table 3. Success rate of legal services (face-to-face) course

Course	Duration of the cycle of studies	Enrolled 1st year 1st time academic year Graduates in 2015/2016	Graduates in the academic year 2017/2018			Total of graduates	Success rate
			N * years + 1	N * years + 2	>=N * years + 3		
Legal services (face-to-face)	3	30	0	0	1	21	66.70%

However, the Legal Services course (e-learning) has a success rate of 36.59%, as it can be seen in Table 4 [10]. The difference of the success rate between the two operating modes of the very same course is 30.11%. In order to reduce this difference we will have to make a different approach.

Table 4. Success rate of legal services e-learning course

Course	Duration of the cycle of studies	Enrolled 1st year 1st time academic year Graduates in 2015/2016	Graduates in the academic year 2016/2017				Total of graduates	Success rate
			N anos *	N * years + 1	N * years + 2	>=N * years + 3		
Legal services (e-learning)	3	41	15	3	0	0	21	36.59%

We emphasize that the data available for the Computer Engineering course, comprises the academic period between 2015/2016 and 2017/2018, with a corresponding success rate of 20% see Table 5 [10].

Table 5. Success rate of computer engineering course

Course	Duration of the cycle of studies	Enrolled 1st year 1st time academic year Graduates in 2014/2015	Graduates in the academic year 2017/2018				Total of graduates	Success rate
			N years *	N * years + 1	N * years + 2	>=N * years + 3		
Computer engineering	3	35	7	3	2	3	15	20.00%

Academic success is defined by the ratio between the number of students approved and the number of students enrolled; however, a significant percentage of students did not perform any of the evaluation's components.

The Table 6 [10] represents the differences between the success rate and failure rate. We can clearly denote that the failure rate of the Legal Services course (e-learning) is higher than the Legal Services course (face-to-face).

Table 6. Success and failure rate regarding all courses

Courses	Success rate	Failure rate
Networks and information systems	21.43%	78,57%
Web technologies and mobile	47.30%	52.70%
Computer engineering	20%	80.00%
Legal services (face-to-face)	66.70%	33.30%
Legal services (e-learning)	36.59%	63.41%

Regarding the CTeSP courses (Table 6): the Web Technologies and Mobile Devices Course is 52.70%. While the Network and Informatics Systems course is 78.57%. The failure rate in the Computer Engineering course is 80%. In this study, this is the higher failure rate recorded.

Dropout Race. We believe that it is a very significant dropout rate, therefore a more detailed analysis should be carried out in order to understand the reason for these values.

We have chosen to display the dropout rate for all courses in a table. Thus, we now have an overview of the dropout rate for each degree. However, we must not forget that the sample differs from course to course. The Computer Engineering course has the largest number of students enrolled. Whereas, the course with the lowest number of students is Computer Networks and Systems.

As can be seen in the Table 7 [10], the dropout rate for the Legal Services (face-to-face) course is 15.49%. We can clearly denote that the failure rate of the course (e-learning) is higher than the Legal Services course (face-to-face).

Regarding the CTeSP courses: the Web Technologies and Mobile Devices Course is 26.32%. While the Network and Informatics Systems course is 21.43% (see Table 7).

The failure rate in the Computer Engineering course is 14.48%. In this study, this is the higher failure rate recorded.

Table 7. Dropout race regarding all courses

Course	School year 2016/2017			School year 2017/2018			School Dropout	
	Total enrolled students	Total graduated students	Students carried over to the following year	Total enrolled students	Students enrolled for the 1st time	Students carried over to the following year	N.°	Rátio
Legal services (face-to -face)	83	12	71	98	38	60	11	15,49%
Legal services (e-learning)	147	21	126	160	57	103	23	18,25%
Computer engineering	160	15	145	183	59	124	21	14,48%
Networks and information systems	14	0	14	27	16	11	3	21,43%
Web technologies and mobile devices	28	9	19	33	19	14	5	26,32%

When correlating the data regarding the school failure rate with the school dropout rate, we observe that they are not directly related. It should be pointed out that this study is not prepared to justify this difference, in other words, there are variables that have not been taken into account, such as employment, family situation, economic factors, inability for the course, among others, that can justify this non-correlation.

We consider these values considerable, so a more detailed study is needed.

3.4 Measures to Combat School Failure

Concerning CTeSP courses, the course coordinators indicated the following measures: Introduction of the PBL methodology which allows greater performance from students. However, there has not yet been enough time to evaluate this new method as we are still only on the second year since this method has been put to practice.

Regarding Legal Services' Course (face-to-face and e-learning), the course's technical-scientific committee (composed by 4 professors of the course) alongside with professors and professionals in the field, created the solicitor's professional profile. Diagnosed their professional activities, as well as a reference to the knowledge, skills and competences that students must acquire and, the subsequent correspondence between the curricular units and the learning results. In addition, the course coordinator introduced the following measures: seminars in all curricular units, broadcasted online, study visits to notaries, study visits to courts, where they can attend trials. It should be underlined that, in addition to the measures mentioned, there was an obligation to be tested by an oral assessment, since this scientific area requires that knowledge is also evaluated in oral expression. Concomitantly, regarding scientific and pedagogical autonomy, the academic staff was called to attention that the teaching methodologies were not to be limited to the expository method but they were to be complemented with the presentation and discussion of case studies in the classroom, with a participatory analysis in legal proceedings, through a guided debate with students and simulation of activities.

Regarding the Computer Engineering course, the Technical-Scientific and Pedagogical Course Commission (composed by 4 teachers) presented some measures to cope with failure, namely: 1) Offering the chance to do the most unsuccessful disciplines in both semesters. 2) Allow the implementation of TP and PL classes, in order to allow the application of active learning methods, to the detriment of expository methods so as to enable students to improve their learning. 3) The implementation of optional subjects in Programming 0 and Mathematics. It should be mentioned that during the academic year of 2018/2019, measures 2 and 3 were implemented. However, we cannot obtain precise results, as only one year has passed since its implementation.

4 Conclusions

Analysing our study, we conclude that age is not necessarily a factor that promotes failure. Since, in the courses studied, the age group with largest significance is identical in all of them. With the exception of the Legal Services course (e-learning), which 85.6% of students are over 28 years old, however this is not the course with the highest failure rate, as a matter of fact, it's the computer engineering one.

Regarding school success, we found out that the most successful course is the Legal Services course (face-to-face), with a percentage value of 66.70%. If we compare it with the same course, but in the e-learning model, we observe that the traditional model of classroom teaching has better results. Thus, what should be improved in distance learning methodologies in order to promote success? To make videoconferencing activities more dynamic, to have differentiated evaluation methods, these and other questions should be introduced in discussion forums.

When comparing the Legal Services Course (face-to-face) with the Computer Engineering Course, we can see that the Legal Services Course has a higher success rate, actually the difference between the success rates is quite significant (46.70%). In order to reduce such difference, it should be implemented new methods such as the logic of problematics, as well as fostering critical thinking, different ways of assessment, regarding the engineering fields, reduce the amount of expository classes.

According to [9] school success may imply a change in the structure of the evaluation, aiming at building knowledge in stages and not a final evaluation, all these factors need further debate.

Regarding school dropout, the data presented differ between 14.48% and 26.32%. However, we have no data allowing us to develop a theory about school dropout. According to the analysed data, what we know is that the failure rate is not directly related to the dropout rate. Therefore, it would be interesting to debate whether school dropout is related to: the inadequacy of the course; economic reasons; exercise of a professional activity, incompatible with school hours. All things considered, these question need answered, therefore maybe more surveys and studies with wider ranges must be made.

Thus, the next study consists of investigating and bring success/failure/school dropout in higher education for debate. With the goal of developing methods and processes to tackle failure and dropout.

References

1. DGES: Acesso ao Ensino Superior (2019). https://www.dges.gov.pt/pt/pagina/cursos-tecnicos-superiores-profissionais?plid=593
2. DRE: Diário da República Eletrónico, March 2018. https://dre.pt/pesquisa/-/search/571937/details/maximized. Accessed 2019
3. Bogdan, R., Biklen, S.K.: Investigação Qualitativa em Educação. Porto Editora (1994)
4. Hébert, M.L., Boutin, G., Goyette, G.: Investigação Qualitativa - Fundamentos e práticas. Instituto Piaget (2013)
5. Pinto, J.M.: Factores de sucesso/insucesso. Sucesso e Insucesso no Ensino Português, Lisboa (2002)
6. Astin, A.W., Oseguera, L.: Pre-college and Institutional Influences on Degree Attainment. Alan Seidman, Westport (2005)
7. Herzog, S.: Measuring determinants of student return vs. dropout/stopout vs. transfer: A first-to-second year analysis of new freshmen. Res. High. Educ. **46**, 883–928 (2005). https://doi.org/10.1007/s11162-005-6933-7
8. de Educação, D.-G.: Direção-Geral da Educação (2019). https://www.dge.mec.pt/cursos-cientifico-humanisticos
9. DGES: Candidatura ao Ensino Superior Público - Colocações 2018 (2018). https://www.dges.gov.pt/coloc/2018/
10. Conselho para a Avaliação e Qualidade: Instituto Politécnico de Beja. https://www.ipbeja.pt/SobreIPBeja/caq/Paginas/Miss%C3%A3o.aspx. Accessed 2019

Use of IT in Project-Based Learning Applied to the Subject Surveying in Civil Engineering

Oihana Mitxelena-Hoyos[1,2], José Lázaro Amaro-Mellado[3,4](✉), and Francisco Martínez-Álvarez[5]

[1] Department of Graphic Expression and Engineering Project, University of the Basque Country, Donostia, Spain
oihana.mitxelena@ehu.eus
[2] Cantabria-Basque Country Division, National Geographic Institute of Spain, Donostia, Spain
[3] Department of Graphic Engineering, University of Seville, Seville, Spain
jamaro@us.es
[4] Andalusia Division, National Geographic Institute of Spain, Seville, Spain
[5] Data Science & Big Data Lab, Pablo de Olavide University, 41013 Seville, Spain
fmaralv@upo.es

Abstract. This work describes the design and implementation of the subject of surveying under the Project-based Learning method. The modernization of teaching-learning and the new requirements of society to our graduates, force us to move towards a new style of higher education. In the case of topography, which is a transversal science closely related to the various skills of the degree, the learning conveyed by a project provides verisimilitude and depth of the knowledge acquired. Given the previous experiences, it is expected that better marks and performance are reached by the students.

Keywords: Project-based learning · Surveying · Civil engineering · Active methodologies

1 Introduction

Problem-based learning is defined by practicing a student-centered approach with an emphasis on student motivation and learning experiences to achieve the skill and knowledge development goals required by numerous institutions and agencies internationally [1].

Since the Public University of the Basque Country (UPV/EHU), and especially the School of Engineering of Gipuzkoa (EIG), wants to modernize the teaching-learning process, this paper presents the design and implementation of a project-based learning (PBL) strategy in the subject *Surveying,* from the Civil Engineering degree. *Surveying* is a compulsory subject in the second year.

According to the academic order established in the syllabus of the degree, it comprises six European Credit Transfer System (ECTS) with a distribution that equals the theory and practice part.

Á. Herrero et al. (Eds.): ICEUTE 2020, AISC 1266, pp. 428–437, 2021.
https://doi.org/10.1007/978-3-030-57799-5_44

The approach of the course is extensive so that the specific skills include:

1. Knowledge of topographic techniques essential in civil engineering in matters of representation of the land and the work as well as the setting out on the ground defined geometries and control of movements of structures or earthworks [SS1].
2. The study of the sciences of Cartography and Geodesy [SS2].
3. Knowledge of techniques and tools for surveying and setting out plans [SS3].

In addition to the specific competences, this subject should collaborate in the achievement of different basic or transversal competences, related to the acquisition of knowledge in an autonomous way, the application of the same in the resolution of problems, interprofessional communication and integration in teams. Such skills are listed below:

1. Solving specific problems of the second-year subjects in a reasoned way and starting from the knowledge acquired in the first year [CS1].
2. Gathering and interpreting relevant data that allows the resolution of problems as well as their justification taking into account technical-scientific aspects [CS2].
3. Communicate correctly and clearly the ideas and opinions and the specific topics related to the subjects in writing [CS3].
4. Correctly and clearly communicate ideas and opinions and specific subject-related topics orally [CS4].
5. Integrate in work teams when the activity requires it [CS5].

Furthermore, the School of Engineering of the University of Seville also teaches the same subject and intends to partially implement this paradigm as well, by the innovation project *Generation of 3D terrain models for land use planning simulations*. This experience has been taken into account in this research.

The initial model used is the one developed by the internationally recognized University of Aalborg, whose study programs have been based on problem-based learning since its inception in 1974, with its pillars being workgroup and collaboration with the professional environment for the development of projects of a high academic level and adapted to the requirements of society and the market at all times.

The rest of the paper is structured as follows. Section 2 reviews relevant and recent papers related to PBL. Section 3 introduces la methodology introduced in this paper. Section 4 reports the results achieved, and it discusses them. Finally, Sect. 5 summarizes the conclusions drawn.

2 Related Works

In the learning results, the use of specific information and communications technology (ICT) and geomatics tools has a specific weight, both for topographic calculation and for modelling and representation in plans and maps. It is also essential the management of networked geoinformation resources, available through the Spatial Data Infrastructures and the assimilation of the principles that govern these resources, such as accessibility, interoperability and reuse of resources. Furthermore, the increasing interest in assessing quality assurance systems in higher education centers is demanding new and sound teaching strategies [2].

Although traditional methods for assessing students' performance still provide satisfactory results [3], the use of ICT tools for teaching contributes to student satisfaction and quality in the management of content and information flows in the subject.

Indeed, machine learning approaches are currently being applied to analyze student achievement scores [4]. Results suggested that several activities were not really relevant and, so, machine learning techniques may be helpful to design new relevant and non-redundant activities for enhancing student knowledge acquisition.

A game-based student response system applied to a multidisciplinary teaching context can be found in [5]. The outcomes showed that this methodology increased the involvement and reduced part of the degree of abstraction that complex contents entail.

Chen and Yang [6] proposed revisited the effects of project-based learning on students' academic achievement by means of a meta-analysis investigating moderators. After an extensive study, they concluded that PBL has a medium to large positive effect on students' academic achievement compared with traditional instruction.

An interesting project-based methodology to lecture on web frameworks applied to the management of health-related data was introduced in [7]. They concluded that the application of flipped teaching along with PBL may have a very positive impact in drop rates for late years.

Recently, online PBL has also been proposed via cloud computing [8]. The authors expected to provide insights about the roles of the instructor and students in online PBL design activities. Along with the study findings, some implications were discussed for using cloud computing effectively in PBL.

The comparison of two systems of PBL applied to Civil Engineering is given in [9], where emphasizes to the need for adequate spaces, facilities and schedules.

3 Methodology

The modernization of teaching by a faithful approach to the requirements of society is the reason why the PBL is chosen as the methodology to be implemented, formulated on a stage in which the subject is developed around, such as set out below.

3.1 Project-Based Learning

The activity of a civil engineer, for the most part, takes place in environments organized by projects, with multidisciplinary tasks that require the collaboration of professionals with different profiles and areas of knowledge, and these requirements are those that society will demand from the students and professionals of the future.

PBL is defined by practicing a student-centered approach with emphasis on student motivation and involvement and learning experiences to achieve the required knowledge and skill development goals. If we consider the differences between problem-based learning and project-based learning, it is feasible in this subject to pose a single global project as a thread of different problems, trigger more in-depth learning and the ability to relate theory to its practical application. This issue is emphasized when choosing a real project adapted to the time and effort dimensions that the subject requires.

The chosen project will be developed in a normative context, with external collaborating agents, as is the case of the National Geographic Institute of Spain (NGIS), with project results that adhere to real technical specifications.

3.2 Proposal for the Design of the Subject *Surveying* by Means of PBL Methodology

The starting point in the PBL methodology is a scenario that poses the "problem" and should serve as an incentive and challenge to begin the learning sub-process [10]. The particular scenario proposed exposes the requirement to define the administrative boundaries of a territory under great urban and environmental pressure. Students will have to investigate the regulatory framework and choose the source of information for the reconstruction of the historical topographic survey of the line according to the current specifications. With the results, a report will be drawn up containing all the actions and a presentation will be prepared with a recreation of the reality in order to visualize the project in the most enlightening way possible.

The driving question is "How far does Zubieta go?" It is attractive, convincing and open to arouse curiosity and stimulate the student to understand its scope.

The approach to each phase of the project is inductive and is preferably solved with active methodologies; the topographic calculation must be taken into account first, in the insertion in the existing cartography with global coordinate system, and therefore the needs of geodesy and cartography in the project, to continue with the adjustment and modelling with design tools in civil engineering; in this case *AutoCAD Civil 3D* 2018 and QGIS 2.18, free and open source software.

The report requested must respond to the current requirements of these actions of jurisdictional demarcation so that different forms of expression of a geographical reality are addressed, 2D plans altimetry profiles, 3D modeling and computer graphics and virtual recreations, in addition to the memory justification of all the work done.

The implementation of this methodology is done on a group of ten students, a number that ensures personalized monitoring during the development of the subject, but small to ensure the extrapolation of results and a significant statistical analysis. Groups of two or three students are formed, taking into account their distribution in other subjects.

The application of PBL reaches 75% of the learning required outcomes.

4 Results and Discussion

Once the PBL has been designed, its first implementation has been carried out during an academic year, verifying that the subject's competencies materialize in the different steps of the actual execution of the project. Regarding the discussion on said implementation, special emphasis is placed on the material and temporal requirements, on the leading role of active methodologies and ICT, and on the impact on evaluation, as explained below.

4.1 Results of the Implementation

The result of the design implemented in academic year 2018–2019 was executed upon the schedule shown in Table 1.

Table 1. Schedule for the academic year 2018–2019.

Sessions	Activity	Learning techniques	Skill #
1	Presentation of the scenario	1-2-4. Discussion. Socrative	SS1 CS2, CS4, CS5
2	Definition and design of the proposal for the historical analysis of a jurisdictional boundary	Determination of the 1-2-4 plan. Discussion	SS3 CS1, CS3, CS4, CS5
3	Calculating the historical field notebook. Radiation	Self-learning card: RADIATION	SS2 CS1, CS5
5	Calculation of a reference point. Bisection	Bisection: puzzle. Self-learning card: BISECTION	SS1, SS2, SS3 CS1, CS2
2	Development of the altimetric profile of the field notebook itinerary	Self-learning card: TRIGONOMETRIC LEVELING	SS2, SS3 CS1, CS3, CS5
4	Writing of the report and preparation of the results.	Seminar	SS3 CS2, CS3, CS4, CS5
1	Google earth smartphone	BYOD	SS1, SS3 CS2, CS3, CS5
4	Field examination. Setting out		SS2 CS1, CS2, CS5
1	Presentation and defense	Results discussion	CS2, CS4, CS5

It must be noted the flexibility derived from the personalization of the project, assuming the phases of the project in a professional environment:

a. Review of legal and technical sources in the definition of the geometry of this geographical relationship between two jurisdictional areas. In this case, it is a boundary document and a field notebook drawn up by classical topography in 1928.
b. Planimetric development of the field notebook, a self-oriented open route that is calculated in a chained way by the radiation method and linking the calculation with a point of the cartography by means of the bisection topographic method.
c. Superimposition of the complete calculation on the cartography, for which it is necessary to present various concepts of geodesy and cartography as well as spatial data infrastructures and data sources and formats obtained from the NGIS download center [11], the cartographic repository of the Provincial Council of Gipuzkoa [12] and the web geoservices of the Basque Government [13] and the NGIS [14].
d. Altimetric development of the trigonometric levelling of the notebook and representation in a longitudinal profile (Fig. 1), for which the use of *AutoCAD Civil 3D* design tools is addressed.
e. Contextual analysis and justified adjustment of the boundary line on the mapping from the literal information of the demarcation act (Fig. 2).

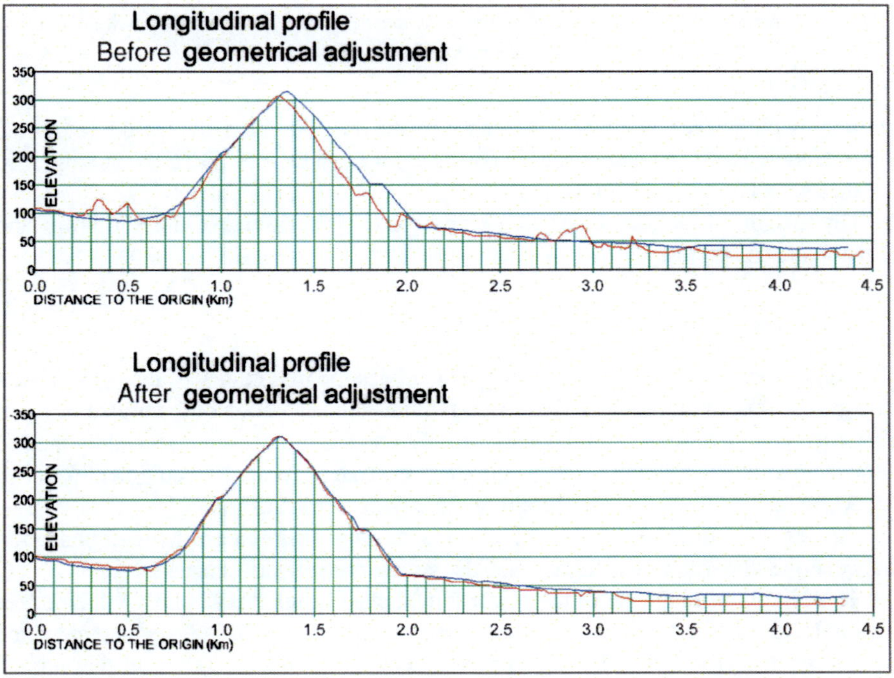

Fig. 1. Evaluation of the adjustment of the longitudinal altimetry profile before and after the fitting of the route, produced by the students.

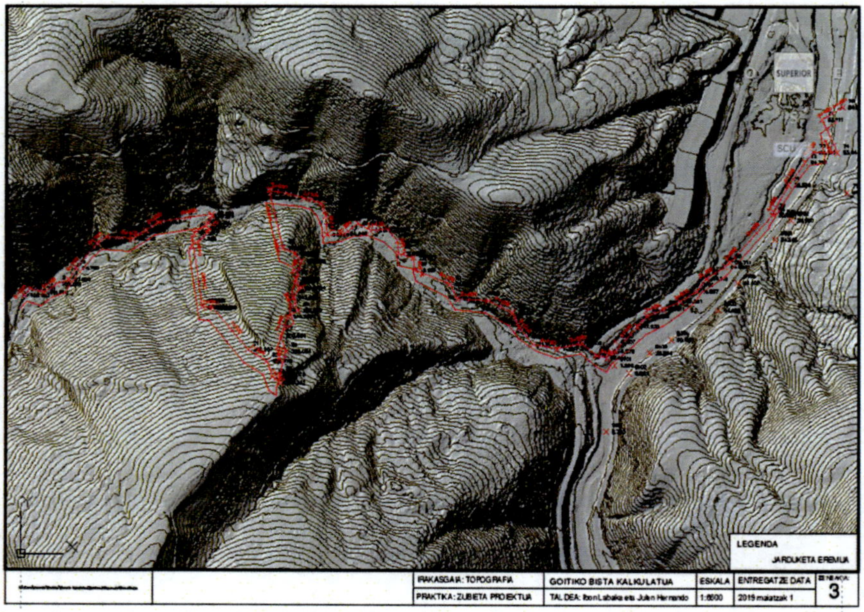

Fig. 2. Presentation of the development work of the field notebook, overlaid with the DTM Lidar and adjusted, carried out by the students.

f. Setting out of the benchmarks that make up the line. For them, they will be exported to Google Earth and Google Maps (Bring Your Own Device –BYOD- paradigm) so that students can access the area.
g. Drafting of the project description report, plans, preparation of the 3D-model and computer graphics.

4.2 Discussion About Resources Required

The application of the PBL methodology underlines the importance of the organization and group work of the students (Fig. 3); it requires the adaptation of the organization of the space and the resources in the classroom. These requirements are solved by the use of a seminar room, a flexible space equipped with some computer equipment for each work group, and with modular furniture that encourages dialogue.

As for the timetable, students have had sessions with freedom of organization between the classroom, the library and the laboratories of the center.

In addition, use is made of the eGela virtual classrooms that the UPV/EHU makes available to each subject in its study plans, based on the Moodle platform which constitutes the repository of basic and complementary documentation in the various formats, such as texts, bibliographic references, and video links. Secondly, it is the communication channel, on which the teacher will make the activity dynamic through announcements, forums for voluntary participation, individual communications and publications on the calendar.

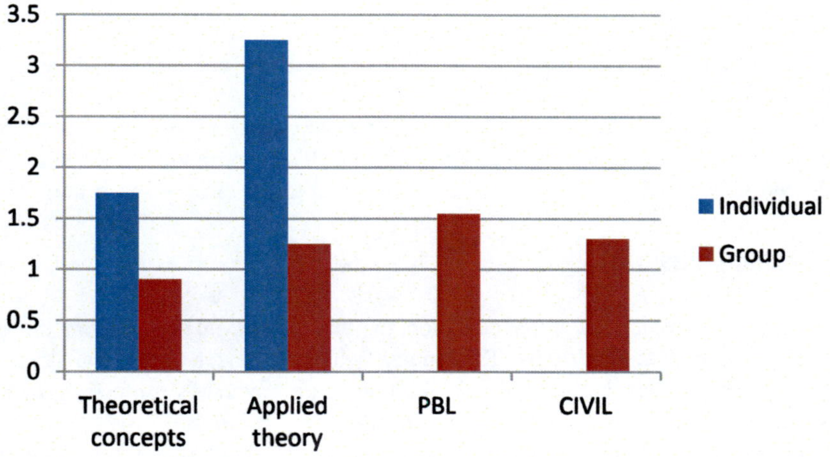

Fig. 3. Distribution of the contribution of each teaching modality.

4.3 Participation: Active Methodologies

The use of active methodologies in solving the problems that make up the project encourages in-depth learning. For this purpose, the use of resources from the recreational and social activity is also considered, such as the integration of *AutoCAD Civil 3D* data in the contents of Google Maps. Some examples of techniques used are the 1-2-4 technique, puzzle, seminar or self-learning cards:

The 1-2-4 technique is used both to enhance the progressive synergy in the search for ideas, and to synthesize and refine them, and is very effective in promoting participation, also takes into account the previous knowledge that students have on a particular topic

Where the puzzle technique has been used, it has been simplified for application in small groups, and is used for the parts of applied theory traditionally taught through classroom practice, and exercise resolution. In contrast, the seminar technique has been used for specific aspects of the project.

Based on the OCW courses, the different teaching units are organized in self-learning sheets for each surveying method. In them the objectives of the method, the basic formulation, the bibliography, list of recommended exercises of gradual complexity, problem solved and proposal of task to be delivered to acquire the knowledge required in the resolution of the PBL are synthesized.

The promotion of student participation in the discussions is reinforced by the use of ICT in the two-way teaching-learning processes, and the interactive and real-time response capacity of questionnaires. In addition to providing a surprise factor, it is an enjoyable way of probing the degree of understanding of the fundamental concepts of any knowledge area.

4.4 Evaluation

The evaluation is formative and based on daily activity, combined with a test of minimum content.

The deliverable activities of the project are accompanied by the rubric corresponding to the learning results, as a self-evaluation. The contribution to the evaluation of the individual and group activities is proportional to the effort of each one.

5 Conclusions

This work presents a PBL implementation for the subject *Surveying* from the civil engineering degree at UPV/EHU. This course, which addresses transversal skills and knowledge for the development of a civil engineering graduate's profession, demonstrates its potential for development through PBL.

During the implementation of PBL, the positive stimulus has been observed through the surveys carried out with students, and the samples of curiosity for the profession and the incidence of surveying activities in civil works.

The execution of the project has hardly deviated from the planned planning, possibly due to the active tutoring of the teacher, necessary because it is one of the first subjects of the degree with a PBL context. Having a small number of students allows for a personalized approach to teaching-learning, but can impoverish and detract from the depth of the discussions generated. It also requires the adaptation of the active techniques used to the characteristics of the group.

The adaptation of the space and the temporal structure of the subject has been critical in the implementation of the PBL, and the support of the virtual classroom, essential, both in the registration of evidence of learning, and in the achievement of a formative evaluation, contributing to transparency, traceability and student satisfaction. Achieving this comprehensive virtual management involves a great effort in content management in the tool and a large workload.

A close line of improvement is the coordination with different subjects of the degree so that this interrelationship affects the involvement of the student and offers a credible framework for the development of varied and realistic projects. In addition, the design of PBLs that address different phases of execution of a work is considered.

Acknowledgments. The authors would like to thank the University of Seville for funding the project *Generation of 3D terrain models for land use planning simulations* framed in the Own Plan for Teaching Innovation (2018) which has supported this research.

References

1. Rodríguez-Mesa, F., Kolmos, A., Guerra, A.: Aprendizaje basado en problemas en ingeniería. Aalborg Universitetsforlag, Teoría y práctica (2017)
2. Rubio-Escudero, C., Martínez-Álvarez, F., Atencia-Gil, E., Troncoso, A.: Implementation of an internal quality assurance system at Pablo de Olavide University of Seville: Improving computer science students skills. In: Martínez-Álvarez, F. et al. (eds.) International Joint Conference CISCS'19-ICEUTE'19. AISC, vol. 951. pp. 340–348. Springer, Cham (2020)

3. Rubio-Escudero, C., Asencio-Cortés, G., Martínez-Álvarez, F., Troncoso, A., Riquelme, J. C.: Impact of auto-evaluation tests as part of the continuous evaluation in programming courses. In: Graña, M. et al. (eds.) International Joint Conference SOC'18-CISCS'18-ICEUTE'18. AISC, vol. 771. pp. 553–561. Springer, Cham (2019)
4. García-Torres, M., Becerra-Alonso, D., Gómez-Vela, F.A., Divina, F., Cobo, I.L., Martínez-Álvarez, F.: Analysis of student achievement scores: A machine learning approach. In: Martínez-Álvarez, F. et al. (eds.) International Joint Conference CISIS 19-ICEUTE 19. AISC., vol. 951. pp. 275–284. Springer, Cham (2020)
5. Amaro-Mellado, J.L., Antón, D., Pérez-Suárez, M., Martínez-Álvarez, F.: Game-based student response system applied to a multidisciplinary teaching context. In: Martínez-Álvarez, F. et al. (eds.) International Joint Conference CISIS 19-ICEUTE 19. AISC, pp. 329–339. Springer, Cham (2020)
6. Chen, C.H., Yang, Y.C.: Revisiting the effects of project-based learning on students' academic achievement: A meta-analysis investigating moderators. Educ. Res. Rev. **26**, 71–81 (2019). https://doi.org/10.1016/j.edurev.2018.11.001
7. Fernández-Cerero, D., Fernández-Montes, A.: Project-based methodology to lecture on web frameworks applied to the management of health-related data. In: Martínez Álvarez, F., Troncoso Lora, A., Sáez Muñoz, J.A., Quintián, H., Corchado, E. (eds.) CISIS/ICEUTE - 2019. AISC, vol. 951, pp. 321–328. Springer, Cham (2020)
8. Çakiroğlu, Ü., Erdemir, T.: Online project based learning via cloud computing: exploring roles of instructor and students. Interact. Learn. Environ. **27**, 547–566 (2019). https://doi.org/10.1080/10494820.2018.1489855
9. Ureña, J.M., Menéndez, J.M., Coronado, J.M.: Project/problem based learning in civil engineering. The ciudad real (Spain) Experience. In: International Conference on Engineering EducationDep, Valencia, Spain (2003)
10. Norman, G.R.: Problem-solving skills, solving problems and problem-based learning. Med. Educ. **22**, 279–286 (1988). https://doi.org/10.1111/j.1365-2923.1988.tb00754.x
11. NGIS: Download Center, http://centrodedescargas.cnig.es
12. Gipuzkoa Provincial Council: Spatial Data Infraestructure of Gipuzkoa. https://b5m.gipuzkoa.eus/web5000
13. GeoEuskadi: Spatial Data Infrastructure of Euskadi. https://www.geo.euskadi.eus
14. NGIS: MTN, 1st edn. http://www.ign.es/wmts/primera-edicion-mtn

Author Index

Á. Herrero et al. (Eds.): ICEUTE 2020, AISC 1266, pp. 439–441, 2021.
https://doi.org/10.1007/978-3-030-57799-5